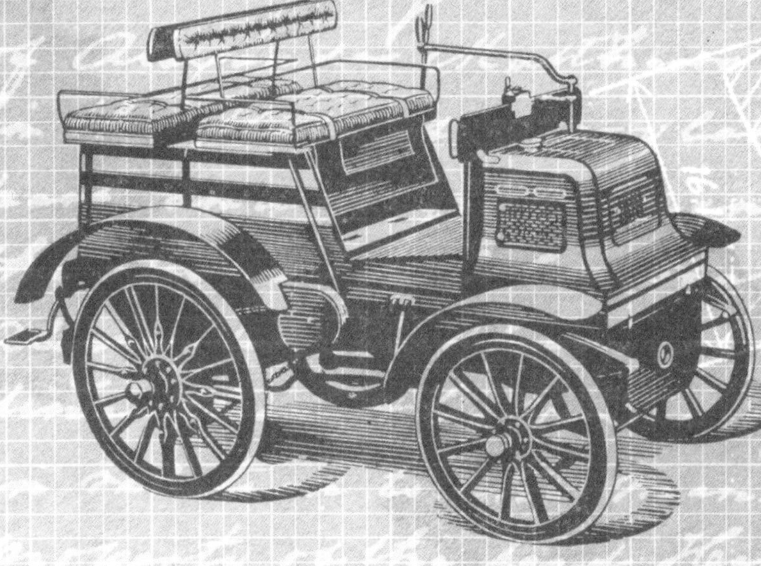

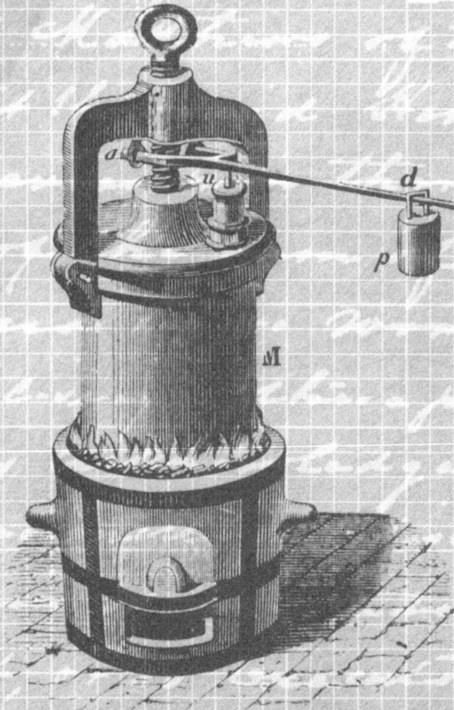
a
u
d
b
p
M

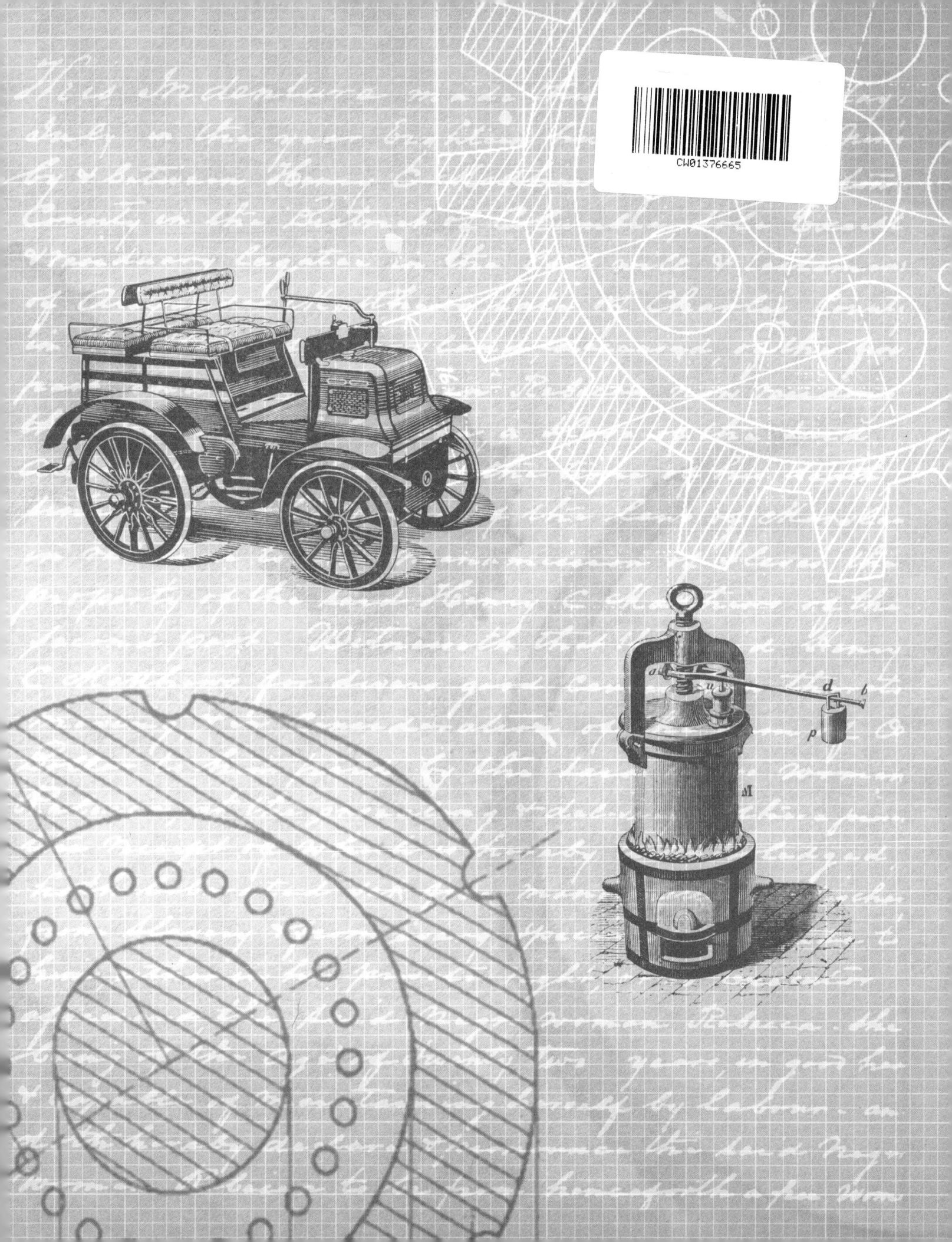

GREAT INVENTIONS

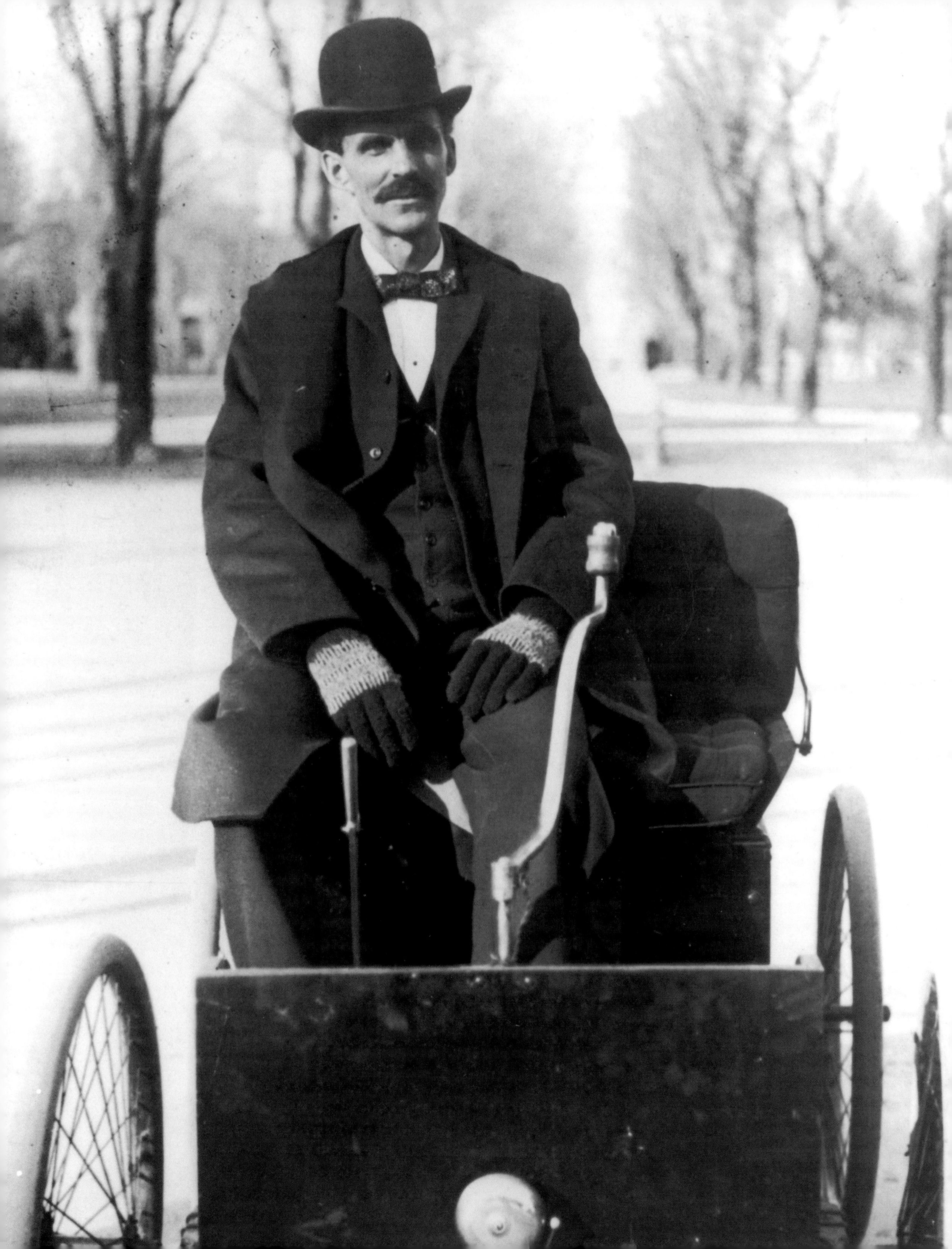

GREAT INVENTIONS
OF OUR TIME

DAVID ELLYARD

NEW HOLLAND

First published in Australia in 2007 by
New Holland Publishers (Australia) Pty Ltd
Sydney • Auckland • London • Cape Town

www.newholland.com.au

1/66 Gibbes Street Chatswood NSW 2067 Australia
218 Lake Road Northcote Auckland New Zealand
86 Edgware Road London W2 2EA United Kingdom
80 McKenzie Street Cape Town 8001 South Africa

Copyright © 2007 in text: David Ellyard
Copyright © 2007 in images: Getty Images and NASA
Copyright © 2007 New Holland Publishers (Australia) Pty Ltd

All rights reserved. No part of this publication may be reproduced, stored in a retrieval system or transmitted, in any form or by any means, electronic, mechanical, photocopying, recording or otherwise, without the prior written permission of the publishers and copyright holders.

National Library of Australia Cataloguing-in-Publication Data:

Ellyard, David, 1942- .
Great inventions of our time.
ISBN 9781741104899 (hbk.).
1. Inventions - History. I. Title.

609

Publisher: Martin Ford
Production: Linda Bottari
Project Editor: Michael McGrath
Editor: Kirsten Chapman
Designer: Simon Rattray
Cover Design: Simon Rattray
Printer: SNP/Leefung Printing Co. Ltd (China)

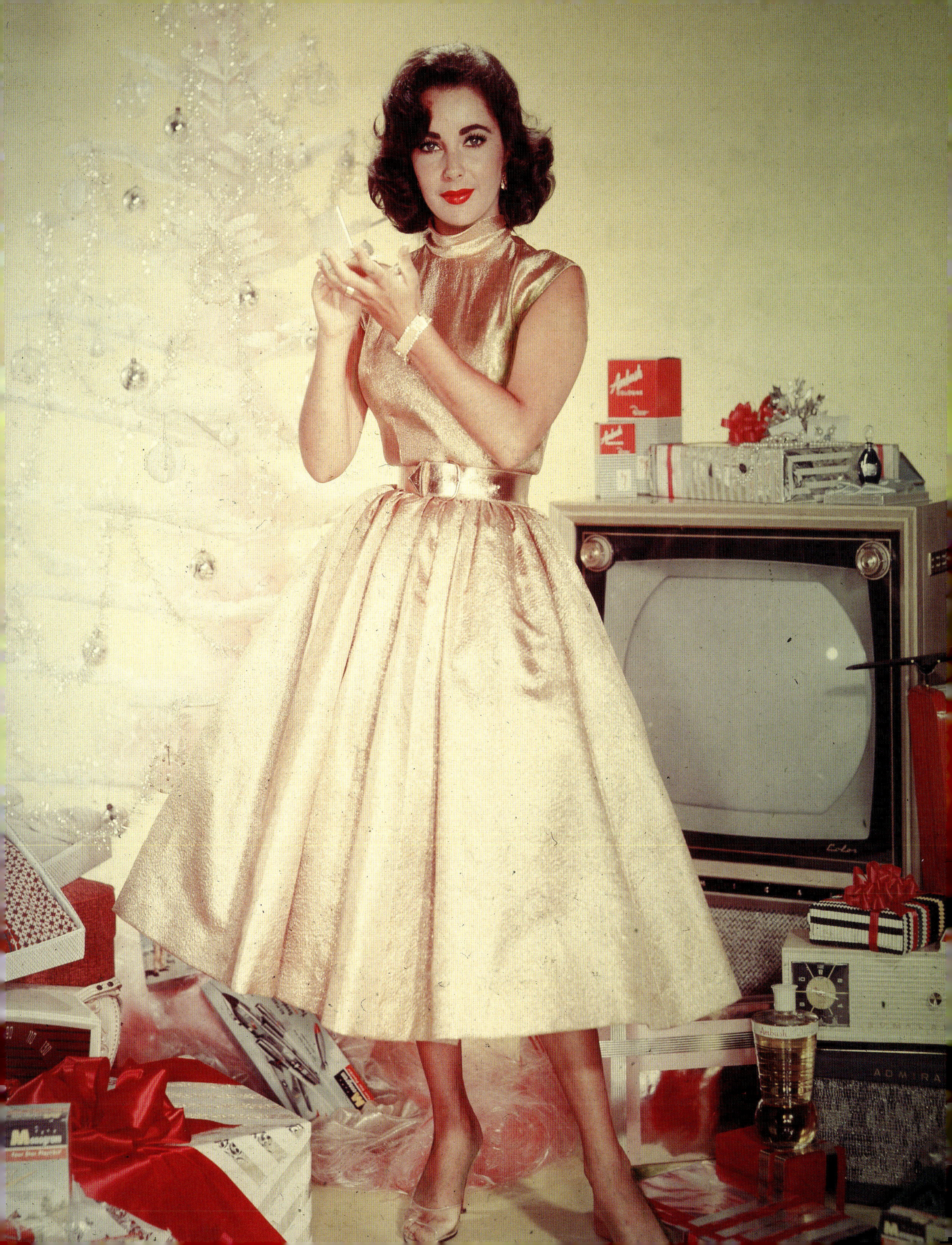

CONTENTS

Introduction 8

Where was technology in 1500? 9

PART 1
Energy meets matter — 14

Chapter 1
Essential energy 16

Chapter 2
From farm to food and fibre 34

Chapter 3
From iron to steel 46

Chapter 4
Materials matter 64

Chapter 5
Putting chemicals to work 70

Chapter 6
Not found in nature 82

PART 2
Information impact — 90

Chapter 7
Pens, paper and printing 92

Chapter 8
Sending a message 100

Chapter 9
Making an image last 114

Chapter 10
Saving sounds 128

Chapter 11
The wireless world 138

Chapter 12
The box in the corner 152

Chapter 13
The computer story 160

28

79

118

168

PART 3
On the move — 174

Chapter 14
Running on iron rails 176

Chapter 15
The world on wheels 184

Chapter 16
Under and over the water 198

Chapter 17
Taking to the air 206

Chapter 18
Time and place 218

PART 4
Healthy and happy — 228

Chapter 19
Medical matters 230

Chapter 20
Looking within 244

Chapter 21
Lifestyle 252

Chapter 22
Having fun 266

Chapter 23
Making music 278

Chapter 24
Lighting the way 284

Information sources 294

About the author 296

Index 297

Introduction

New technologies, the offspring of human inventiveness and persistence, continue to penetrate every aspect of our lives. Better ways of meeting our needs for food, clothing, shelter, transport, communications, entertainment and protection of our health and environment have transformed our standard of living and quality of life today when compared with that of previous generations.

We now find it all but impossible to imagine life without, for example, anaesthetics, antiseptics and antibiotics, or the capacity for almost instantaneous global communication through the mobile phone and the Internet. Yet those vital medical technologies did not exist until the 19th century and the communications capacities not until the late 20th.

Invention is a profoundly human phenomenon, driven by human desires for a better life, and empowered by human minds, hearts and hands. This book explores the background and impact of dozens of the most important inventions to emerge in Western society in the last 500 years.

Here you will meet the creative spirits who have seized an opportunity to create some new device or process, enriching the lives of their fellows and often their own lives, although no invention comes without some negative impact. Some of the names you will know well, others will be new.

Like the science that increasingly underpins the creation of new technology, invention is cumulative. It is often impossible to say exactly who was responsible for some particular invention. In general, many people will have contributed to it, perhaps over several centuries. This 'building on the past' also ensures that the pace of technological change increases over time, leading to the seemingly relentless emergence of new devices and processes, often at breathtaking speeds, that so characterises our lives today.

A word about the selection of inventions that feature in this book: given the immense number and diversity of technologies past and present, many more are not included than are. I sought to cover the wide spectrum of human needs, to cover the major developments over the half millennium and to tell intriguing or inspiring stories of human inventiveness and endeavour.

You will find here many things you expect, but not all. The laser, for example, does not have its own entry, since its value does not lie in the device itself, which was for some years a solution in search of a problem, Rather it comes from diverse applications which were later found for the laser, such as those in communications, which are discussed here.

The earliest inventions chronicled in this book, such as the first watches, date from around 1500, though invention is as old as our race or even older. Around that time the pace of both scientific discovery and technological advancement increased markedly, driven factors that include increasing prosperity, the emergence of early industries, the discoveries made by intrepid explorers and the growing challenge to ancient ideas, represented in part by the Protestant Reformation.

That impetus is by no means exhausted. In coming decades, we will both know more and be able to do more, making the lives of future generations even more different from today than today is from yesterday. Whether you welcome the prospect or fear it is for you to decide.

Where was technology in 1500?

The earliest inventions chronicled in this book are from around 1500, but finding better ways to meet everyday needs is a quest as old as the human race and even older. From the most ancient times, we had to provide ourselves with shelter, clothing, warmth, food, means of communication and defence against attack—human and animal. The most venerable of our technologies emerged thousands of years ago, and we continue to make use of them: the wheel and its many applications; the plough and other farming implements; the use of fire for cooking, heating, the firing of pottery and bricks and the smelting of metal ores; the working of copper, bronze and iron into tools, armour and weapons; materials like glass.

While much of the technology that dominates our lives now is modern, developed within the last few centuries and within the nations of Western Europe and its former colonies, significant inventions have long-ago origins in other regions of our planet, notably in Asia. An immense array of technology originated in the ancient civilisations of China and the Middle East, which flourished while the lights were out in Europe following the fall of Rome, though the pace of innovation in these places slackened markedly just as it was picking up in Europe. Significant Chinese inventions which came overland or by sea to Europe—aided by exploration, trade or conquest—include paper, the magnetic compass, sandpaper, water-driven clocks, wallpaper, earthquake-detecting apparatus, noodles, paper money, the toothbrush and playing cards.

Around 1000 CE, the Chinese perfected the formula for gunpowder and, soon after, printing with movable type, but it was centuries before those reached Europe. The Arab scholar Alhazen knew about lenses for focusing light, but spectacles were not used in Europe

The 15th century saw the pace of invention quicken, spurred by the Renaissance, rising prosperity and the patronage of able minds—like Leonardo da Vinci.

until the 13th century. Omar Khayyam's 11th-century calendar was much more accurate than any used in Europe for many years. However, Islamic science and technology would soon be in decline, suppressed by strict religious teachings that opposed progress.

Not every new idea was imported. The particular requirements of Western music demanded some way of writing it down, rather than each new set of players or singers having to learn it by heart. Around 1036, Guido D'Arezzo devised a notation much as we have it today, with the sounds marked as dots on a set of horizontal lines—four to begin with—as well as the do-re-mi system for memorising.

Twelfth-century Europe created some home-grown initiatives in invention and engineering. The Dutch led the way in reclaiming flooded land with dams, dykes, canals and wind-driven pumps; low-lying regions of Germany and elsewhere sought Dutch skills and technology. The first mills to make paper, another Chinese invention, appeared in Spain, built by Moors who had carried the innovation west, though most writing was still done on parchment.

Late in the 12th century, the grand new architectural style known as 'Gothic'—initially a criticism—began to take hold, though even this had Middle Eastern origins, with the concept returning home with the Crusaders. Masons began to experiment with flying buttresses to support ever-higher roofs and spires, raised to the glory of God; these remain impressive achievements today. Domestic housing was more modest, though increasingly windows were filled with glass, chimneys vented the smoke from hearths, and tiles replaced thatch in large cities like London to reduce the risk of fire.

In the 13th century the magnetic compass, of such value to navigators, was improved—perhaps by Peter Peregrinus who described it—with the needle on a pivot around which it could swing freely. Previously the needle had been floated on a mattress of straw, which was not much good if the sea was rough. European ships began to use the sternpost rudder in place of a sweep oar to control their course. In time this Chinese innovation would enable much larger ships to be built. Ships also began to sail in convoy to provide greater mutual protection.

The spinning wheel, invented in India, came into use in Europe, significantly speeding the production of fibres for clothing from wool and flax, and foretelling the later large-scale mechanisation of textile production. The production of silk (another import from China where the silk worm had been domesticated thousands of years earlier) was speeded by mechanical devices that wound silk onto reels, invented probably in Bologna around 1270.

One of the most important imports was the system of numbers we use today—variously called Indian or Arabic—introduced by the Italian Leonardo Fibonacci, replacing the clumsy Roman numerals and making possible the calculations that would increasingly back up science and technology. In an early commercial use, accountants in Italy invented double-entry bookkeeping around 1270.

English natural philosopher Roger Bacon—who already knew the secret of gunpowder—wrote of the magnifying glass, and by the end of the century lenses in frames formed the first spectacles in Europe. Again China seems to have had them first. Brandy was first distilled in France around 1300; chemical distillation apparatus was another legacy from the Middle East.

Introduction: Where was technology in 1500?

The 14th century saw the arrival of gunpowder on the battlefields of Europe. Invented in China and used in rockets and firearms against the Mongols, it soon found employment in cannon, rockets and small-arms, with lead or iron shot fired from iron tubes held against the shoulder—the Chinese had used bamboo. It was a century or more before small-arms replaced the longbow and the crossbow, as they lacked the range, accuracy and rate of fire.

Though sundials were still the most common timepieces, clocks began to chime the hours and quarters from church towers and royal palaces in England and elsewhere—the word 'clock' comes from the French for 'bell'. Not until much later did such clocks have faces to show the time.

Among other innovations were the first locks for canals, devised by the Dutch and enabling a canal boat to go up or down the side of a hill; the re-invention of the sawmill; the form of navigation chart known as a 'portolano', and blast furnaces with water-powered bellows for the production of iron.

The 15th century saw the pace of invention quicken, spurred by the Renaissance, rising prosperity and the patronage of able minds—like Leonardo da Vinci—by the many competing courts.

This was the century of printing, at least in Europe. Printing with movable type, which could be rearranged to form words and sentences, had been known for centuries in China and Korea. German goldsmith Johann Gutenberg is usually credited as the first printer in Europe, but others may have beaten him by a few years. The innovation spread rapidly across Europe, a major development in itself, and speeded the growth of technology in other fields, enabling news of developments to spread much more quickly.

Left: Windmills used to grind grain or pump irrigation water and ploughs powered by powerful domesticated animals represent the advanced labour-saving devices on the eve of the technological revolution we now call industrialisation.

The 15th century also began the great age of European exploration, with first Portuguese sailors and later those of Spain and other nations daring to voyage further and further along coastlines then across open oceans. They were initially looking for routes to the Spice Islands in the east, as overland routes became blocked by the Saracens, but discovered much unknown territory along the way, such as the unsuspected Americas. These voyages drew on new methods of shipbuilding such as the 'carrack', with flush rather than overlapping exterior planking. Navigation was aided by better map-making and the early quadrant used to find latitude.

Musical instruments with keyboards first appeared—the clavichord, in which metal bars struck strings, and the air-powered organ. Instrumental music could now become more complex, with the fingers of both hands employed. The new oil paints, pioneered in Flanders, gave new life and depth to portraits and religious works, largely replacing the frescos painted in wet plaster.

Crafts advanced. Glassmakers found that by adding what we now call manganese to the melt, they could remove all colour. The first primitive lathes to turn wooden objects came into use, probably in Nuremberg, driven by a cord moved back and forth by a bow. The carpenter's brace for drilling holes was another innovation of the time.

Medicine remained primitive, though mercury was now used to treat venereal diseases and gold had become the preferred material for filling teeth, in place of the wax and gum used in medieval times. Dentistry, like surgery, was still mostly done by barbers.

Right: Until the 15th century, heating, cooking and metal work all relied on the naked flame, usually from burning timber or vegetable oils for light. Almost all work required large amounts of manpower. Strength and stamina were essential assets for survival.

The dawn of the modern world

So, as we draw near 1500, we can see how the lives of the people of Europe were constrained by the technologies at their command. Moving about was still done essentially on foot or on horseback, unless you were rich or powerful enough to be carried by sedan-chair. Heavy loads went by cart, aided by the horse-collar and shoulder harness, a genuine European invention some 600 years old.

There was no faster way to send a written message than to give it to a man on a horse. City streets were usually unpaved, dusty when dry, muddy when wet; country roads were often merely tracks. Water travel was powered by oars or relied on the vagaries of the wind. Nobody travelled by air, except in their imagination.

The only sources of energy to do most work were human or animal muscles, though moving water or the wind could be harnessed to specific tasks, such as grinding grain or pumping water. Rich footwear was made from leather, poorer footwear from wood. Fibres for clothing were mostly wool and linen, with spinning and weaving still done by hand and therefore slow, even with the aid of the spinning wheel and the loom. Cotton, later to become so valuable and versatile, was little known in Europe, since its origins were far away in India.

Save for candles, oil lamps and flares, it was dark when the sun went down, before the moon came up. Coal was only just coming into use, so cooking and home heating depended on wood. Preserving food still relied on ancient techniques such as salting, drying, smoking and pickling, or chilling if you had access to ice. Water was commonly polluted and unsafe to drink in major centres of population. New beverages such as tea and coffee had yet to arrive from distant lands. Most people drank ale with their meals. Soap was a luxury and not very effective when available, so cleanliness had low priority; the link between dirt and disease wasn't understood for another 300 years.

> **The only sources of energy to do most work were human or animal muscles ... Nobody travelled by air, except in their imagination.**

The only explosive, military or civil, was gunpowder; that would be true until the 19th century. The dominant metal was iron, especially for weapons and armour, but it was too expensive to allow a farmer to have an iron plough. The crudeness of its production from iron ore, using charcoal from the rapidly disappearing forests, left it brittle unless extensively reworked by heating and hammering, which again added to the cost. The other metals known were precious gold and silver, liquid mercury, copper and tin, which could be made into bronze, and soft, unreactive lead, much used in plumbing – the word comes from *plumbum*, the Latin for lead. Zinc was still unknown, so there was as yet no brass.

The practice of medicine was constrained by an almost total lack of the essential knowledge of what was needed to preserve and restore health, mixed with erroneous philosophies which promoted hazardous therapies like bloodletting. External symptoms were the only clue as to what was happening within the body. There was no stethoscope, no clinical thermometer. Lacking the 'germ theory' of disease, and the antiseptic techniques that would ultimately flow from that, doctors were commonly unable to control infection. Anaesthesia for surgical procedures was possible only with the aid of narcotics or alcohol, and surgery generally was rudimentary and risky.

Every one of these varied aspects of life would be transformed in the 500 years to come.

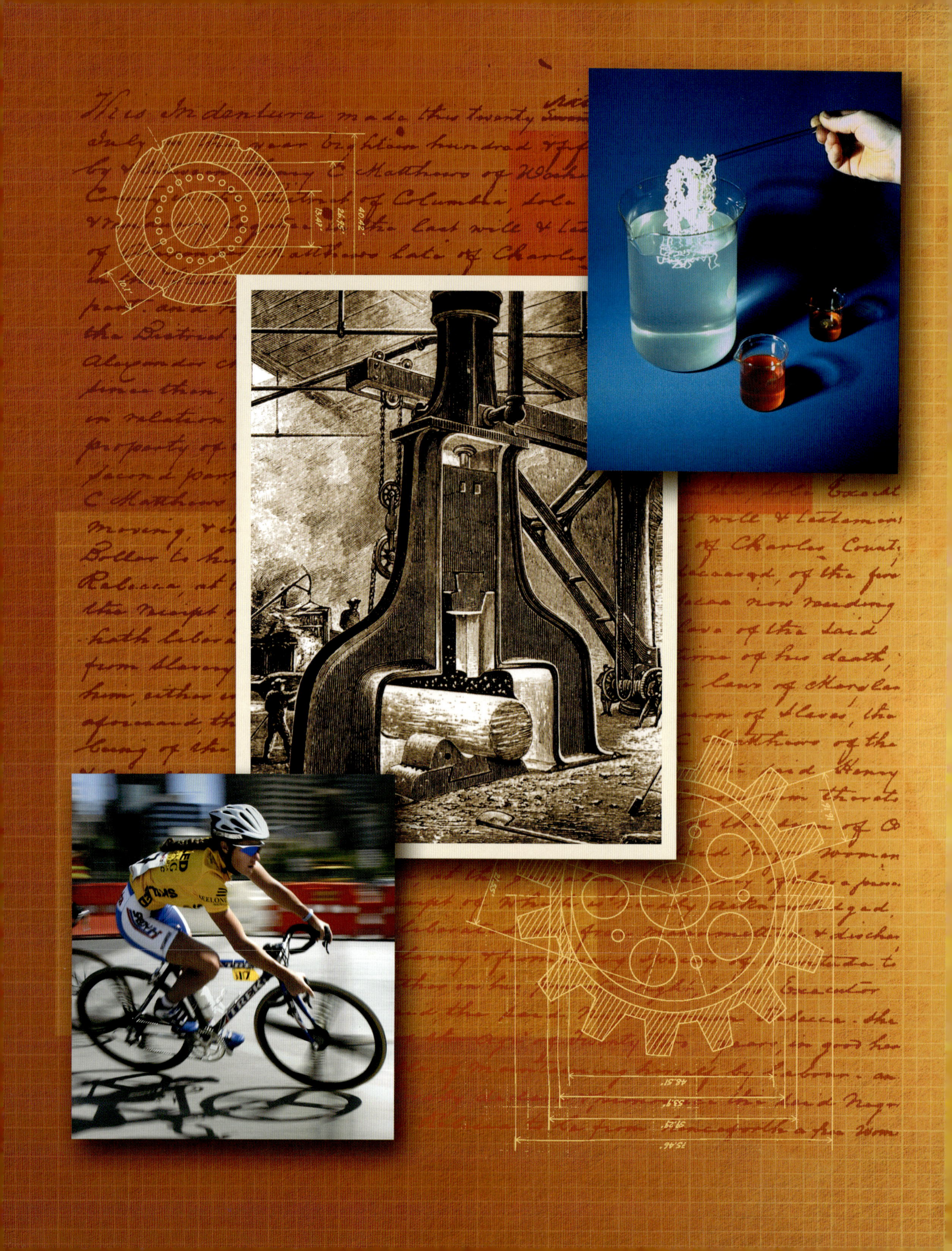

01
PART ONE:

Energy meets matter

CHAPTER 1:
Essential energy

It is not easy to say just what energy is. But we certainly can say what energy does. It makes things happen. Nothing changes without the exchange of energy in one form or another. Making an object move, constructing a building, transforming one material into another—all these need energy.

For most of human history, the energy we needed to run our society came from the movement of muscles—whether human or of other animals—and therefore ultimately from the food we and they ate. We also knew the transforming power of fire—in heating, cooking, baking pottery and releasing metals from their ores—and quite early on had techniques to harvest some of the energy in wind and running water.

The creation of the modern world flows mostly from our success in finding better ways to harness energy, from steam, from electricity, from liquid fuels, from the atom, and from even more efficient ways to capture the primeval energy of the sun. As energy has become cheaper and more abundant, our lives have been transformed. Of course, the energy released has not always been for peaceful or constructive purposes.

What happened when?

1698	Thomas Savery's The Miners' Friend pumps water from mines using steam
1712	Thomas Newcomen refines steam pump design and they become widespread
1776	First Boulton and Watt steam engines installed in factories to drive machinery
1783	Luigi Galvani investigates 'animal electricity'
1800	Alessandro Volta develops the voltaic pile, essentially a battery
1812	Richard Trevithick introduces efficient, high-pressure Cornish engines
1820	Hans Oersted discovers electromagnetism
1821	Michael Faraday designs and trials an electric motor
1843	Robert Groves produces first fuel cell, using hydrogen and oxygen to make electricity
1854	Benjamin Silliman distils petroleum to find useful vapours
1859	Edwin Drake establishes first successful oil well
1860's	August Mouchout develops a steam engine powered by sunlight
1883	Charles Fritz turns sun's rays into electricity
1884	Charles Parsons designs a steam turbine
1893	Rudolf Diesel uses heavy fuel oil to run an engine
1932	The atom is split at Cavendish Laboratory, Cambridge, England
1942	First nuclear reactor created in Chicago, United States
1954	Russell Ohr develops solar cells able to convert sunlight to electricity
1955	First nuclear-powered submarine, USS *Nautilus*, launched
1956	Calder Hall, England, nuclear reactor starts operating commercially
1959	Francis Bacon develops a fuel cell that can produce 5kw of power
1961	Apollo space missions begin using the fuel cell

Chapter 1: Essential energy

Harnessing the power of steam
Thomas Savery and Thomas Newcomen

The world's first practical steam engine did not drive any machinery; it pumped water from coal and tin mines. Driving ever deeper in search of deposits, miners reached the water table, a mostly constant level of water permeating low-lying rocks. Keeping mines adequately dry was a great labour, cutting profits severely. A mine might need 500 horses to drive pumps. The inventor of the steam-driven pump, Englishman Thomas Savery, called his invention The Miners' Friend. His pamphlet *An Engine to Raise Water by Fire* appeared in 1702 but he had been running one successfully since 1698.

It worked but not very well. The height to which the water could be raised depended on the pressure of the steam. That was limited by the strength of soldered joints in the boiler and cylinder. A 20-metre vertical lift was about the limit. More than that needed several pumps in series.

Savery's model was overtaken in 1712 by a design from his business partner Thomas Newcomen. This separated the pump from the engine. The cylinder, open at the top, now contained a piston. Steam pressure pushed the piston up. When the steam was condensed, unbalanced air pressure and gravity forced it down again. The movements of the piston rocked a beam to drive the pump.

Newcomen's engine was soon driving pumps all over Britain and Europe. At the time of his death, a hundred Newcomen engines were in service. Users had to tolerate their slowness and wastefulness for more than 50 years—less than 1 per cent of the energy in the coal used to raise the steam did any useful work.

Savery's steam pump: how did it work?

Three pipes fitted with one-way valves were connected to a large metal cylinder. The cylinder was filled with steam through one of these pipes connected to a boiler. A spray of cold water condensed the steam, creating a partial vacuum inside the cylinder. This would cause outside air pressure—which was greater than that inside the cylinder—to push water up from the mine through the second pipe and into the cylinder, much like water sucked up through a straw. A steam blast drove the water out again, upwards through the third pipe, leaving the cylinder full of steam. The process was then repeated.

Above: An early steam engine, designed for pumping water, designed by English blacksmith and inventor, Thomas Newcomen.

More energy from steam
James Watt and Matthew Boulton

Many will say James Watt invented the steam engine, but steam-enabled 'atmospheric engines' had been pumping water from mines for many years before the Scots-born Watt first took an interest. These worked, but were horribly inefficient.

Asked to repair a model Newcomen engine in 1769, Watt saw a major reason for the waste. The cylinder and piston had to be cooled during each stroke to condense the steam and, when they were heated again, they absorbed some of the heat of the next steam blast. Watt's innovation was to build a separate condenser so the cylinder could be kept hot. Science had played a role here; studies by Watt's countryman Joseph Black measured the energy needed to heat the cylinder each time, energy that essentially went to waste.

Watt had some other very useful ideas, including a 'governor', which kept the engine running at a steady pace, and 'sun and planet' gearwheels, which let the engine drive rotating machinery, like the large-scale spinning and weaving equipment coming into use. Watt also had a business partner, Birmingham iron-maker Matthew Boulton. By the 1780s, the pair enjoyed a good living manufacturing the engines that drove the early Industrial Revolution.

Watt's engines still could not pull their own weight continuously, so there was no chance of a reliable self-moving steam engine, a 'locomotive', at that time. However, progress demanded the use of higher-pressure steam to drive the piston, without any help from air pressure. This meant higher-temperature working as well, extracting more energy from the steam. Over the following years, efficiency climbed steeply; steam engines became lighter and more compact.

There were hazards too—early boilers could not withstand the increased steam pressure and often burst with catastrophic results. One precaution was the safety valve, invented a century before by Frenchman Denis Papin for his 'steam digester' (which has become the 'pressure cooker'); this released steam if the pressure rose too high. Acceptable safety came only with much experience, better design, materials and manufacture and, ultimately, government regulation.

Another name to note in relation to steam was Englishman Richard Trevithick; his high-pressure 'Cornish engines' were successful and popular in factories and mines. An additional refinement was the introduction of 'double acting' engines, in which steam pushed alternately on both sides of the piston. This paved the way for the first steam-powered vehicles and boats.

Electricity makes magnetism
Alessandro Volta, Hans Oersted and Michael Faraday

Through the late 18th century, the courts and drawing rooms of Europe were entertained by tricks played with 'electrical machines', generating through friction what we now call static electricity. Participants could give each other electric shocks while kissing, set fire to flammable liquids with sparks, and watch paper and feathers fly through the air.

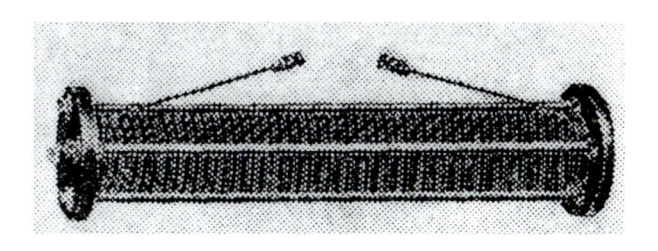

Above: In this early 'battery', dozens of interleaved discs of metal and salt-soaked paper interact chemically to generate a strong current between the ends of the wires.

It was all good fun, and the early scientists gained some insight into the nature of the phenomenon, but there seemed little use for such charges, since they could not be sustained. One spark, one shock and they were gone. A continuous flow of electric charge, an 'electric current', might be useful but no-one know how to produce one.

A device to produce such a current was discovered in a curious way. The Italian anatomist Luigi Galvani found that if he put two pieces of different metals, such as copper and iron, in contact with animal body fluids it could cause sudden movement in apparently lifeless tissues. Galvani put it down to 'animal electricity', some sort of 'life force', but a few years later his countryman Alessandro Volta hit on the real explanation of the workings of an electrical 'cell'—it was probably called this as it resembled the biological cells of plants and animals.

Volta discovered that if you connected a number of electric cells, you could increase the current. This became known as a 'voltaic pile' or what we would today call a battery— the analogy was to an artillery battery, a number of guns side-by-side firing at once. To make matters confusing, we would now usually define a battery as a single cell, but let that pass.

Volta's batteries were soon the rage of scientific Europe, and researchers rushed about to find new things to do with the current. The inventor was summoned to demonstrate the device before Napoleon, who had a good feel for the importance of such discoveries.

In 1820 Danish physicist Hans Oersted did a simple experiment and made a striking discovery. He brought a compass needle (which is a small magnet) near a wire carrying an electric current from a battery. The needle moved, meaning that the wire was now acting like a magnet. Electricity and magnetism, both long studied but thought unrelated, were closely linked.

Pondering Oersted's discovery, English physicist Michael Faraday, at London's Royal Institution, noted that the shift in the compass needle when the current flowed indicated that the newly created magnetic field ran around the wire in a circle. Could the effect be used to push a current carrying wire round and round a magnet, as a sort of 'electric motor'? He made one in 1821. It was not very practical, needing the wire to drag through a bowl of mercury, but it worked. Unfortunately Faraday published this work without due acknowledgement to colleagues. The resulting fuss caused him to bypass work on electricity and magnetism for ten years.

By the time Faraday came back to the field in 1831, various combinations of magnets and coils of current-carrying wire were being investigated, and the modern electric motor slowly evolved. These would ultimately find an immense array of uses, driving machinery in factories, appliances in the home, power tools in the workshop, electric trains, trams and even cars. Similar principles would soon be used to generate electricity supplies, hastening the coming of the 'electric age'.

Crude oil challenges coal
Benjamin Silliman and Edwin Drake

In the mid 19th century, coal was king, burnt in vast amounts to make steam to drive factories and locomotives. Coke (a sort of distilled coal) was used to release iron from its ores in blast furnaces. Coal gas lit homes, shops, factories and streets. Coal tar was beginning to yield its multitude of chemicals to make dyes, drugs and synthetic materials.

However, a challenger was emerging, in the form of petroleum, the flammable 'oil from rock', collected for millennia as it seeped from the ground. Researchers had begun to wonder what petroleum really was. American geologist and opponent of slavery Benjamin Silliman was the first to supply an answer in 1854. Encouraged by New York lawyer George Bissell, who suspected petroleum might contain something useful for lighting, he 'distilled' petroleum from a seep in Pennsylvania. Heating it first gently, then more strongly, he drove off a succession of vapours with ever higher boiling points. When cooled, these vapours condensed to form liquids.

Petroleum proved to be a mixture of such liquids, from those that were light and quick to evaporate, through to some so dense and tarry they were almost solid. The lighter fractions had no immediate use, but a slightly heavier one, dubbed 'paraffin' or 'kerosene', was a good substitute for the expensive whale and seal oil used in lamps. It was also a solvent for greasy substances that would not dissolve in water. Some of

The electric cell: how does it work?

To make a simple 'wet' electric cell you need strips of two different metals, say copper and iron, sitting in water with dissolved salt or acid. A chemical reaction between the iron and the solution allows tiny particles of electricity (electrons) to accumulate on the iron, which slowly dissolves. These particles will flow to the copper if the two metals are connected by a wire. The flowing electrons form an electric current.

Chapter 1: Essential energy

Above: Stokers in a 19th century factory feed the furnaces with broken coal, so supplying heat to steam-generating boilers. The French called them 'heat-men' or 'chauffeurs'.

the denser liquids were good lubricants, able to keep machinery running smoothly and iron free from rust. Collectively the products made by distilling crude oil were worth having. Holes were soon being drilled into the ground in search of deposits.

The first in North America, pioneered by Edwin Drake—and dubbed 'Drake's Folly' by sceptics—hit oil only 20 metres down in Pennsylvania. Soon dozens of drill rigs and pumps pocked the landscape in promising areas, pipelines and special ships were built to transport the 'crude oil', 'refineries' were established to process it into useful products.

The age of oil made a quiet beginning. In time a vital use would be found for the mix of more volatile

Collectively the products made by distilling crude oil were worth having. Holes were soon being drilled into the ground in search of deposits.

fractions, which were usually thrown away or remained as a contaminant in kerosene, causing lamps to explode and houses to burn. Renamed 'gasoline' in America and 'petrol' elsewhere, this would set the world on wheels in the century to come.

Still more power from steam
Charles Parsons

The turbine (literally 'something that rotates') has a long history and many contributors, although some of these merely dreamed and designed, rather than built. In 1500 Leonardo da Vinci sketched one powered by air rising up a chimney above a fire. Fan-like blades, spinning in the airstream, turned meat roasting on a spit. A century later another Italian, Giovanni Branca, drew plans for a boiler directing steam onto a turbine like a horizontal waterwheel. This was to turn machinery, specifically a 'stamping mill' to crush rocks. We are not sure if either of these were ever more than an idea.

Real progress took a long time. The key issue was efficiency, getting the most value from the burning coal. To absorb an acceptable fraction of the energy in the steam, the turbine had to spin very fast, many thousands of times a minute. Until the late 19th century, no metal had been found that could withstand both those stresses and the heat of the high-pressure steam. Many good minds were onto the problem, including ingenious Dutchman Carl de Laval, inventor of the cream separator.

Irish engineer Charles Parsons was the son of the Earl of Rosse, builder of a great telescope known as the *Leviathan of Parsonstown*. In 1884 he designed a turbine with a series of angled blades one behind the other on one shaft. Steam bounced from one blade to another as it passed through, delivering energy at each blow, the blades getting larger as the steam pressure fell. Turning 300 times a second, the first such turbine generated 10 horsepower. It was a brilliant concept, making the turbine efficient enough to compete with, and later surpass, steam engines with pistons.

Parson's early interest was in ships. A demonstration craft, the *Turbina*, zipped through the water in 1894 at an unheard-of 60 kilometres an hour. Over coming decades, steam turbines went into many naval vessels and commercial liners, including famous Cunarders like the *Lusitania*, replacing conventional steam engines until themselves replaced later by diesel engines and gas turbines.

Parsons initially planned the turbine to make electricity and designed a compatible high-speed generator for the task. Today virtually all coal-fired power stations use steam turbines descended from Parson's first models, though improvements in design mean they can turn rather more slowly.

A man and his engine
Rudolf Diesel

Rudolf Diesel remains well known today through the widely used engine that bears his name and its distinctive fuel. However, he orginally trained as a refrigeration engineer in Munich, where his exam results were the highest ever achieved. He had a mixture of motives in seeking new sources of power. Steam engines, though much improved by James Watt and others, remained unacceptably inefficient, turning perhaps 15 per cent of the energy from coal into useful work. They were also too big and costly for small factories and workshops. More efficient engines would be smaller and might even be able to power vehicles.

Etienne Lenoir and Nicolaus Otto had already created 'internal combustion engines', in which the fuel was burnt inside the cylinder rather than outside, as in a steam engine. Seeking greater efficiency in a 'rational heat engine', Diesel proposed injecting fuel into the cylinder and igniting it, not with a spark or flame, but by the high temperature generated as the piston compressed air in the cylinder. By working through a much wider range of temperatures and pressures, his engines would exploit more of the energy of the burning fuel. He took fourteen years to turn these ideas—much disputed at the time—into a patent.

Diesel's first engine, set running in 1893, was a failure. The intended fuel was coal dust, piled up in vast heaps around the German Ruhr as waste; funding came from industry giants like steel magnate Baron Von Krupp. The coal dust-fuelled engine exploded during trials, nearly killing Diesel.

A second model, also operational in 1893, used heavy fuel oil, less refined and therefore cheaper than gasoline. With his 1895 model he achieved unprecedented fuel efficiency, opening up many applications. Diesel became famous. Within a few decades, diesel engines were powering factories, mills,

Below: Holiday-makers swim close to a petrochemical site in Fos-Sur-Mer, France. The use of coal and oil as useful energy sources has transformed the natural environment.

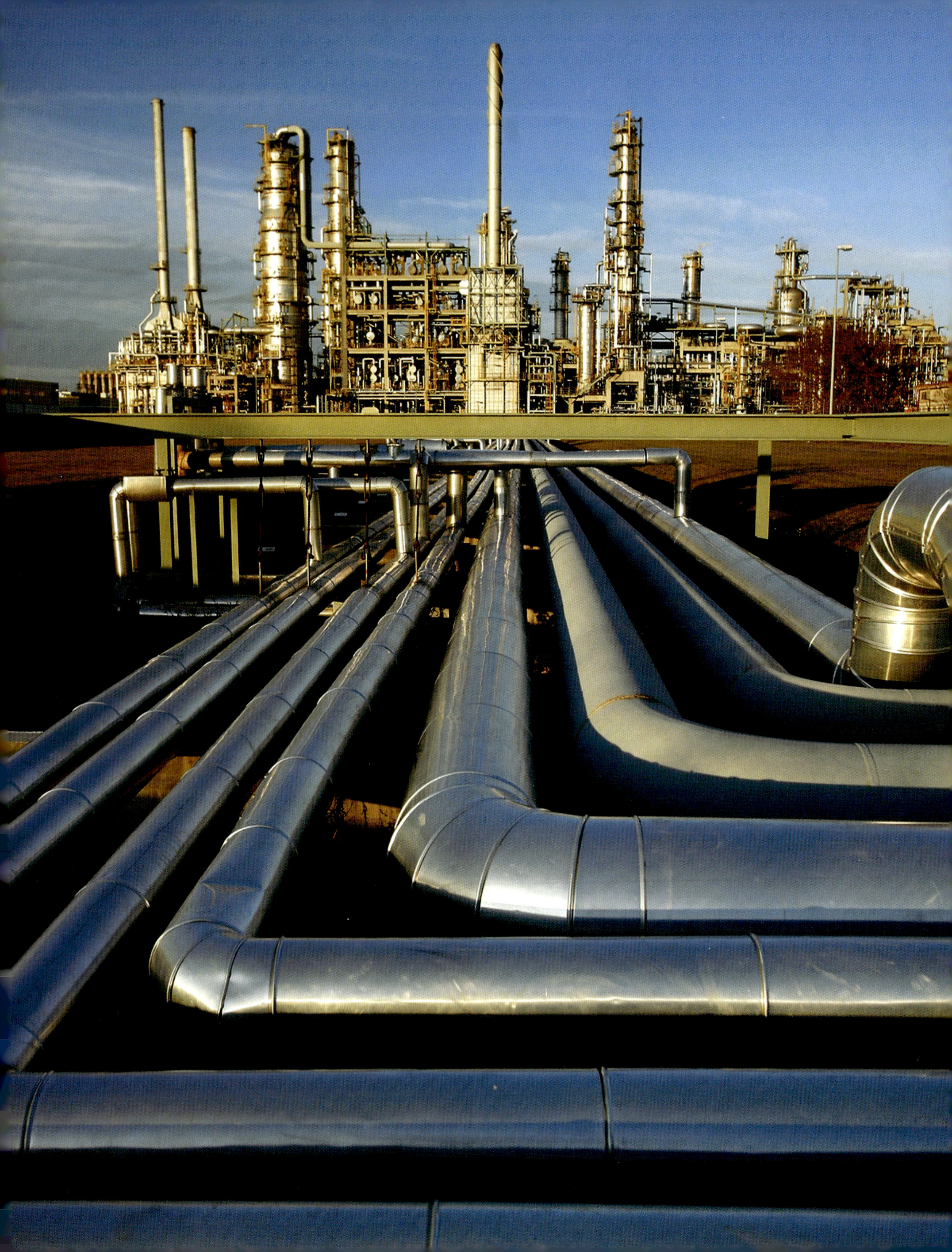

water pumps, automobiles, trucks, boats, trains and international shipping.

Diesel engines have traditionally produced smoky exhaust fumes, posing risks to health and the environment, and poor acceleration, making them less popular for automobiles than petrol-driven engines. But more refined technology, especially in fuel injection, and better quality fuels have largely solved these problems, letting the inherent benefits of Rudolph Diesel's creation shine through.

The energy withing the atom
Enrico Fermi and the Manhattan Project

In terms of political and military consequences, no scientific discovery in history outranks nuclear fission. 'Atomic' or 'nuclear' energy was not invented; the potential had always been there. What we invented were ways to harness atomic energy, releasing it on demand, gradually for peaceful purposes or violently for military ones.

The key discoveries were made through 1938 and 1939, on the very eve of global conflict, by scientists in half a dozen countries, including Germany. Nuclear fission brought the prospect of nuclear power, but there was uncertainty. The energy release might or might not be explosive, able to power a bomb.

Scientists in Britain uncovered the crucial factor in 1940. Uranium has two isotopes (forms): the rare U235 and the abundant U238. Research showed that neutrons split only U235. To make an atomic bomb, the uranium must be enriched (processed) to greatly increase the proportion of U235.

Opposite: A network of pipes delivers petroleum into a refinery and carries away the host of useful products distilled from the 'crude'. Those products, such as petrol, diesel, kerosene and lubricating oils, keep modern civilisation running.

Stranger than fiction

Diesel should have made a fortune from his inventions, but had no head for business. He disappeared from a Channel steamer in 1915, most likely falling overboard, although some suspect he was driven to suicide by financial worries.

There was another way. Years before Italian physicists, led by Enrico Fermi, sought to make artificially elements that were heavier than uranium, called 'transuranics', by bombarding uranium with neutrons. Fermi's experiments continued in the United States after war broke out in Europe, with the help of many Jewish scientists who, like Fermi himself, had fled persecution. Just as fission was being discovered, researchers learnt that one of the newly found transuranics, the ominously sounding plutonium, made by adding neutrons to U238, would also split when hit with neutrons, perhaps even more violently than uranium.

The military implications were clear. If enough plutonium could be created, it too could be made into a weapon. (Indeed, the bomb that destroyed Nagasaki was powered by plutonium.) There was another benefit. Since the plutonium was manufactured from the 'useless' U238, which made up 99 per cent of natural uranium, the amount of potentially available fissile material rose 100-fold.

All that was discovery. Now came the invention, on a scale and with a sense of urgency never seen before. The vast Manhattan Project, charged with creating war-winning weapons from uranium and plutonium, involved top-secret laboratories in a dozen places across the United States. It ultimately employed hundreds of thousands of people, including many of the brightest minds of the day, solving an immense array of technical problems on the way to making the 'gadget', so-called for the sake of security.

In Chicago, a Fermi-led team built the first nuclear 'pile' on a university squash court. The name recalled the first electric piles made by Alessandro Volta. The reactor (our term today) was also a literal pile; pieces of uranium distributed among blocks of graphite, which acted as the 'moderator' to slow the neutrons down so they would be more easily absorbed by U238, creating plutonium. The gradual build-up of the reaction was controlled by metal rods, which absorbed neutrons and were slowly withdrawn.

On 2 December 1942, the pile went 'critical'; initiating the first man-made chain reaction. The way was now open to make plutonium, a few milligrams at a time. There could be enough for a bomb, given enough time and enough reactors.

The other route to an atomic bomb, enriching the uranium mixture in U235, was being pursued elsewhere, with various methods being trialled for this difficult and delicate task. A lot of new physics and chemistry was learned in a hurry, but gathering the fissile material was excruciatingly slow. By July 1945 there was still only enough available for three bombs. One would use uranium, two would use plutonium; one of those was needed for a test. With the war in Europe over—the whole project had been driven mostly by the fear of Hitler having the bomb—Japan was the target. The Hiroshima bomb was dropped on 6 August 1945, the Nagasaki bomb three days later. The Pacific War ended soon after, although its outcome was already settled.

Nuclear fission: how does it work?

When the cores or 'nuclei' of uranium atoms are bombarded with tiny, uncharged particles called neutrons, some of them split in two, releasing energy, a few neutrons and smaller nuclei, which are strongly radioactive. The freshly released neutrons can cause other atoms to split, starting a self-sustaining chain reaction, with ongoing release of energy. This splitting is known as fission. Compared with a conventional explosive like dynamite), the energy liberated is immense.

Left: To energise modern society, electricity is carried in immense volumes by networks of power lines. By greatly increasing the voltage (essentially the 'electrical pressure') of the supply—using transformers—electricity can be sent over long distances with little loss.

The vast Manhattan Project... employed hundreds of thousands of people... solving an immense array of technical problems on the way to making the 'gadget'.

In the decades ahead, the fission bomb would give way to the fusion bomb and be married to the rocket to make the fearsome Intercontinental Ballistic Missile and Fermi's first reactor would be followed by hundreds more.

Electricity from the atom
Hyman Rickover and Walter Zinn

When the first nuclear reactor went 'critical' in Chicago during World War II, no-one was really interested in the energy it released. The objective was to 'breed' the newly-discovered plutonium for use in nuclear weapons. The plutonium was created from U238, the more abundant of the two uranium isotopes and not itself a useful source of power.

At war's end, priorities shifted a little. Many reactors were still built purely to produce plutonium, as the Cold War 'arms race' got underway, but others were designed to make use of the heat generated by carefully controlled nuclear reactions, say to boil water, make steam and generate electricity. The nuclear reactor was intended to replace a furnace burning oil or coal.

The earliest interest was from the military; US Admiral Hyman Rickover spearheaded the development of compact nuclear power plants for submarines, with the electricity driving motors connected to the propellers. Submarines no longer had to surface to charge their batteries, so became less vulnerable to detection and attack, and could sail long distances without refuelling. The first operational nuclear powered submarine carried a famous name; the USS *Nautilus*, launched in 1955, demonstrated its prowess by sailing under the icecap at the North Pole in 1958.

The world's navies still operate several hundred nuclear submarines, though many were scrapped with the ending of the Cold War. During that period they played a vital role in the balance of terror, remaining submerged out of sight for weeks or months, ready to launch missiles with nuclear warheads. Only a few types of surface ships proved suitable for nuclear reactors, very large vessels such as aircraft carriers and specialised vessels like ice-breakers, which require lots of power and need to stay at sea without refuelling for long periods.

From the early 1950s, engineers began scaling-up reactors built on similar principles so they could make large amounts of electricity for sale to homes, factories and offices. A reactor in the Soviet Union started supplying electricity for 2000 homes in 1954, but the first big—for the day—commercial reactor opened for business at Calder Hall, Cumbria, in Britain in 1956. This would supply electricity to the grid for

Stranger than fiction

Safety measures have improved since the first nuclear reactor. Had anything gone wrong with the pile, a man stood by with an axe ready to cut the rope supporting the control rods. They would have dropped back into the reactor and halted the reaction. Shutting down a reactor is still called scramming. SCRAM is reputed to stand for Safety Control Rod Axe Man.

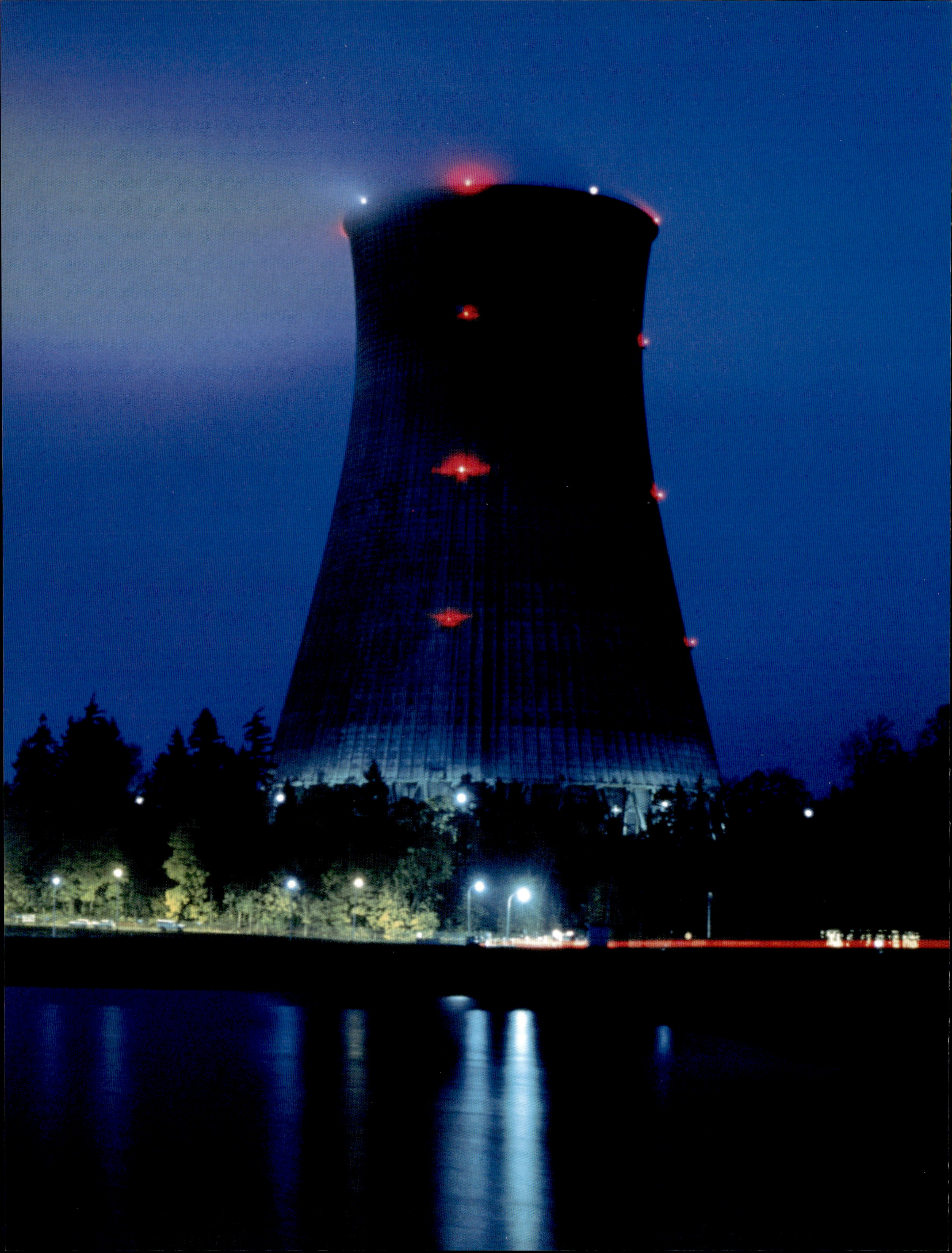

Chapter 1: Essential energy

Some early proponents had believed that nuclear electricity would prove 'too cheap to meter'.

47 years until it closed in 2003. Each of the four reactors at Calder Hall could power a town of 40,000 people; many of the reactors to follow would be ten times as large.

The United States opened its first power reactor at Shippingport, Pennsylvania, in 1957. Over following decades many hundreds of power reactors were built in many countries, with France and Japan in particular coming to rely on electricity from reactors because of a shortage of coal and gas. In recent years safety and cost issues, with public alarm at accidents at Three Mile Island, Chernobyl and elsewhere, have eroded both enthusiasm and construction rates. Hopes had been high to start with.

Some early proponents had believed that nuclear electricity would prove 'too cheap to meter', a goal always far from reality.

The link between nuclear electricity and plutonium was always close. American physicist Walter Zinn was the first—with his colleagues—to produce electricity from atomic energy. Zinn had been part of the Fermi-led team in Chicago in 1942. Indeed, it was he who withdrew the first control rod to set the reactor running.

He was also a key figure in the early design of reactors for nuclear submarines. In 1951 his Experimental Breeder Reactor One (EBR-1) was turned on in the deserts of Idaho, producing enough electricity to light four 200-watt lamps and later the whole building in which it sat. That was not much power of course; indeed the main purpose of EBR-1 was to prove that by converting U238 into plutonium while extracting energy from fissile U235, a breeder reactor could create more fuel than it consumed.

The reactor certainly did that, and for many years there were confident hopes that breeder reactors would solve the world's energy problems. By making plutonium from the non-fissile U235, they would in

Right: An employee monitors the level of radioactivity at the Chernobyl nuclear plant which blew up on 26 April 1986, spewing a radioactive cloud that contaminated much of Europe. The destroyed reactor is entombed in an immense concrete and steel structure known as the 'Envelope' or 'Sarcophagus'.

Opposite: The immense concrete towers that dominate many nuclear power plants do not generate any energy. They simply allow the water from the turbines to cool and are also a feature of many coal-fired power stations.

What's that?

Albert Einstein's famous equation ($E = mc^2$) sums up a profound and powerful idea, one thought revolutionary when Einstein proposed it in 1905. Matter and energy are interchangeable. Matter can be transformed into energy and vice versa. m stands for mass (in kilograms), E is energy in joules and c^2 is the speed of light multiplied by itself. Since c is very large (300 million metres per second) converting a small of amount of matter can release an immense amount of energy.

effect increase the energy available in natural uranium 100-fold. But high costs and technical delays have seen that promise dwindle away and no breeder reactor today provides commercial electricity, though research continues. There was always another issue: unease at the prospect of a 'plutonium economy', an energy generating system that depended on producing and exploiting large amounts of plutonium. Plutonium is not only highly toxic, with lethal doses measured in millions of a gram; it can be subverted into nuclear weapons. The risk always seemed too high.

Harnessing what drives the sun
Mark Oliphant, Edward Teller, Stanley Ulam and Andrei Sakharov

For centuries we wondered what keeps our sun shining. In airless space it cannot actually be on fire; if it releases heat simply by shrinking under gravity, it would have gone out long ago. We now know that the source of its vast and enduring energy is nuclear fusion. Atoms of the lightest and most abundant element hydrogen blend together to form atoms of the second lightest element, helium. Mass is lost in the transformation, re-emerging as energy, as Einstein allowed for in his famous equation 'E equals MC squared' ($E = mc^2$).

Hydrogen comes in several isotopes (forms): ordinary hydrogen, deuterium and tritium, which is radioactive. While the sun runs on the common ordinary hydrogen, researchers in their laboratories have had more success in making deuterium fuse. In the 1930s Australian physicist Mark Oliphant, working at the Cavendish laboratory, Cambridge, figured out the key features of the reaction. He found that deuterium atoms—or strictly speaking their nuclei—could be fused in two ways, one making tritium, which he discovered, the other a previously unknown isotope of helium. He tried to extract energy from this reaction, by firing one beam of deuterium nuclei into another, but the process took much more energy than it created, so he let it drop. Anyway, nuclear physics at that time was not supposed to be useful.

The first success in getting energy on command from hydrogen fusion was much more violent, the explosion of the first 'hydrogen bomb' by the United States in 1952. In the context of the increasingly chilly Cold War, the hunt was on for ever-more powerful weapons of mass destruction, much bigger than the atom bombs detonated in World War II. Leading the effort to create such a 'super bomb' was the Hungarian-born American physicist Edward Teller. With his colleague Stanley Ulam, he invented a way to make such a bomb work, a configuration still called the Teller-Ulam design. For a long time, this was a closely guarded secret, though nations like Britain, France, China and Russia were able to work it out for themselves.

Of the thousands of 'thermonuclear' warheads the world clings to today, nearly all use just deuterium, turned from a gas into a manageable solid by chemically reacting it with the metal lithium. Fusion will take place freely only at extremely high temperatures, many millions of degrees such as are found in the core of the sun. To set a hydrogen bomb going, it is

Chapter 1: Essential energy

Above: This immense machine is a turbine for harnessing the energy of the wind. The turning blades drive a generator (under the raised lid), producing electricity whenever the wind blows, but releasing almost no pollution in the process.

necessary first to explode a fission bomb, using plutonium or uranium, to create the tremendous heat necessary and to compress the deuterium.

These weapons have unparalleled destructive power, released in a split second. A World War II atomic bomb released as much energy as 10,000–20,000 tonnes of TNT; hydrogen bombs liberated hundreds of times more power. Within a few years hydrogen bombs were small enough to be carried by aircraft and missiles, preparing the ground for the 40-year balance-of-power standoff between the United States, the Soviet Union and their allies.

While the development of nuclear fusion for military ends proceeded briskly through the 1950s, Oliphant's 1930s efforts at making controllable energy from the same reaction were followed up much more slowly. There was not the same urgency fuelled by national security and the technical challenge was enormous. How could enough deuterium be heated to the necessary millions of degrees and held in some sort of container, able to compress it until the fusion reactions were set going? Technical engineering on this scale had never been faced before. Two basic approaches were tried. In 'inertial confinement' pellets of deuterium would be bombarded by laser beams of enormous power, so quickly compressed that the fusion would get under way before the pellets blew themselves apart. 'Magnetic confinement' used very

Progress remains relatively slow even today, despite the fuel cell's advantages. It can use any sort of fuel and does not waste energy.

powerful magnetic fields shaped like cages or bottles to confine, squeeze and heat the gas, now electrically charged because of its temperature.

Doughnut-shaped machines following the second path were called 'tokamaks' by the Russians, who set the early pace through the 1960s under the leadership of Andrei Sakharov, who was also the godfather of the Soviet H-bomb. Those remain the front-runners, as researchers approach the much-sought 'break-even' point, at which more energy is generated from fusion than the machine requires to heat and control the deuterium. With plenty of cheap fuel—though relatively rare, deuterium is abundant in the sea—added to the environmental benefits—fusion does not generally produce radioactive wastes or greenhouse gases—fusion is appealing for large-scale central power plants. But affordable electricity from fusion is still at least several decades away.

The promise of the fuel cell
William Groves and Francis Bacon

Many people know how close the *Apollo 13* came to disaster. The spacecraft relied on 'fuel cells' for its power, combining hydrogen and oxygen to make electricity—as well as drinking water. An explosion in one of the oxygen tanks almost ended the mission.

Their use in late 20th century spacecraft might suggest that fuel cells are a new idea, but the concept is as old as electricity. From 1800 we knew that passing an electric current through water breaks it up into hydrogen and oxygen. How then to make hydrogen and oxygen combine chemically and generate electricity? Around 1840 Englishman William Groves tried it, with sealed gas bottles and platinum electrodes. Current flowed, water formed, but very little and very slowly. Fifty years later, two German chemists tried combining air and coal gas. Success was again limited. The new internal combustion engine), which could burn coal gas, plus the realisation of how difficult the problem was, put further research on the backburner.

Interest revived in the 1930s. British chemist Francis Bacon, descended from his more famous 17th century namesake, used cheaper materials, such as nickel instead of platinum, and certainly made current flow. But he took until 1959 to make a hydrogen-oxygen cell produce useful power, for example 5 kilowatts to run a welding machine. The cells needed gases at very high pressure and quite high temperature, and efficiency remained low; much of the gas did not combine.

Other inventors and manufacturers were now tinkering, but it took a decision by the American space agency NASA to use Bacon-style fuel cell technology in its spacecraft to drive the technology forward. Some predicted a rapid uptake of fuel cells, but progress remains relatively slow even today, despite the fuel cell's advantages. It can use any sort of fuel and does not waste energy. At best, a coal-fired power station converts 50 per cent of the energy in the burning coal into electricity. In a fuel cell, it should be almost 100 per cent.

Work continues on many different designs, now encouraged by rising oil prices. Some schools and hospitals have large stationary fuel cells in place, and major automobile manufacturers have plans. As technical problems are solved and costs come down, fuel cells will be big players in the drive for clean, affordable power.

Running on the sun?
August Mouchout and Russell Ohr

Nearly all the energy we use comes ultimately from the sun. Burning coal, oil and natural gas liberates energy trapped long ago in plants by photosynthesis. The sun's heat evaporates water to fall later as rain and drive hydroelectric systems. Only over the last century or so have we invented ways to capture solar energy more directly.

Sunlight is absorbed by what it falls on, such as the ground beneath our feet, and liberates heat. The quest to harness that to do useful work is at least a century old, with French academic August Mouchout one of the pioneers. In the 1860s, concerned his country was relying on coal that would run out one day, Mouchout built a steam engine that ran on sunlight. Mirrors tracking the movement of the sun concentrated its energy onto a boiler that turned water into steam. It worked, but not well enough to convince bureaucrats it could replace coal, especially once coal became cheaper. Mouchout went back to his teaching.

Today solar-powered steam engines are a rarity, but many buildings collect the sun's energy in roof-top devices that heat water to around 60 degrees Celsius for space heating and hot water supply. Such use is doubling worldwide every five years.

A more versatile approach is generating electricity, which can provide heat, light and mechanical power. A key discovery early in the 19th century was the 'photoelectric effect'. When sunlight struck certain newly-discovered elements known as 'semiconductors', such as selenium, an electric current was generated. The process was not very efficient. In the 1880s only 1 per cent of the energy of sunlight striking a 'photocell' ended up as electrical energy. At this stage the process was mostly a laboratory curiosity, though it was used in 'light meters' to measure how strong light was.

Interest revived in the 1950s, following the invention of the transistor, which also used semiconductors like silicon and germanium. By 1954 Russell Ohr from the Bell Telephone Laboratories had invented solar cells that converted 4 per cent of sunlight into electricity, and patents to prove it. Better technology since then has pushed the conversion rate to 25 per cent in the laboratory and 15 per cent in commercially-available devices. The initially high costs are coming down, and global production of photovoltaics rose eightfold between 2000 and 2005. Into the 21st century, solar cells are increasingly attractive as a source of renewable, non-polluting energy.

Below: Ranks of 'solar cells', made from glass thinly coated with silicon, turn sunlight into electrical current collected by embedded wires, though only when the sun shines.

CHAPTER 2:
From farm to food and fibre

Working the land to produce crops for food and fibres has long been a human preoccupation. The first agriculture, perhaps 10,000 years ago, was a stimulus to the growth of the other arts and technologies that make up modern civilisation. The Industrial Revolution with its great demand for labour was only possible because farming had become much more productive, freeing farm workers to migrate to the industrial cities. That added productivity was in turn the consequence of mechanisation.

In the 20th century and up to today, technology has continued to make its impact, particularly through better ways to stop food 'going off'.

Seeds in the ground
Jethro Tull

Around the turn of the 18th century, crops like wheat, barley or oats were usually planted by hand. Seeds were broadcast wastefully, flung out onto ploughed land, much as had been done since biblical times. The seeds did not spread evenly, too few here, too many there. They were not automatically covered by soil, so many would fail to germinate; others would be eaten by birds.

The man usually credited with finding a better way, Englishman Jethro Tull, trained as a lawyer, but could not afford to start a practice. So he took to managing the family farm in fertile Berkshire—he named it Prosperity Farm, perhaps more in anticipation than actuality. An early invention of his, or perhaps reinvention, around 1701, was the 'seed drill', a machine on wheels pulled behind a horse. This planted three parallel rows of seeds in furrows, at regular intervals and the right depth, covering the seeds as it went. According to some reports he made his first working model using parts from an old pipe organ.

This was obviously a good idea, reducing seed wastage and increasing productivity—up to eight times, Tull claimed. Since the growing crops were in neat rows, it was much easier to keep the ground between them free of weeds, using a horse-drawn hoe, another

Below: An early example of farm mechanisation: a small petrol engine drives a threshing machine, separating the grain from the straw. The grain is bagged and the straw discarded. The thresher would most likely still have needed a horse to pull it into position.

Farmers then were a conservative lot, and Tull's ideas seemed radical.

of his inventions. But the seed drill was slow to take on, especially in England, though it was more popular in France and the American colonies. Farmers then were a conservative lot, and Tull's ideas seemed radical.

Some of his ideas were wrong. He did not like using animal manure to keep fields fertile, since it often contained the seeds of weeds, and he recommended repeatedly breaking up the soil into tiny particles to 'release the nutrients'. This drawn-out and ineffective process often delayed planting and made late growing crops vulnerable to diseases. His advocacy of such ideas may have made others slow to take up his good ones, like the seed drill, his improvements to the plough, and the use of horses instead of oxen.

Tull was poor, ignored and even ridiculed most of his life, but his inventions, popularised by his writings, won through in the long run. The increased capacity of the land to produce food made possible the growth of the big industrial towns.

Above: An early 'combine harvester', self powered and designed to cut the cereal stalks, thresh off the grains and bundle the straw, which is dropped off behind. With such machines. farmers could now cover very large acreages.

Threshing the grain
Andrew Meikle

Andrew Meikle's gravestone in Preston Kirk, near Edinburgh, where he died at the age of 92, tells us he was descended from 'a long line of ingenious mechanics'. Certainly his father had invented a 'winnowing machine' to separate the grains of, say, wheat, from their stems, though the common hostility to machines at this time—around 1710—meant that it had not been well received.

Andrew did rather better. His 'threshing machine'— or 'thrashing machine'—did more than separate the wheat grains from the straw; it actively shook the ears of wheat and liberated the grains, a task done by hand until then. His was not the first such machine, but previous attempts had had limited success, apparently— or so Meikle thought—because they tried to rub the grain from the ear, as a human hand would do. His machine beat the ears of wheat against a drum to set the grains free, much as the fibres of flax (the basis of linen) were broken from their plant stems.

Meikle build his first model around 1778, patented it in 1788 and starting manufacture soon after. At that time—and indeed for most of his life—he was employed as a millwright at Heston Mill, where wind power drove the grinding of wheat into flour. The owner of the mill was John Rennie. He supported Meikle, putting the new technology to work in mills other than his own. It is said that Meikle's inspiration contributed to Rennie later becoming a leading civil engineer.

Mechanical threshing had come to stay, adding to the boost in agricultural productivity from the use of machines. The early machines were powered by the wind, but they could also be driven by water, by the newfangled steam engines or, if all else failed, by horses.

As a millwright, Meikle made other improvements, most noticeably replacing the traditional canvas sails of the windmill with a series of shutters that could be opened by levers. The mill could now be closed down quickly in the event of damaging strong winds.

Meikle's inventions did not make him rich or even comfortable. Late in life his lack of means caused the former President of the Board of Agriculture, realising what this pioneering 'agricultural engineer' had done, to raise £1500 to support him.

Cotton seeds and the Civil War
Eli Whitney

Eli Whitney's 'cotton gin' changed history. It transformed productivity and profits on the cotton plantations in the American South. Escalating production necessitated the continued—and growing—use of slaves to work the fields. The American Civil War followed almost inevitably.

To allow spinning and weaving, cotton growers had to separate the fibres from the small, sticky cottonseeds. These could be pulled out by hand, but a worker would struggle to clean half a kilo of cotton in a day. Existing machines to do the job, such as the East Indian *charka* which pulled the cotton between rollers, did not work with the short-fibre cotton grown in the American South away from the coast.

Eli Whitney had been fascinated by machinery since childhood. A law graduate from Yale and in need of money to pay off debts, he became a tutor on the Georgia plantation of Catherine Greene. Seeing the need for a machine to clean cotton, he quickly devised his 'gin' (short for 'engine', meaning a mechanical device). Cotton fibres, snagged on a set of rotating spikes or hooks, were pulled through narrow slots, leaving the seeds behind.

It was simple and very effective. Cleaning rates rose 50-fold, costs fell dramatically, cotton production boomed, doubling every decade. So did the demand for slave labour. By the mid-19th century, American cotton would make up 75 per cent of the world supply, and more than half the nation's exports.

Whitney patented the gin in 1794. According to legend, the concept had come from seeing a cat claw at a chicken through the slatted walls of its coop, grabbing nothing but feathers. The original idea may have come from Whitney's employer. She certainly gave him financial and moral support. He needed it; Whitney and his business partners struggled to protect their patent, which was routinely infringed, and to make any money from the innovation.

Stranger than fiction

In one way the cotton gin hurt the South. Cotton growing became so profitable that industry and cities saw little investment. Nearly all population and manufacturing growth went to the northern states, leaving the South badly under-resourced when the Civil War came and perhaps ensuring its defeat. Ironically, Whitney would make more money from his methods for the mass production of guns, such as those used in the war, than he ever did from the cotton gin.

Above: A symbol of 'industrial scale' farming, this huge header drops the threshed grain in a steady stream into a truck alongside, leaving the stubble standing. Such machines cut costs, but require enormous investments and encourage the growth of huge farms owned by international 'agri-businesses'.

Bringing in the harvest
Cyrus McCormick, John Ridley and Hugh McKay

In 1831, 90 per cent of Americans lived 'on the land'. Today the United States meets its needs for food and fibre through the labours of only 2 per cent of the population. That transformation is due to mechanisation.

One machine, conceived in 1831, ultimately increased tenfold the productivity of farmers growing grains like wheat, rye or barley. The slowness of harvesting such crops had long limited how much a farmer planted. Using a model left incomplete by his father, Virginian Cyrus McCormick took only six weeks to design, build and test a 'reaping machine'. By replacing men on foot using scythes and rakes with a horse-drawn scythe that piled the cut grain onto platforms, McCormick enabled one farmhand to do the work of five.

Local farmers were wary of innovation; some

McCormick's easy credit, talent for publicity, ready availability of spare parts and guaranteed productivity—'15 acres a day!'—won them over.

thought the new iron ploughs 'poisoned the soil'. A decade passed before McCormick's easy credit, talent for publicity, ready availability of spare parts and guaranteed productivity—'15 acres a day'—won them over. He sold two machines in 1840, seven in 1842, and in 1844 he sold 50 and had to build a factory in Chicago to meet the demand. In 1851 his machine won a Gold Medal at the Great London Exhibition.

McCormick continued to innovate. His later models cut, bundled and bound the sheaves of grain, ready for transport to the thresher. They needed just one man to drive the horse, or later the tractor. Despite inevitable court challenges to his patents, competition from other manufacturers and the loss of the factory by fire, he won out in the end, founding the International Harvester Corporation.

McCormick's machine cut the crop with the heads of grain intact; threshing took place elsewhere. Another harvesting strategy 'stripped' or 'headed' the crop, removing the ears of grain while leaving the stalks standing. Australian innovators John Ridley and Hugh McKay made major contributions here. Ridley's 'stripper', devised in 1843 to counter a labour shortage during a good harvest, held the heads in combs and beat off the grain. He sold tens of thousands of these machines. Forty years later McKay's 'combine harvester' stripped the heads, then threshed, winnowed and bagged the grain. The Sunshine Harvester factory outside Melbourne became the largest of its kind in the southern hemisphere.

The 'Devil's Rope'
Joseph Glidden

Anyone could have invented 'barbed wire'; putting spikes on smooth fencing wire was an obvious thing to do. Many designs were tried during the late 19th century, when demand was growing, especially out on in the vast American prairies, where, according to some accounts, fencing cost as much as livestock. However, fame and fortune went to the man who made the idea work, devising machinery to make the stuff by the kilometre.

At an a Illinois county fair in 1873, 60-year-old rancher Joseph Glidden saw a new form of fencing, a wooden rail with protruding metal spikes hanging behind a smooth wire fence. Glidden thought he could do better by fixing the spikes to the wire. His first lengths of hand-made barbed wire surrounded his wife's vegetable patch, but he already had thoughts on mechanised manufacture. His friends Jacob Haish and Isaac Elswood, having seen the same demonstration, had similar plans. Glidden secured a patent in 1874, and then spent several years in court battling Haish, Elswood and others over ownership of the intellectual property. He emerged victorious as the 'father of barbed wire'.

Stranger than fiction

Combine harvesters today are usually self-propelled and may be guided by global positioning systems (see page 226) in large fields, but their basic operating principles remain unchanged.

The success story

Glidden, a teacher from New Hampshire before turning to ranching, made a fortune from royalties on his invention, dying in 1906 aged 93 as one of the richest men in America. Sales boomed, as barbed wire remade the West. Elswood was making his own pile from manufacturing the wire; output from his factories went from 1000 tonnes in 1876 to 25,000 tonnes in 1879.

Success was not inevitable. 'Free-rangers' were against any sort of land enclosure, and religious groups called barbed wire 'the Devil's Rope' because it could injure animals encountering it for the first time. Texas ranchers, a large potential market, doubted the flimsy-looking fence would contain their powerful long-horn cattle. A public demonstration in 1881 changed many minds. Wild cattle were driven into a barbed wire enclosure. Agitated by the onlookers, the cattle repeatedly charged the fences, but the wire held and the animals suffered little harm. Soon barbed wire fences were criss-crossing the West and lined the rapidly-spreading railroad tracks to stop farm animals straying onto the line.

Barbed wire had a more macabre application a few decades later, erected along hundreds of kilometres of trench lines in the Great War. Many soldiers on both sides met their deaths tangled in the 'Devils Rope'.

Making food last longer
Nicholas Appert and Peter Durand

Over the millennia, many ways have been found to help to store food for long periods—drying, smoking, salting, pickling, impregnating with sugar, the use of spices or chemicals, and chilling in ice or snow. All of these methods work for the same reason. They prevent, or at least slow, the activity of micro-organisms (yeasts, fungi, bacteria), which, as the French scientist Louis Pasteur would later show, cause the chemical changes we call 'going off' or 'going bad'.

By the early 19th century, it was well known that if a meat broth, for example, was first boiled and then sealed from contact with the air, it would 'keep' and be safe to eat for several weeks. Again it was Louis Pasteur who showed why; the boiling killed the micro-organisms, and isolation of the food prevented further contamination.

A major step forward in food preservation, and indeed in convenience, came early in the 19th century with the first sealing of cooked foods into metal containers. French pastrycook Nicholas Appert had already pioneered the idea of storing food in glass jars with cork stoppers in 1807, securing a prize offered by Napoleon, who was concerned about providing safe food to his troops in far-flung battlefields. London merchant Peter Durand took the idea further around 1810, using cans of iron coated with a thin layer of tin to minimise rusting—hence the name 'tin cans' or simply 'tins'.

Canned food reached the London shops around 1830, but for decades it was used mostly by the military and explorers. Indeed, inadequate preparation of canned foods by an unscrupulous supplier helped precipitate the disaster that overtook the Franklin expedition in search of the Northwest Passage in the 1840s.

Getting the food out of the can and into the pan was a separate challenge. Early tins were made of thick, heavy iron, weighing half a kilogram even when empty. You needed a hammer and chisel to open them, which limited their appeal to the householder. By the 1860s tins of lighter, thinner steel with a rim around the top made possible the can opener, beginning with the bullhead type around 1885, with a lever-like body and a point to pierce the can. Much

Chapter 2: From farm to food and fibre

ingenuity has been expended on alternative designs in the century or more since.

The world's favourite snack?
George Crum and Herman Lay

When it comes to popular snacks, the potato chip or 'crisp' must be up there with the leaders. Potatoes have been a staple in many European countries since they were brought from the New World centuries ago, and are served in an immense range of styles. Thomas Jefferson, later US President, came to enjoy the French style while on duty as ambassador there in the late 18th century. He brought the recipe home and served the thick slices of fried potato—different to today's hot chips or 'French fries'—to his guests, many of whom had previously thought potatoes only suitable for feeding to pigs.

In 1853 the part-Native American George Crum was chef at the Sun Moon restaurant at the Saratoga

Below: On an early food production line, workers fill steel cans with fresh beans. Once the lids had been soldered on, the cans were heated to cook the contents. The customer then merely had to open the can, reheat and eat.

Above: The world's favourite snack and some avid young consumers. Potato chips are symbolic of processed food, dependent on modern technology, including new packaging materials to keep the food fresh.

Springs resort in upstate New York. He had hot chips on the menu, but one diner, reputedly the millionaire banker and New York social icon Cornelius Vanderbilt, did not care for them. Too thick, he declared, and sent them back. Crum prepared a second, thinner serving, but the guest was still not satisfied. Irate, Crum cut them so finely that they went crisp when fried, too thin and hard to be speared with a fork. He expected the diners to be angry, but they were delighted, proclaiming the browned, paper-thin titbits delicious and demanding more.

Crum was onto a winner. He made his invention a specialty of the house, calling them 'potato crunches', and was soon packaging them for sale as Saratoga Chips. He ultimately set up his own restaurant nearby; Vanderbilt was one of his backers.

Some distance remained to be covered before a New England dinnertime delicacy could be counted a global gastronomic phenomenon. Three key events all occurred in the 1920s: the invention of machines to peel and slice the potatoes, previously done laboriously be hand, the first use of waxed paper bags to keep the crisps crisp—plastic film not yet being available—and the intervention of Herman Lay. A travelling salesman from the north, he peddled potato chips to storekeepers throughout the American South from the boot of his car, building a business and linking his own name indelibly with the salty snacks, especially once he merged his company with Frito, a Dallas-based firm that made corn chips. Frito-Lay is still the largest maker of potato chips in the United States.

He expected the diners to be angry, but they were delighted, proclaiming the browned, paper-thin titbits delicious and demanding more.

The genesis of pasteurisation
Louis Pasteur

Louis Pasteur carries one of the greatest names and reputations in the history of science. Through the late 19th century, the French biologist won the battle on behalf of his 'germ theory' against its rivals. He maintained and conclusively proved that plant and animal diseases are caused by microscopic living things, which became known as 'micro-organisms' or 'bacteria'. The same is true of the fermentation of wine and beer, the souring of milk, the decay and corruption of all kinds of once-living materials and even the digestion of foods in the intestines of animals. All these are the consequences of chemical changes induced by micro-organisms as they release the energy they need to live.

This new understanding, which included the revelation that many micro-organisms do not need oxygen to live, had profound impacts on technology. It transformed medicine by generating the new 'aseptic' surgery of Joseph Lister, which saved many lives by reducing deadly infection in surgical wounds. It led Pasteur himself to develop 'vaccination' against many diseases. In industry it enabled the makers of products based on fermentation, such as alcohol, to enhance quality by ensuring that only desirable microbes were present, since each micro-organism catalyses only one particular reaction.

Pasteur's name is also recalled when we refer to pasteurisation, widely used for a century to make foods last longer and to stop fruit juices and dairy products going sour. Its origins lay in Pasteur's discovery that the responsible bacteria are sensitive to heat and certain chemicals, and so can be killed or incapacitated. In 1864 Pasteur was asked by Napoleon III to investigate diseases threatening ruin to the wine industry. He found that heating the wine to 55 degrees Celsius for a few minutes disabled the bacteria and prevented the disease. Similar treatments to beer and milk followed soon after and 'pasteurisation' entered the language.

Generally, the higher temperature used, the less heating time is needed. Pasteurisation of milk now takes place continuously rather than in batches. Milk is judged pasteurised after being heated to 62 degrees Celsius for 20 seconds, but it is not completely sterile and still requires refrigeration.

UHT: what's that?

Ultrahigh temperature (UHT) pasteurisation, used on fruit juices and 'long-life' milk which require no refrigeration or preservatives, involves heating the liquid to over the boiling point, but under pressure so that boiling does not actually occur. This can affect flavour, and all pasteurisation reduces the nutritional quality of foods.

Separating the cream from the milk
Gustav de Laval

Oil and water don't mix. Farmers have used that fact for centuries to separate oily cream from watery milk. The cream is also 'lighter' (less dense) than the milk, floating to the top if given enough time, skimmed off the milk with a spoon or a special shallow dish—hence the term 'skim milk'. It was slow and not very efficient, a lot of cream remaining mixed with the milk.

Swedish inventor Gustav de Laval heard that a German brewer had spun a barrel of milk to make the cream separate faster. When the German showed no interest in collaboration, Laval experimented alone. In his first machine, a number of buckets were chained to a vertical shaft. Spinning the shaft made the buckets swung out horizontally. The heavier milk particles were thrown outwards and ended up at the bottom of the bucket, leaving the cream behind, just as happened under gravity, but much faster. This 'centrifuge' (to use a later term) produced a form of 'artificial gravity'.

By 1878 Laval had a machine that separated continuously. Whole milk went in at 150 litres an hour; cream ('butter fat') and skim milk emerged from separate spouts. Dairy industry productivity rose, costs fell, the machines became very popular. Modern cream separators use the same principles but with greater efficiency. The separation of the high-value buttermilk from the less valuable skim milk is all but total.

We find centrifuges (the name means 'flying from the centre') everywhere nowadays. Your spin drier is one, producing about 2000 times the force of gravity. Industry uses similar machines to separate solids from liquids, such as impurities from oils. High-speed

The success story

Though never rich, Laval was a prolific and much acknowledged inventor, making vital improvements to the steam turbine (page 22), building a factory to make the new incandescent light globes, experimenting with milking machines and showing great interest in early aviation.

Left: Not all technology outwits nature. Pasteurised and homogenised milk in glass bottles, sealed with aluminium lids, was a common sight before the advent of cardboard packaging and plastic bottles. The birds now have to find their treats elsewhere.

centrifuges can separate the isotopes of uranium for power generation and weapons manufacture, purely on the basis of their slight mass difference. In medicine, centrifuges separate blood cells from plasma, the liquid in which they are carried. 'Ultracentrifuges', which spin glass containers thousands of times a second with jets of air to produce a million times normal gravity, separate tiny particles of different proteins on the basis of their mass, a crucial capability in modern biotechnology.

Frozen fast for flavour
Clarence Birdseye

The English philosopher Francis Bacon reportedly died in 1626 from a chill caught while stuffing a chicken with snow as an experiment in food preservation. But not until the 1920s did anyone devise a commercially-viable food freezing process that retained food flavour and texture. His surname is still seen in many supermarket freezer cabinets: Clarence Birdseye.

Brooklyn-born Birdseye always loved the outdoors. He hunted and fished, paying his way through college by catching rats for sale to a laboratory, spending five years in Labrador trapping foxes for fur. There he observed the traditional Eskimo method for preserving fish. Thrown onto the ice, fish froze almost instantly in the frigid air. Birdseye noted that the freezing took place so quickly that only very small ice crystals formed, tissues were less damaged and, when thawed, the fish tasted much better than if frozen more slowly. Living in a remote location, Birdseye froze vegetables like cabbage to give his family fresh food.

Back home, he applied what he had seen in Labrador to large-scale food preservation. In 1923, using a fan, tubs of brine and cakes of ice—total cost $7—he experimented in his kitchen, quickly freezing rabbit meat and fish fillets and packing them into waxed boxes ready for defrosting and cooking without further preparation. His innovation combined

Above: Snap-frozen broccoli, ready for warming and eating, typical of the many foods now available: essentially fresh though preserved. Thanks to the inventiveness of Clarence Birdseye our diets now have much more variety.

convenience with flavour. Turning that process into a viable business took time and even made Birdseye broke at one point. In March 1930 a range of frozen meat, fish, vegetables and fruits went on sale in 18 stores in Springfield, Massachusetts. Sales were slow initially, prices high and shop displays unattractive.

By the mid-1930s the idea began to catch on. Birdseye leased freezer cabinets to retailers who could not afford to buy them during the Depression. By 1944 he was leasing insulated railroad cars to send frozen food around the country. The amount of space in the new-fangled home refrigerators devoted to frozen foods grew as householders bought more. Frozen food is a major industry today, relying on blasts of extremely cold air for quick freezing (which is often referred to as 'snap' freezing).

Birdseye's inventiveness went beyond frozen foods, securing him 300 patents. He was proudest of an improved process to convert sugar cane waste into paper pulp in twelve minutes instead of nine hours. He attributed his success not to outstanding intelligence or sales skill but to intense curiosity combined with a willingness to take risks.

CHAPTER 3:
Iron and steam

The Industrial Revolution, which transformed Britain in the early 19th century and the rest of the planet in the decades following, was built on two interlocking technologies, iron and steam. Neither was new; iron-making was nearly 3000 years old, and the ancient Greeks had toyed with the motive force of steam. The new world arose from their interaction. Better iron-making allowed the force of steam to be more effectively harnessed; steam power drove much of the iron-making machinery. In time iron gave way to steel, and primitive 'atmospheric' steam engines to better ones, including high-pressure engines and turbines.

Harnessing the power in the air
Evangelista Torricelli and Otto von Guericke

Practical people took a long time to realise that air has weight and to make use of the fact. Popular wisdom, following Aristotle, said that air was intrinsically light and always wanted to rise. He had also argued that a 'vacuum' (a space empty of air or anything else) was impossible. Not so.

Italian Evangelista Torricelli, secretary and assistant to the ageing genius Galileo, made the key discoveries in 1643. In a classic experiment, he filled a glass tube, closed at one end, with mercury and inverted it into a basin of mercury. Some of the mercury ran out, but not all. The remaining column of mercury, about 750 millimetres tall, was apparently held in place by the weight of the air pushing down on the mercury in the basin. That same force would support a column of water up to 10 metres high, which answered an old puzzle faced by the builders of fountains and waterworks. Why could water not be sucked up to a height of more than 10 metres?

Furthermore, there was now a space between the top of the mercury and the top of the tube, which clearly contained no air. Here then was Aristotle's 'impossible' vacuum.

Air pressure is massive, equivalent to one kilogram weight on every square centimetre of any surface. After reading about Torricelli's work, the German engineer Otto von Guericke, later mayor of Magdeburg, staged a striking demonstration in front of the German Emperor in 1854. He had devised a pump to pull air from an enclosed space, similar to a pump used to raise water but made as airtight as possible. He removed the air

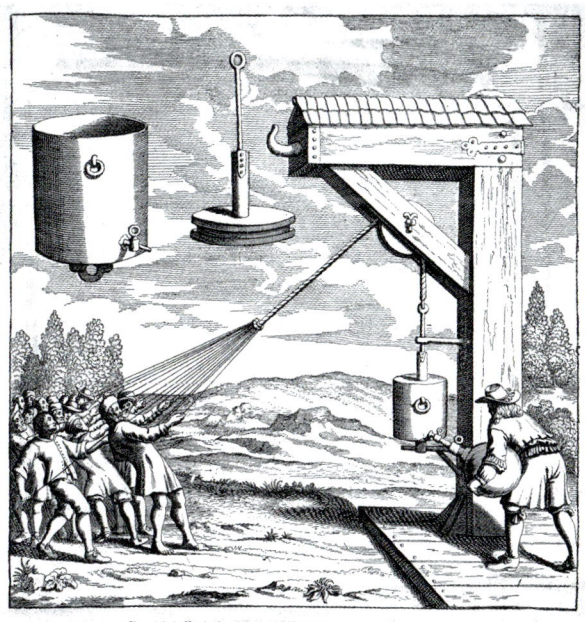

Above: A depiction of the German scientist Otto von Guericke of Magdeburg experimenting with the nature of vacuum using a hollow cylinder and piston.

from the inside a sphere of metal, made in two pieces. With a partial vacuum inside, the imbalance with the air pressure outside was so great that two teams of eight horses could not pull the 'Magdeburg hemispheres' apart.

Von Guericke wondered if imbalanced air pressure could do useful work. In 1672 he removed most of the air from one end of a cylinder containing a tightly fitting piston. The combined strength of twenty men could not stop the external air pressure from pushing the piston into the partial vacuum. Interesting, but such an 'atmospheric engine' was not yet practical. How could he make the vacuum without an air pump and much labour? There was a way.

The imbalance with the air pressure outside was so great that two teams of eight horses could not pull the 'Magdeburg hemispheres' apart.

Forgotten pioneer of steam
Denis Papin

The name of Denis Papin is little remembered nowadays, but this inventive 17th century Frenchman has at least one legacy still active, the 'pressure cooker' or, as he called it, the 'steam digester'. This was—and still is—a pot with a tightly fitting lid, which allowed internal steam pressure to build up when the water-filled pot was set boiling on the stove.

Papin had originally trained as a doctor but turned to science and invention early on. In Paris he helped the eminent Dutchman Christiaan Huygens, builder of the first pendulum clock, with investigations into air pressure, but also with unsuccessful attempts to harness the force of gunpowder, the only explosive known at the time. Hot, rapidly expanding gases from the burning powder were supposed to drive a piston in a cylinder, but the iron of the day could barely withstand the explosion, and the experiments went nowhere.

Papin was one of the many godfathers of the steam engine, if not actually a parent. Like any number of researchers at the time, he dreamt of using the force of steam, perhaps combined with air pressure, to drive a piston back and forth in a cylinder and so do some useful work, particularly pumping water out of mines or raising water to supply fountains.

In his design, a vertical cylinder closed at one end had a tightly-fitting piston with water trapped below it. The water was heated by an external fire, boiling into steam which then drove up the piston. The exterior of the cylinder was cooled with water, the steam condensed to leave a partial vacuum, and unbalanced air pressure—helped by gravity—drove the piston down again. The cycle was then repeated. The movements of the piston could drive a pump.

So the tube was combined boiler, working cylinder and condenser, neither very practical nor efficient. We don't know if Papin ever built such a machine, though

The pressure cooker: how does it work?

Water boils at a higher temperature as the pressure rises, and the combination of higher temperature and steam pressure was very effective in cooking tough meat and even softening bones. If the pressure rose too high, a valve on top opened to let out the excess steam, the first 'safety valve' as later used on steam engines.

Above: Smoke rises from the furnaces of Coalbrookdale, an early iron works in the still-rural English Midlands, built near supplies of iron ore and coal. Pollution from a vast multitude of such smokestacks would later make this region known as the Black Country.

he did work on the idea for years. We know of it mostly from a long series of letters he wrote to the great German mathematician Gottfried Leibniz, with whom he had worked.

The masters of iron
Abraham Darby and John Wilkinson

In 1781 the first bridge made entirely of iron opened for traffic across the River Severn in western England, close to Coalbrookdale. The arched bridge, 15 metres long and 14 metres high, had taken three years to assemble. Nothing like it had ever been built before.

Nearby, in the midst of a rural landscape, stood the state-of-the-art iron works run by Abraham Darby, a third generation ironmaster, with a father and grandfather also called Abraham. Here all the iron in the bridge had been cast and shaped. Darby's partner in this enterprise was an ironmaster of equal repute, John Wilkinson, whose name lives on in Wilkinson's Sword razor blades. Together they

Stranger than fiction

In Papin's letters to Leibniz, he also wrote of a small carriage he had constructed, driven by the force of steam. If this was actually ever built, it would have preceded Nicolas Cugnot's 'steam dray' by half a century.

symbolised iron, the dominant material of the looming industrial age.

In 1709 Darby's grandfather had pioneered the use of coke, obtained from coal, in place of wood-derived charcoal for the release of molten iron from its ores, in furnaces powered by bellows. Suitable wood had become expensive, with the wholesale felling of trees to meet the demand for iron weapons, necessitated by the regular convulsion of Europe by war. By the end of the century, ironmasters would be making quality iron in large amounts using coke, and more of it was for peaceful pursuits—building, transport and manufacturing.

The iron in the bridge and other applications at the time was cast iron—the raw output of the furnaces, run into moulds of sand to cool and harden. It was strong enough for many uses but brittle. Reheating and hammering could strengthen it, but no one had dared to make a substantial structure from the stuff, for fear of catastrophic collapse. A few decades later, the tougher 'wrought iron' was developed and iron was everywhere.

This second Iron Age was also the Age of Steam. Wilkinson played a key role in this with his 'boring machine', which cut the cylinders of steam engines from a block of iron. Making these adequately round and smooth internally was beyond previously established techniques, such as casting, then chiselling and filing. Wilkinson's machine was very robust, holding the iron block firmly in a cradle of oak beams. Water power drove the boring bar round and round. So well did the machine perform that Wilkinson's friends, the steam engine makers James Watt and Matthew Boulton, relied on his borer for 25 years.

The founder of the 'iron aristocracy'
Henry Cort

As an example of an industry pioneer who enriched others and the nation as a whole, yet received almost nothing in return, it is hard to go past Henry Cort. Son of a Lancashire brick-maker, he set up business in London aged 25 as a buying agent for the British Royal Navy, sourcing, among other things, increasing amounts of iron for weapons, anchors and chains. Cort became aware of how poor British iron was compared with imports from nations like Russia and Sweden—knowing how much Britain needed its iron at this time, Russia had had no hesitation in raising the price two and threefold.

After ten years of working with the Navy, and having accumulated a little capital, Cort left London to set up his own iron-making business on the shores of Portsmouth harbour, close to the naval base. Here he began to experiment with new techniques to make stronger, tougher iron, able to compete with imports.

He first developed a process to roll out iron bars from the crude 'pig iron' which came from the furnace,

Coke: what's that?

The fuel coke is created by baking coal in an airless oven in very high temperatures—up to 1000 degrees Celsius. This drives off the more volatile constituents found in the natural product—namely water, coal tar and coal gas—leaving the combination of carbon and ash that comprises coke. With its high carbon content, coke burns smokelessly and more efficiently than coal.

Opposite: The great 19th-century engineer, Isembard Brunel, whose creations were made possible by iron and steam. Here he is pictured against the iron work and anchor chains of his colossal steamship, the *Great Eastern*.

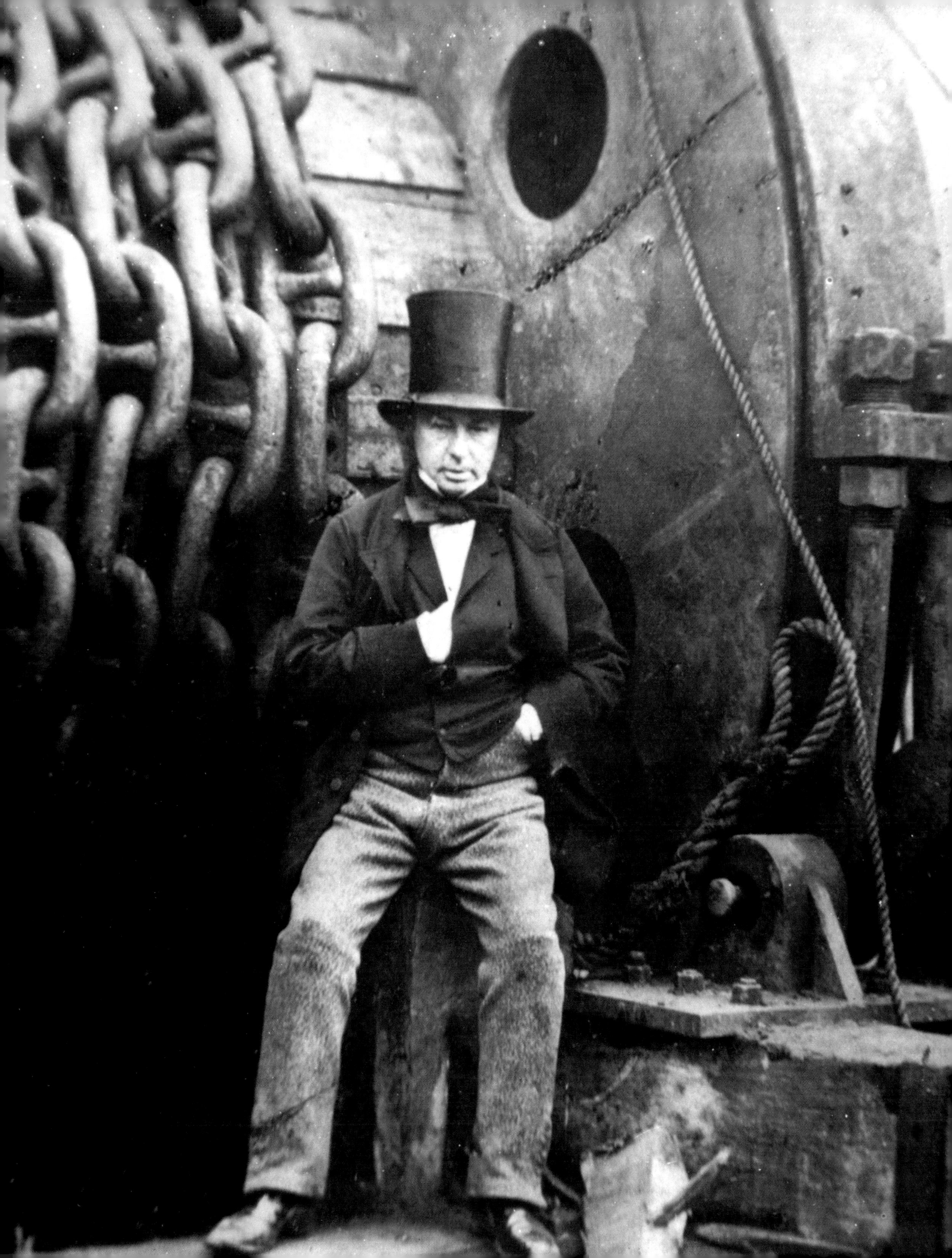

Above: A painting by Charles Graham printed in *Harper's Weekly* around 1890, depicting workers processing pig iron in a Birmingham, Alabama factory.

rather than hammer them, which was much slower and more labour intensive. His process, using grooved rollers powered by the new steam engine, squeezed out many impurities and spread traces of carbon left from the smelting process evenly through the iron, increasing its strength and reducing its brittleness.

Cort's second innovation converted the crude cast iron into 'wrought iron', much more useful and durable, by a process called 'puddling'. These innovations were not Cort's alone; he built on the ideas of others, including the Grange brothers, John Roebuck and Peter Onions, but his own contributions were vital and their outcomes of immense importance. In 1786 Lord Sheffield proclaimed that the new processes meant much more to the nation than the possession of the American colonies—then in the process of being lost—as they would give Britain command of the iron trade 'with its vast advantages to navigation'. The following year tests on iron produced by Cort's methods found it superior to imported iron and imports promptly ceased.

Some figures tell the story. By 1820, 8000 of Cort's wrought iron furnaces were running in Britain, producing 400,000 tonnes annually, as opposed to a mere 90,000 tonnes in 1780. By 1900 production totalled 4 million tonnes, more than all other European nations combined. The 19th-century preeminence of British industry—'the workshop of the world'—was secured, and many ironmasters made fortunes.

Opposite: Wrought iron was not only very useful; it could be beautiful, as in these elaborately-decorated iron gates, still in use a century or more after they were crafted.

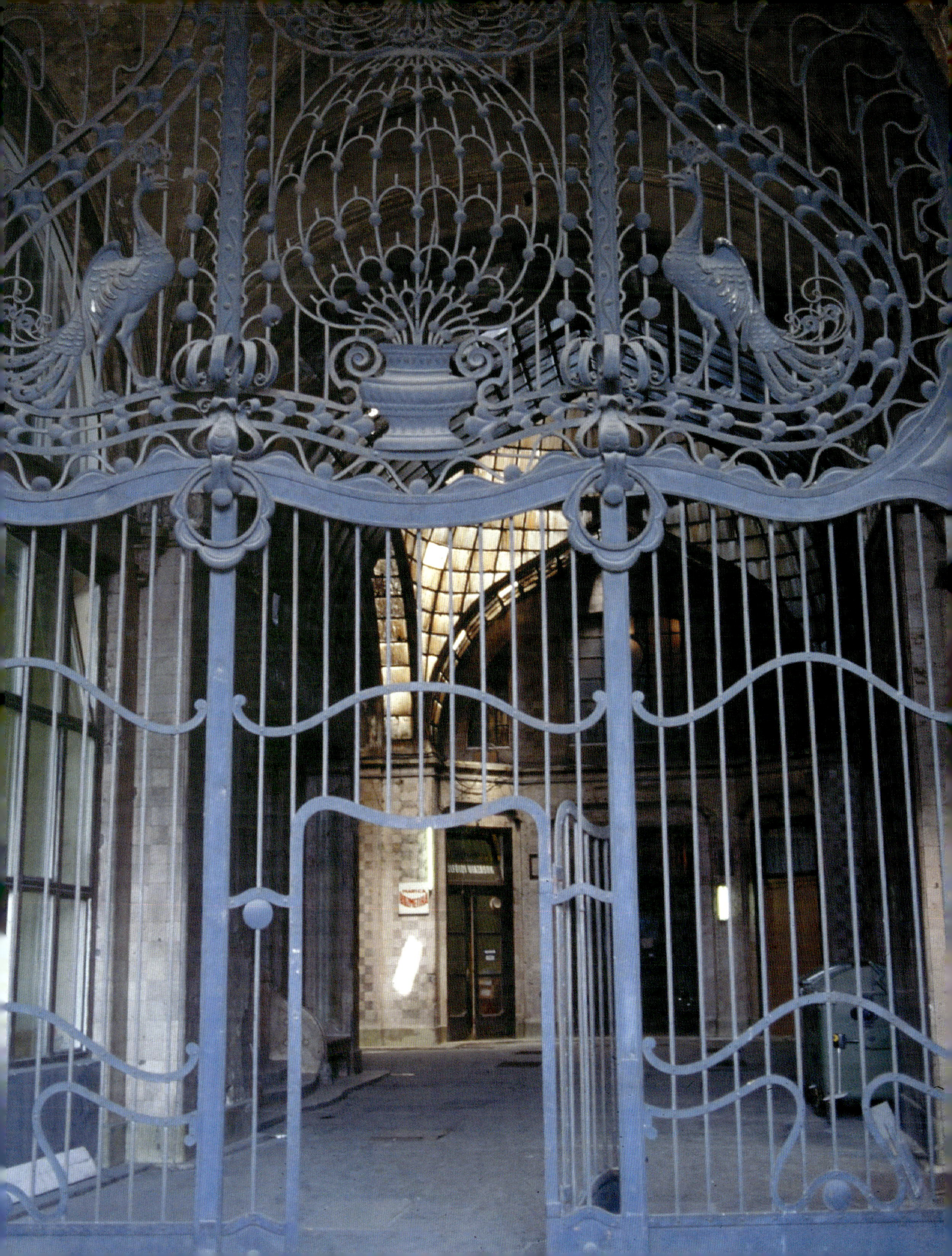

Typical of these—and one of the early movers—was Richard Crawshay, an enterprising Yorkshireman, who had ridden to London at the age of fifteen to seek his fortune—the journey took a fortnight over the appalling roads of the time. He found employment in an ironmonger's shop, and specialised in the selling of 'flat irons' to washerwomen. He inherited the business when the owner died and soon expanded his interests into iron-making, taking a lease over deposits of iron ore and coal, and a small ironworks, in south Wales. Profits from the store in London were ploughed into the new venture.

In 1787 Crawshay struggled to make 10 tonnes of iron bar a week, and of poor quality at that. Needing to build new furnaces, he decided to adopt Cort's methods, and with such success that a quarter of a century later he could report making 200 tonnes a week, and had built a canal to get his produce to port in Cardiff.

Cort himself did not prosper. Early on he had entered into a partnership with Samuel Jellicoe, whose father Adam was Deputy Paymaster for the Royal Navy. The elder Jellicoe financed the iron-making experiments, holding Cort's patents on his two processes as guarantee for the debt.

Adam Jellicoe was in debt himself when he died in 1789. To settle the estate, Cort's debt was called in; he could not pay. His patents, now of vital importance to the Navy, were taken over by the Government, in part because some of the funds Jellicoe had advanced to Cort were moneys entrusted to him to pay the wages of officers and seamen. As no royalties had been paid to him by the many ironmasters who had made use of his inventions—partly because they disputed the patents—Cort was ruined.

Forced to leave his business in the hands of the younger Jellicoe, Cort was ultimately reduced to seeking a government pension to support his wife and twelve children. Later to be described as 'the author of the iron aristocracy', he died in 1800, broken in health and fortune.

Puddling: how does it work?

A vat of molten iron, heated from above, was constantly stirred by a worker with a long iron bar. Many of the carbon impurities were burned off and the pure iron was formed into a ball on the end of the bar, ready to be hammered into a 'shingle' and then rolled.

Steam drives the hammer
James Nasmyth

The great contribution made by Scots-born James Nasmyth to heavy engineering was the 'steam hammer'. It was able to shape very large objects from one piece of hot wrought iron, such as ships' anchors or, specifically, the one-metre diameter drive shaft for the proposed paddle-wheels on Brunel's *Great Britain*—though the shaft was never needed due to changes in the ship's design. Existing 'tilt hammers', essentially a hammer on a long arm pivoted at the lower end, could not be raised high enough to exert sufficient force to forge such big pieces, which therefore had to be made in small sections and welded together, time-consuming and costly.

Nasmyth's insight was to use steam pressure to raise a massive iron block, which could weigh a tonne or more, much higher than was previously possible. Steam also boosted the fall of the hammer under gravity to deliver an immense blow to the iron being banged into shape—and to make a thunderous noise. He also built a pile driver using the same idea.

Opposite: Iron and steam combine in this massive steam hammer, here seen shaping a large axle, perhaps for a steamship. The iron is being pushed through by a dozen or more workers. Even the industrial age needed human muscle.

Chapter 3: Iron and steam

The steam hammer, devised in 1837, transformed heavy industry: costs were halved, quality improved greatly, previously impossible projects became feasible. Many different products could be formed under the hammer, from iron plates for the sides of warships to parts of steam engines to large cannon.

Nasmyth was tardy in patenting his steam hammer; he did so only after seeing a replica in operation in a rival's factory in France. Yet he made a good living and was able to retire at 48, devoting himself to astronomy and other interests 'I now have enough of this world's goods,' he said. 'Let the young have their chance.'

> **The steam hammer transformed heavy industry. Costs were halved, quality improved greatly, previously impossible projects became feasible.**

CHAPTER 4:
Materials matter

To build something useful we need stuff to make it from. Traditional materials are derived from nature, such as stone or timber, usually with some human intervention, baking clay to make bricks, for example, or smelting iron from its ores. Glass is another ancient material, made from molten sand and other minerals, but greatly enhanced by invention over the centuries. The development of some materials relied heavily on innovation elsewhere. Advances in chemistry gave us useable rubber and we could not have aluminium until we could make electricity in large amounts.

Chapter 4: Materials matter

What happened when?

1575	Artificial (soft paste) porcelain created in Italy to rival Chinese imports
1676	George Ravenscroft develops flint glass (lead glass)
1709	Johann Pottger creates hard paste, Dresden china
1750	Bone china developed
1759	Eddystone Lighthouse, built using hydraulic cement, finished
1823	Charles Macintosh patents waterproof material
1824	Joseph Aspidin patents Portland cement
1830	Elasticised female underwear reaches the market
1841	Charles Goodyear produces vulcanised rubber
1856	Henry Bessemer describes his steel-making process to an impressed British Association
1867	Joseph Monier patents reinforced concrete
1886	Charles Hall and Paul Herault independently develop a means to produce aluminium
1892	Monier uses his complete building system based on reinforced concrete to build an apartment block
1909	Duralumin invented accidentally
1912	Harry Brearley invents stainless steel
1915	Pyrex (borosilicate glass) introduced
1931	Glass bricks and fibre-glass come into use
1952	Alistair Pilkington develops a method of producing large sheets of glass cheaply
1957	Glass ceramics appear on the market

The secret of porcelain
Johann Bottger

In pottery, we can go far beyond technology and functionality into the realms of artistic creation; elaborate forms, brilliant decorative colours and patterns, gilding and other refinements combine to generate works of art for everyday use. In 18th-century Europe, the pottery most sought after for cups, plates and other crockery was white, semi-translucent and durable but delicate in appearance. It was called 'china' from its place of origin or 'porcelain', since it resembled a certain type of seashell called porcellana.

Chinese artisans had perfected the process over many hundreds of years, using a unique mix of pure white kaolin clay and the mineral petuntse, found only in China. When baked in the oven, the petuntse melts and fuses with the kaolin, producing the unique qualities of 'hard paste' or 'true' porcelain. The Chinese guarded the secret for centuries.

Early explorers and traders returned home with samples of china. With the growing popularity of new beverages from the Americas and the East, such as tea, coffee and chocolate, demand for porcelain cups and saucers grew rapidly among the well-to-do. European manufacturers tried to emulate 'china' but could not, lacking the right minerals.

Around 1575 experiments in Italy led to 'artificial' or 'soft paste' porcelain, made by firing mixtures of kaolin and local glassy materials. It remained slightly porous, but had a creamy colour which some preferred to the bright white of true porcelain. Over coming centuries, the manufacture of soft-paste porcelain spread across Europe, with still-famous factories founded at Sevres and Limoges,

> **With the growing popularity of new beverages ... such as tea, coffee and chocolate, demand for porcelain cups and saucers grew rapidly among the well-to-do.**

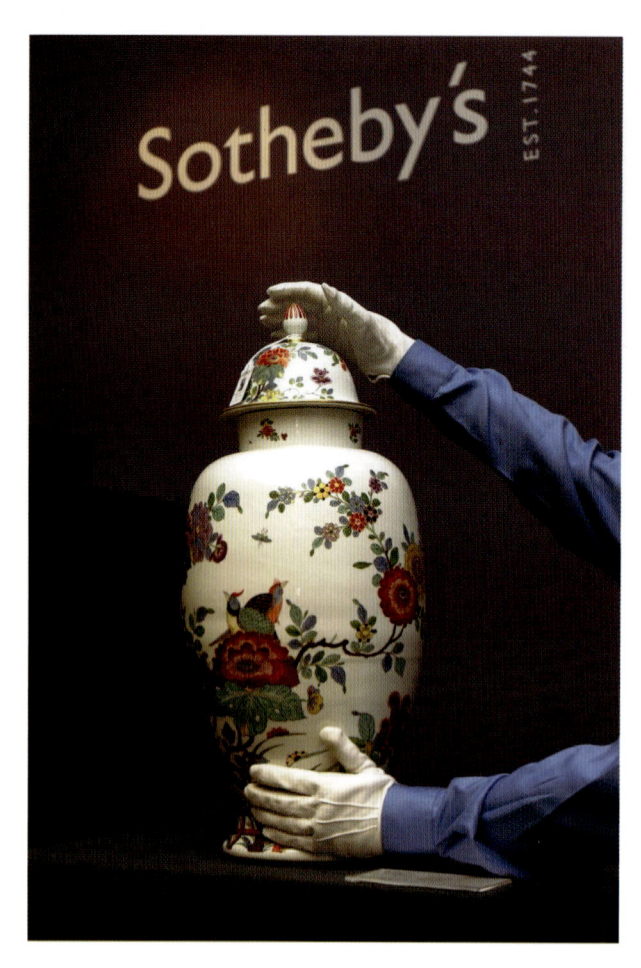

Above: Despite modern synthetics, porcelain remains one of the great inventions. Originally from China, its spread into Europe sparked an outburst of creativity, generating remarkable and highly-prized objects of art.

near deposits of high quality kaolin, and also in England, Germany and Italy.

But it was not yet the 'real thing'. In soft paste, the outer 'glaze' is distinguishable from the inner 'body' of the piece; in 'hard paste', they cannot be separated. The secret was not known in Europe until 1709. German chemist Johann Bottger realised the Chinese were using a mixture of minerals and methodically tested samples from all over Saxony to find what he needed. His factory at Meissen was famous for decades; 'Dresden china' remains well known. However, Bottger died aged only 37.

Around 1750, a new form of porcelain, 'bone china', was developed. Crushed and calcined (strongly heated) ox bones were added to the soft paste mix. The result was harder than soft paste, though not as durable as true porcelain, with an appealing translucence. Bone china became particularly popular in England.

Advancing glass
George Ravenscroft and others

Glass is an ancient invention, much refined by generations of trial and error—versatile, once expensive, now cheap, with some drawbacks, such as its potential to break into jagged fragments. Progress in glassmaking has paralleled mechanisation in other industries. For example, the skilled human glass blower was replaced long ago by mechanical bellows for routine or low-value products like bottles.

The first real advance on ancient glass technology came in 1676, when Englishman George Ravenscroft added some lead oxide to the glass mix. The result was remarkable. The emerging glass was relatively soft to cut, heavy in the hand, rang musically when struck, and brilliantly clear with flashes of light and colour due to its higher 'refractive index'. The product was known as 'flint glass', since a few years earlier an improved

Glass: what's that?

Scientifically, glass consists of the mineral silica (silicon dioxide, the main constituent of sand) and various other minerals in small amounts, such as potash, soda or lime. These react with each other when melted at temperatures of nearly 2000 degrees Celsius. Once cooled and allowed to harden, you have glass.

Chapter 4: Materials matter   59

Above: While everyday glass objects like bottles, jars and light globes are now blown by machines by the identical million, there is still a role in glass-making for the skilled artisan who can produce something exquisite and unique.

glass had been produced by mixing in crushed flint (a mineral composed almost entirely of silica). We now call it lead glass or 'crystal'.

Glass now has many forms, mostly developed in the last century: glass blocks or 'bricks', able to replace ordinary bricks in building, and 'fibreglass', masses of glass fibres thinner than a human hair, both came into use in 1931. The latter is as resistant to chemical attack and weathering as ordinary glass, and is used for heat-resistant installations. Implanted in a matrix it forms a 'composite' in all sorts of building applications, including boats and aeroplane components. Fibreglass composites combine great strength with light weight.

Objects made from 'boro-silicate' glass (glass containing boric acid), such as Pyrex, are much more resistant to shock and heat than those made from 'soda' (ordinary) glass. Glass that has been 'tempered' (carefully treated by heating) forms 'safety glass', which is strong and forms into rounded fragments rather than hazardous shards when it breaks. Since 1957 'glass ceramics', for example CorningWare, have been created with millions of tiny crystals where ordinary glass has none. They are therefore tougher and more heat-resistant, able, for example, to go directly from freezer to microwave to table.

The reinvention of cement
John Smeaton and Joseph Aspdin

Eddystone Lighthouse warns ships of treacherous rocks off the Cornwall coast. The first lighthouses there, built in 1698 and 1709, were wooden structures, soon destroyed by storm or fire. Yet the third lighthouse, finished in 1759, operated for nearly 130 years. When erosion of the underlying rocks required its demolition, the lower levels were so firmly bonded they could not be taken apart. They remain in situ to this day.

One secret to that strength was 'hydraulic cement', that is, cement that sets underwater, even salt water. Its inventor, or rather its 're-inventor', was John Smeaton. The ancient Romans had known that 'quick lime' (made by heating limestone), mixed with larger amounts of a particular crushed volcanic ash, made a cement that did not need to be dry to harden. This knowledge was lost with the fall of Rome and not rediscovered for 1500 years.

Smeaton, already eminent in civil engineering, had been recommended by the Royal Society to build a new lighthouse at Eddystone. To storm-proof the 20-metre tower, he employed three new ideas: a conical shape—modelled on an oak tree; interlocking blocks of stone; and his rediscovered hydraulic cement. Drawing on records of Roman experience, he had experimented with various mixes of lime and clay in search of the ideal formula. The durability of the lighthouse demonstrates that he found it.

Others took Smeaton's innovation forwards in search of a consistent product. In 1824 Leeds bricklayer and stonemason Joseph Aspdin patented a form of hydraulic cement called 'Portland cement', so named because it resembled the colour of stone quarried at Portland on the English South Coast—and when set was as hard.

Aspdin crushed and heated together precise amounts of limestone and clay, grinding the result into cement. A similar process is used today; concrete (cement mixed with sand or gravel) is the world's most widely used building material. Early successes with Portland cement included repairing the first tunnel under the Thames in 1828 and cementing the millions of bricks that formed the new London sewers in the 1860s.

Above: John Smeaton's remarkably durable Eddystone Lighthouse off the coast of Cornwall (the third on the site), a tribute to the fortitude of hydraulic cement, which he rediscovered.

This knowledge was lost with the fall of Rome and not rediscovered for 1500 years.

The taming of rubber
Charles McIntosh and Charles Goodyear

A raincoat is often called a 'mac' thanks to Charles McIntosh, Scottish chemist and inventor of the rubberised fabric originally used in waterproof garments. The 'macintosh' got its start in experiments with naphtha, a liquid waste from the heating of coal to make coke and coal gas.

McIntosh found that naphtha dissolved 'india rubber', made from the sap of certain South American trees. He spread the solution between two pieces of fabric. On drying, the pieces were bound together by a layer of rubber, becoming waterproof. McIntosh received a patent in 1823 and opened a factory in Manchester soon after—that factory remained operational for more than 150 years. Earlier, in 1820, McIntosh's inventive partner Thomas Hancock had patented 'elastic', strips of rubber attached to cloth. Elasticised female undergarments reached the market in 1830, replacing wire and whalebone 'stays' and definitely increasing comfort. Elastic-sided boots appeared in 1837.

Rubber remained troublesome, hard when cold and sticky when hot. American hardware merchant

Below: Once dubbed the 'weeping wood', the tropical rubber tree yields a useful gum, long used locally. European chemistry transformed natural rubber into a powerful and versatile material, now little used … thanks to synthetics.

Left: A giant condom was placed over the Obelisk of Luxor in Paris's Place de la Concorde, in 1993 by the French anti-AIDS group Act Up and Italian clothes maker Benetton to mark World AIDS day.

Charles Goodyear wanted to make rubber more stable and therefore more useful. He experimented for a decade with combinations of chemicals, but the answer came accidentally when he dropped a mixture of raw rubber and sulphur onto a hot stove. When the smoke cleared, he found the mixture had charred but not melted, meaning that 'vulcanised' rubber was heat-resistant.

By 1841 Goodyear could make thin, uniform sheets of rubber that stayed firm and elastic, hot or cold. With little interest in his invention, he tried to sell the idea to McIntosh and Hancock, but Hancock quickly worked out the process for himself and secured a British patent before Goodyear did. Though Goodyear's invention enriched others, he died a poor man, mostly from costly fights against patent infringements.

Another major discovery was by synthetic materials pioneer Alexander Parkes; rubber could be 'cold-vulcanised' by dipping in a solution of sulphur chloride. Natural rubber is little used nowadays, having been replaced in many applications by 20th century synthetic rubber and materials like polythene. But in its day it was a revelation.

Steel replaces iron
Henry Bessemer

Steel-making is ancient, dating back many centuries for high-quality goods like swords and tools. Production was labour-intensive; the 'Hundred Refinings' method used by the Chinese suggests its complexity and consequent cost. Despite its advantages, steel could replace iron more widely only when it could be cheaply mass-produced.

A multitude of uses

The improved vulcanised rubber was versatile. The emerging electrical industry found it good for insulating wires. Vehicles used rubber shock absorbers. Inflatable rubber tyres found ready markets in bicycles and then motorcars. Cold-vulcanised rubber could make very thin rubber sheets for balloons, condoms and the like.

Chapter 4: Materials matter

Elasticised female undergarments reached the market in 1830, replacing wire and whalebone 'stays' and definitely increasing comfort.

That transformation was mostly the work of Englishman Henry Bessemer, though he drew ideas from American William Kelly—there was argument later over patents. Bessemer conquered the troublesome brittleness of iron by reducing its carbon content, blowing air through molten iron in a large pear-shaped vessel he called a 'converter'. Much of the carbon burnt away, in turn releasing heat to keep the iron molten and speed the refining. When Bessemer described the process before the British Association in 1856, iron experts remarked they had never heard of anything 'so striking and impressive'. It still took four years of effort, and some failures, to make the process acceptably reliable in producing steel. He opened his first plant in Sheffield in 1860.

Bessemer steel soon overtook wrought iron, used for many decades to support the Industrial Revolution. Steel matched wrought iron for most purposes and could be made in larger quantities for half the cost.

Below: Skilled and diligent craftsmen have been making precious objects from steel, such as this samurai sword, for thousands of years. In the 19th century, Henry Bessemer and his converter turned steel into an everyday material.

The success story

Famous for his converter, Bessemer was also active in the sugar industry, steamboats and big telescopes. Honours included a knighthood and the Fellowship of the Royal Society and he earned £5 million for his steel process alone. He already had a fortune. When still young, he had powdered brass and sold it as an additive to make 'gold' paint, much in demand for gilded decorations.

In 1850, 50,000 tonnes of steel were made in Britain by a slow and costly process. Thirty years later, production of steel was over one million tonnes, with Bessemer converters making 80 per cent of that.

Military needs drove demand. The Crimean War of the 1860s showed that steel outperformed iron in cannon and armour-plating; steel for guns had been Bessemer's initial goal. Steel was attractive also to builders of ocean-going ships and railways—steel rails were more durable than wrought iron ones. Steel girders soon supported buildings, including the new 'skyscrapers'.

In time Bessemer's method was superseded, initially by the 'open hearth' process developed in Germany, but Bessemer converters were running in Britain until the 1970s. Bessemer had enabled steel to replace iron and that was what mattered.

Below: The advent of cheap, strong steel transformed construction across the planet. Enterprise, like the East Sea Bridge in Shanghai, China, would be impossible without it. Likewise the skyscrapers which now dominate our great cities.

Cement and iron
Joseph Monier and Francois Hennebique

By the mid-19th century cement was a familiar popular building material, increasingly replacing wood and stone. But it had problems. Cement—when mixed with sand and gravel to produce concrete—has great compressive strength (it bears loads well and is hard to crush), but it is weak under tension (stretch it, and it cracks and crumbles easily).

This challenge was addressed by many, most notably by French commercial gardener Joseph Monier. He pioneered the use of steel mesh embedded in concrete tubs and basins, patented the idea in 1867 and displayed it that year in the Great Paris Exposition. Monier had no technical training in these matters, but seemed instinctively to know what he was doing. The embedded wire took the tensile stresses; the cement handled the compressive forces, the two components functioning as a unit. The range of uses to which cement and concrete could be put was vastly expanded.

Among those who admired Monier's innovation on display at the Paris Exposition was engineer Francois Hennebique. Quick to the possibilities, he set to work devising ways to apply the technology to buildings, using steel rods for reinforcement. By 1879 he was making whole floor slabs from reinforced concrete, and by 1892 had devised and patented a complete building system, using reinforced concrete floors, beams and columns, connected by metal stirrups. He used this to construct an apartment building in Paris, setting a pattern for building construction that continues largely to the present day.

Above: Steel-reinforced concrete has both high tensile and high compression strength, making it an ideal modern building material. In extreme cases it can be used for protection from bullets and bombs, like these German observtion posts, erected in Normandy during World War II.

A multitude of uses

In the decade after he patented it, Monier extended his concept of 'reinforced cement'—and the number of his patents—to drainage and sewer pipes, water tanks, reservoirs, panels for building facades, foot and vehicle bridges, tunnels under rivers and bays, the sleepers or ties supporting railway tracks and concrete beams for structural strength within buildings. A bridge built with reinforced concrete opened in San Francisco's Golden Gate Park in 1889—it is still in use. Within a few more decades, reinforced concrete would be forming the huge locks in the Panama Canal.

Aluminium becomes abundant
Paul Herault and Charles Hall

Aluminium is not an invention; it has been abundant in rocks, soil and gemstones since the earth formed. From the late 18th century, chemists suspected it existed and named it after the compound alum, already in use in papermaking, textile dyeing, and medicine. They produced small impure samples, but the costly process made aluminium a semi-precious metal. Extracting it in larger, purer quantities for industrial use required major innovation, and depended on cheap and abundant electricity, available only in the late 19th century.

Aluminium is so tightly bound to oxygen in the plentiful compound alumina that heating and chemical attack will not release it. The invention of electrolysis by English chemist Humphry Davy provided a way of prising aluminium out, using an electric current. A process to do this cheaply did not arrive until 1886, courtesy of Charles Hall in the United States and Paul Herault in France, working independently and both aged only 22 at the time. Starting with the common ore bauxite, they firstly cleaned away impurities such as iron and silica, then dissolved the purified alumina in molten cryolite, a

Below: The first successful extraction of aluminium from its ore coincided with the widespread availability of electricity. The new material became cheap enough to be used for many purposes, such as the sculpture 'The Awakening' in America's capital, Washington, DC.

Brearley had to have samples made with the new steel by the local cutler Mosley's at his own expense.

compound containing fluorine. The melt was placed in large carbon-lined vats. Heavy doses of electrical current were passed through, liberating molten aluminium, which collected at the bottom of the vats ready to be tapped off. This is still the basic method of production today.

With a unique combination of strength, lightness and resistance to corrosion, aluminium was soon in widespread use, rivalling steel in many applications. The famous statue of Eros in London's Piccadilly Circus was cast in aluminium in 1893. Aluminium's properties are enhanced by alloying (mixing) it with other elements. Duralumin, invented accidentally in 1909, contains copper and magnesium as well as aluminium. As strong as mild steel, but weighing only one third as much, duralumin made possible the first airships and radically transformed the early aircraft industry. Military demand for aluminium during World War II saw housewives urged to donate their pots and pans to the war effort.

We now find aluminium everywhere, in food containers, foil wrappings, electrical conductors, car engines and fittings, furniture and lightweight girders. It is almost as ubiquitous as steel. The huge amounts of electricity required to make it, however, poses a challenge for the future.

Steel without stain
Harry Brearley

The invention of stainless steel by British chemist Harry Brearley in 1912 shows how an invention can end up being used for very different purposes than those originally intended. At the time Brearley was in charge of a laboratory set up by two of the major steelmakers in Sheffield in the United Kingdom, Firths and Browns, to solve various problems the industry faced. Such collaborative research was a new idea.

A small-arms manufacturer had reported that steel barrels wore away quickly from the effects of the hot gases released. After many trials, Brearley found that adding about 10 per cent of the metal chromium to the steel greatly reduced wear. Problem solved. Needing to etch samples of the steel for study under the microscope, he found that the chromium-enriched steel also resisted attack by acids, including common food acids like vinegar and lemon juice. Brearley was quick to see the potential for cutlery manufacture, a major business in Sheffield. Much cutlery was electroplated with chromium or nickel already and so resisted corrosion, but plated steel would not take a good edge. In consequence, knife blades had to be made from mild carbon steel, which rusted quickly unless less well cared for. Knife blades of the new 'rustless' steel would be largely immune to staining and other corrosion.

Sheffield cutlery makers were conservative. Brearley had to have samples made with the new steel by the local cutler Mosley's at his own expense. The manager at Mosley's suggested calling the new material 'stainless steel'. Slowly its advantages were accepted. The major German steel manufacturer Krupps obtained a similar result using nickel; the steel produced was softer, more easily worked and even more resistant to acids. The outbreak of war halted production, but in the 1920s, the use of chromium and nickel together was pioneered and remains the basis of stainless steel today.

Brearley soon fell out with Firths. They had made use of his services to run the laboratory, but claimed

Above: The ill-fated DeLorean DMC-12 was made with a durable, paint-free, stainless steel body and designed to break the 'built-in obsolescence' cycle.

ownership of the invention because he worked for them. Brearley joined rival company Browns to continue development of stainless steel. Having left school at the age of twelve to wash bottles, Brearley was largely self-taught in steel matters, but his expertise was recognised in 1920 by the Iron and Steel Institute, which awarded him the Gold Medal named after Henry Bessemer.

Cities of glass
Alistair Pilkington

We commonly associate glass with windows, but a pane big enough to let in significant light was not easy to make. So window glass has traditionally been expensive. The number of windows in a house indicated how rich the owner was—and made a 'window tax' a good money earner for the authorities in times past. Large windows, say for churches, could be made only by joining many small pieces of glass together. Arranging the fragments of glass, some of them coloured, to form a decoration or a religious image made a virtue of necessity.

Craftsmen and glass firms devised various techniques to make larger pieces of window glass, including pouring molten glass into a mould, letting it set then drawing out a hollow cylinder of glass from the melt (just as glass tubing is made, with air blown in to stop the tube collapsing). The cylinder was then cut down one side and flattened out using heat and pressure.

From around 1900 glassmakers could pull out a flat sheet in a continuous ribbon, much as you might draw out a long strand of toffee. This process was faster and cheaper but often left visible ripples and other imperfections in the glass, however carefully it was done. Light passing through it was distorted, acceptable in small domestic windows but not in the larger sheets needed for, say, a shop front. The sheet had to be made smooth by grinding and polishing both sides, a laborious process and wasteful of glass, to produce the much prized 'plate glass'.

Window glass is now made by a process conceived in 1952 by Alistair Pilkington of the famous English firm, which had been making glass for 150 years. A decade of development was needed to perfect the process, which involves floating the molten glass on a layer of molten tin. Natural forces within the glass produce a continuous ribbon of highly uniform thickness, which doesn't need grinding.

Plate glass could now be made relatively cheaply in very large pieces, leading to its growing popularity as a building material and generating the 'glass towers' look of modern cities. New building techniques, using cores of steel and concrete to support buildings from within, allows the glass walls to be simply hung on the outside, since they do not carry any weight.

Opposite: The 'cities of glass' look is now almost universal in developed and developing countries. There is a downside, however. Glass transmits heat easily, leading to much wastage of energy to heat and cool the buildings.

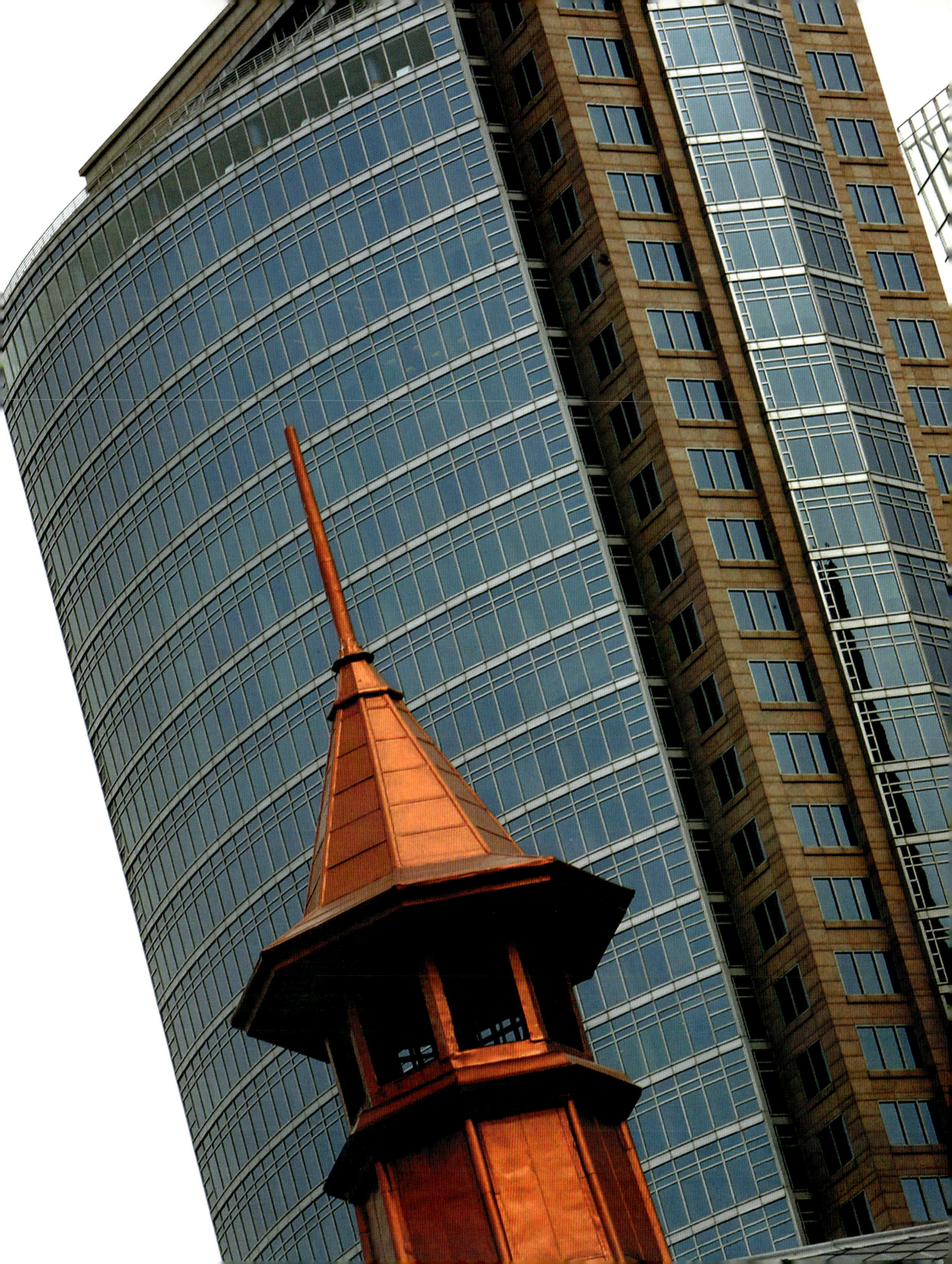

CHAPTER 5:
Putting chemicals to work

Chemistry is a difficult and obscure subject for most people, but we gain value from its practical applications in everything we do. Chemical manipulations liberate energy, sometimes violently, release materials from their sources, and convert natural products like rubber into something much more useful.

Here is just a sample of what chemistry has made possible over the last few hundred years, thanks to inventive minds and assiduous entrepreneurs. Here are the stories behind the mass production of vital chemicals such as sulphuric acid, ammonia, 'soda ash' and chlorine. You will find a lot more in other sections of this book, dealing with issues like the control of pain or disease. Not all the stories have happy endings, either for the inventor or for the environment.

Chapter 5: Putting chemicals to work

Chemicals old and new
Charles Tennant and John Roebuck

Chemistry came of age in the late 18th century. Before then it was just a motley collection of barely understood recipes. Brilliant discoveries in England, Scotland and Sweden culminated in the new chemical theories of the French genius Antoine Lavoisier. We began to understand how and why chemicals react as they do, and how these reactions could be controlled and directed. At the same time, chemistry became more useful. Ways were found to make large amounts of useful chemicals in industries like the manufacture of textiles, glass and paper, often replacing substances derived from nature that had been used for centuries.

Take the case of chlorine, a green gas released from hydrochloric acid, discovered by the industrious Swedish chemist Karl Scheele in 1774. He soon found that chlorine bleached the colours from cloth or other natural materials. However, chlorine is nasty stuff to handle, poisonous and corrosive, and therefore not safe for routine fabric whitening and stain removal. Its use was limited.

It was the Scotsman Charles Tennant who found a way to render chlorine safe to handle, and therefore genuinely useful. He passed chlorine over calcium hydroxide (lime), making solid 'bleaching powder', which could be stored until needed. In water, bleaching powder releases chlorine. Suddenly dazzling white cottons and linens were easily available and the textile industry, already mechanising at a fast rate, was given a tremendous boost.

The traditional way to bleach cloth was slow and laborious, involving repeated steeping of the fabric in sour milk (a mild acid) and exposure to sunlight for weeks at a time. The Dutch were good at this and had dominated the business. Now anyone could do it, thanks to bleaching powder. The same process whitened paper, removing natural colours from the vegetable fibres.

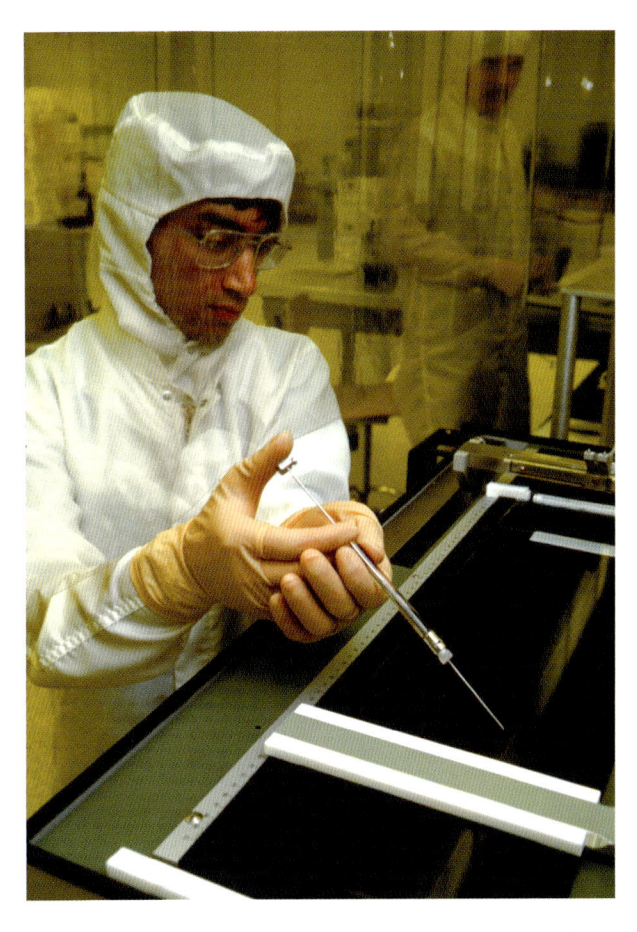

Above: Today's polymer scientist has come a long way from the pioneers of the chemical industry, employing advanced techniques, highly controlled working conditions and strict safety standards. The knowledge they have, however, is built on the foundation of experimentation and leaps of faith taken by the early chemical explorers.

Sulphuric acid became the most important industrial chemical ... Until the mid 20th century, a nation's level of industrialisation could be measured by its production of the stuff.

Another chemical coming into vogue was sulphuric acid. Diluted it was a satisfactory bleach; in concentrated form it could refine gold and silver. Sulphuric acid had been made in small amounts for hundreds of years, for example by burning sulphur and mixing the resulting gases in water in glass vessels.

Around 1760 English chemist John Roebuck began to run this process in large chambers made from lead—immune to the corrosive acid—and produced the acid much more cheaply and in greater quantity than ever before. Sulphuric acid became the most important industrial chemical, so diverse were its uses. Until the late 20th century, a nation's level of industrialisation could be measured by its production of the stuff.

The quest for the 'alkali'
Nicholas Leblanc and Ernest Solvay

Late in the 18th century, industries making glass, soap, textiles and paper were beginning to boom, driving up demand for raw materials. Chief among these were the 'alkalis', particularly 'soda ash' (aka 'washing soda'), made at the time by boiling down seaweeds such as Scottish kelp or barilla from the Canary Islands. Supplies were unreliable, quality was variable. More abundant and reliable sources were urgently needed.

The French, at war with much of Europe and unable to import soda ash, were in particular need. In 1783 King Louis XVI had his Academy of Science offer a prize to anyone who could 'make the alkali' (that is, soda ash) by decomposing ordinary sea salt using the 'simplest and most economical method'.

The starting point was already known. German doctor Johann Glauber had found it a century before, heating sulphuric acid with sodium chloride (sea salt) to make hydrochloric acid. This left a residue, which he called Glauber's Salt, another popular name was 'salt cake' and formally it is sodium sulphate. It was much used in medicine. Various people thought soda ash

Above: The 19th century saw an upsurge in concern for cleanliness, since we learned that 'dirt brings disease'. Soap became more popular and more sophisticated, putting pressure on the manufacturers of chemicals such as alkalis.

(sodium carbonate) could be extracted from the salt cake, but no-one could do it cheaply and easily enough to meet demand.

Nicholas Leblanc, a doctor employed by the Duke of Orleans, came up with his process in 1791. He roasted the salt cake with charcoal (carbon) and chalk (calcium carbonate). Off came carbon dioxide, leaving a mix of the desired sodium carbonate and calcium sulphate waste; the two could be separated by washing. The problem appeared solved. Leblanc's patron built a big plant outside Paris and production began. Demand was high because of war.

Though the outcome for Leblanc was eventually tragic, his process of making the industrially important soda ash was popular for a while. Plants sprang up all over Europe, especially in Britain, to supply booming industries making glass, paper and soap. Hundreds of thousands of tonnes of 'alkali' were produced every year. However, the Leblanc method was an environmental disaster. Vast amounts of hydrochloric acid gas, judged worthless, were let off into the atmosphere. The resulting 'acid rain' destroyed vegetation and corroded buildings, iron work and clothing. Land around Leblanc plants was laid waste; property owners sued.

Governments responded by passing the first ever environmental control legislation. The officials of the Alkali Inspectorate were empowered to demand that the corrosive gas be absorbed in water rather than released to the air. This made hydrochloric acid, which was quietly poured into nearby streams, causing further environmental damage. In time ways were found to add value to this waste. The acid could be converted to chlorine and then to bleaching powder, for which there was a market.

The other waste, a 'black ash' of mixed calcium sulphate and unused coal, was equally noxious and disfiguring. Huge dumps, stinking of rotten eggs, built up on vacant land. These wastes, as much as the smoke pouring from chimneys, made the English Midlands the 'black country'.

Belgian industrialist Ernest Solvay invented a better process in 1861. It still started with salt. Brine was saturated with ammonia and then with carbon dioxide, forming sodium bicarbonate and ammonium chloride. When heated, these compounds yielded the soda ash and the ammonia was recovered for reuse. The products were cheaper to make, the wastes relatively harmless. The new process rapidly took over and the last Leblanc plant closed in the 1920s. The Solvay family made a vast fortune.

Coal tar yields its wonders
August Wilhelm von Hofmann and William Perkin

Coal tar does not look very promising; it's a thick, black, smelly, oily mass left behind when coal is strongly heated ('destructively distilled') in a closed container to make coke, for smelting iron ore, and coal gas, for lighting and heating. Chemists soon guessed that coal tar is not a single substance but a mixture of compounds, some of which could be liberated by gentler, 'non-destructive' distillation. If the heating was done carefully and the right temperatures chosen, the various compounds could be driven off and condensed one by one.

An early discovery was the compound aniline and its close relative phenol. German Friedlieb Runge found

Stranger than fiction

When a plant was built to manufacture soda ash using Leblanc's process, it looked set to make him a fortune. Larger events, however, overcame him. The plant was soon appropriated by revolutionaries, his patron, the Duke, beheaded and Leblanc forced to reveal the secret of his process. Others quickly put it to work. He never got compensation for the takeover of the plant, and by the time he resumed production his competitors were too far ahead. The Academy of Science had been dissolved, so he never got the prize offered for developing the process. Bankrupt and despairing, Leblanc finally shot himself. At least his descendants finally got some reward. In 1855 Napoleon III made a payment to them in lieu of the prize. A statue of Leblanc now stands in Paris.

He mopped the mess up with his wife's cotton apron and hung it up again. When the apron dried, it suddenly vanished in a puff of smoke, completely consumed.

those in 1834. Phenol would later be called carbolic acid and valued for its ability to kill the micro-organisms that could spread disease. It played a major role in the early growth of antiseptic surgery

Guessing how the colourless, oily aniline was put together, English chemists thought it might be similar to quinine, the only known defence against malaria, then a scourge of the British Empire. In 1845 the leading German chemist August Wilhelm von Hofmann was imported to run the newly established Royal School of Chemistry in London, with the major task of somehow converting aniline into quinine.

In 1856 William Perkin, a laboratory assistant still in his teens, reacted aniline with potassium dichromate and washed the result in alcohol. He found not quinine, but a beautiful purple dye, which he called 'Perkins mauve' or 'mauveine'. It was the second synthetic dye, following the excellent aniline black that Runge had patented in 1834, and the source of Perkin's personal fortune. He dropped out of the college to set up a dye-making business with his father's money. Mauveine became immensely fashionable and popular for women's clothing. Perkin prospered. Biology benefited too. Perkins mauve proved very useful in staining specimens to reveal previously hidden details.

The synthetic dye industry was born. It grew rapidly, especially in Germany once Hofmann returned to his homeland. Synthetic dyes in many other colours were soon developed and, together with drugs and other useful products made from coal tar, would be the wellspring of German industrial and economic strength in the 20th century. There was a downside: many compounds distilled from coal tar would prove to be cancer-causing.

Bigger, safer bangs
Christian Schonbein and Alfred Nobel

Nineteenth-century miners, hunting for coal and metal ores, and engineers, building canals, bridges and railroads, often needed to blast and break up rock. They had only one explosive to work with, the venerable gunpowder, a mixture of sulphur, carbon (charcoal) and nitre (saltpetre). The last of these supplied the oxygen in which the others burned very rapidly, liberating the very hot, high-pressure gases that generated the explosive force. As its name implies, the first—and still main—use of the explosive was in guns, both cannon and small arms.

The first alternative explosive was invented, apparently by accident, by the Swiss chemist Christian Schonbein, who also discovered ozone. In 1845, the story goes, he was experimenting at home and spilt a mixture of nitric and sulphuric acids. He mopped the mess up with his wife's cotton apron and hung it up again. When the apron dried, it suddenly vanished in a puff of smoke, completely consumed.

The mixture of acids had added small clusters of atoms called nitrate groups to the cellulose in the cotton, making nitrocellulose. These groups supplied oxygen for rapid burning just like the saltpetre (potassium nitrate) in the gunpowder. The new explosive was first used in warfare, hence its name 'guncotton'. Its white smoke did not obscure the battlefield, unlike the black smoke of gunpowder.

Opposite: Setting off an explosion at a construction site. Pushing down the handle sent an electric current through a long wire to the charge. This meant the man with the plunger could be at a safe distance, even with a cigarette.

Chapter 5: Putting chemicals to work 75

Above: Chemistry in conflict. Soldiers in the Great War blinded by the newly-invented mustard gas—designed to incapacitate rather than to kill—are led along by more fortunate comrades. Mustard gas later brought a benefit.

Chemists were soon adding nitrate groups to other flammable substances, seeking new explosives. The most powerful and hazardous was nitroglycerine, discovered in 1847 by the Italian chemist Ascanio Sobrero. This was too unstable for use in war; even peacetime use in civil engineering required very careful handling. The slightest shock could set it off; carelessness cost many lives.

The Nobel family from Sweden was a major manufacturer of the explosive, and they later employed Sobrero. When his brother was killed in a factory explosion, Alfred Nobel worked to make nitroglycerine safer. In 1867 he soaked an absorbent mineral in liquid nitroglycerine and produced a stable material, able to be moulded into sticks for easy handling—dynamite. Combining high explosive force with acceptable safety, it became tremendously popular in both peace and war. Together with gelignite, a mixture of nitroglycerine and guncotton, the explosive funded the Nobel Prizes, first awarded in 1901.

The hunt for better explosives went on. Early in the 20th century, three nitrate groups were added to molecules of the hydrocarbon toluene, so making tri-nitro-toluene (TNT).

Ammonia in war and peace
Fritz Haber and Carl Bosch

When the German chemist Fritz Haber won the Nobel Prize in 1918, many people were alarmed. Haber had promoted a new and terrible weapon for use in the trenches of the Western Front—poison gas, initially chlorine, first used at Ypres in 1915, then mustard gas after 1917. His wife, herself a chemist, had already reportedly committed suicide over the matter.

Haber's Nobel Prize had nothing to do with poison gas, but still had a lot to do with war. The award acknowledged his invention of a process to make the vital chemical ammonia synthetically, by combining nitrogen and hydrogen under the right conditions of temperature and pressure, with a catalyst to greatly speed the process. The ammonia, oxidised to become nitric acid, could then be used in explosives and fertilisers, particularly through the still much-used compound ammonium nitrate.

Stranger than fiction

Fritz Haber had a broad range of interests and achievements. He even attempted, unsuccessfully, to recover gold from seawater to help Germany to pay its war reparations. He died in Switzerland in 1934, exiled from Germany because of his Jewish background.

Chapter 5: Putting chemicals to work

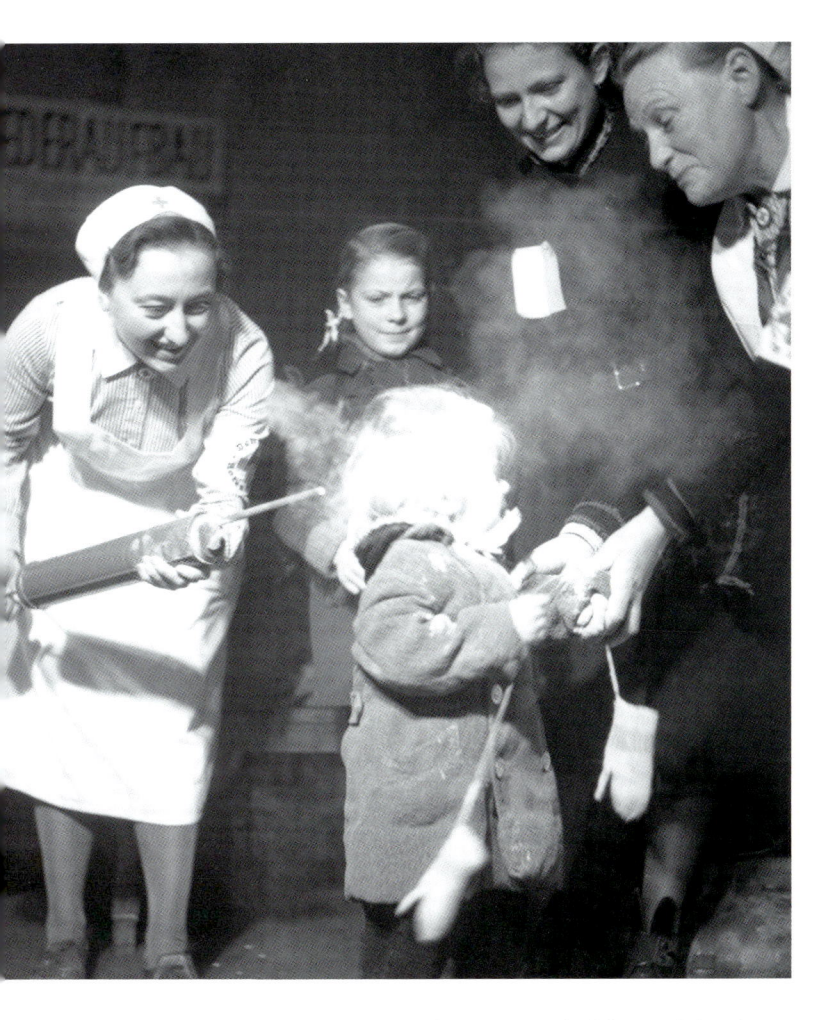

Above: A child cries as she is sprayed with DDT delousing powder in a German school in 1945, shortly after the end of World War II.

The invention of the Haber-Bosch process may have helped to precipitate the Great War, by encouraging the German government to think it could be self-sufficient in ammonia and its products, and so sustain its industry through war and the expected trade blockades. At least it probably prolonged the war, by maintaining German supplies that would otherwise have been exhausted.

With modifications, however, the Haber-Bosch process still produces ammonia for synthetic fertilisers today. So its impact in peace is perhaps even greater than in war, empowering, for example, the Green Revolution. In wider industry the use of high pressures and catalysts to speed up chemical processes, as Haber did, has helped the refining of petroleum and the production of oil from coal.

DDT: a wonder chemical?
Paul Muller

Our ambivalence about the insecticide DDT pits its unprecedented success in controlling serious diseases carried by insects against its proven harm to wildlife, as set out in biologist Rachel Carson's influential 1962 book *Silent Spring*. Dichloro-diphenyl-trichloro-ethane (DDT) was first synthesised in 1874 by Othmar Zeidler, a Viennese pharmacist, but he did not realise its properties. When rediscovered in 1938 by Swiss chemist Paul Muller, working for the Geigy company, its lethal impact was soon recognised. Ready victims included mosquitoes, fleas and other insects carrying malaria, yellow fever, typhus, bubonic plague, river blindness and elephantiasis, as well as agricultural pests such potato beetles, cotton boll worms and coddling moths.

DDT was patented in 1940 and reached the market as a dust in 1941. Its use was greatly stimulated by the outbreak of the Pacific War and the need to control pests in areas of conflict, especially in the tropics. It was routinely dusted on soldiers to control lice that spread typhus.

Germany, like most industrial nations, depended on imports of the mineral saltpetre from Chile and on by-products from the making of coal gas. Fears were growing that the rising demand would see supplies of the first run out, with the latter unable to make up the difference. The atmosphere, of course, could provide an almost unlimited supply of nitrogen, but some 'fixing' process was needed to form it into useful compounds. Haber discovered it, and another leading chemist Carl Bosch, from the company BASF, refined and upgraded the process for major industrial use.

The success story?

The results of DDT's widespread use are striking. Spraying with DDT virtually eliminated malaria from twenty countries where it had been endemic. Death rates in India, for example, fell from one million a year to less than 5000, extending average lifespan from 32 to 47 years. The incidence of malaria in the United States dropped from 250,000 a year in the 1930s to less than ten a year in the 1950s. For his invention Muller was awarded the Nobel Prize for Medicine in 1948.

DDT appeared an ideal insecticide: inexpensive to make, long-lasting in effect, so it did not need frequent re-application, and apparently harmless to plants and animals. It seemed to deserve the 'wonder chemical' tag. However, Carson's book, highlighting previously unsuspected damage to bird reproduction—hence the potential for a 'silent spring'—together with evidence from laboratory rats and mice, saw it banned in a number of countries and damned as one of the worst environmental pollutants. It remains in use in the developing world, where nothing matches its cost-effectiveness, particularly against malaria.

DDT dissolves readily in fat and so builds up in human body tissues, but the threat posed remains uncertain, unlike the millions of lives undoubtedly saved. The shortcomings of DDT, which included the appearance of insects resistant to its effects, have driven sophisticated research into alternatives over the 60-plus years since its first use.

Opposite: The drug LSD produced powerful hallucinations in users but failed to deliver any useful therapeutic benefits, other than helping in the treatment of long-term alcoholics.

The invention of LSD
Albert Hoffman

The 1960s saw 'flower power', the anti-Vietnam War protests and the emergence of the Beatles and the Rolling Stones. It also saw the rise to popularity of LSD, also called 'acid', a mind-altering chemical that spawned the cultural phenomena of psychedelia. The name most associated at the time with LSD use was Timothy Leary, a Harvard University psychiatrist, but the man who first discovered, even invented, LSD was Swiss researcher Albert Hoffman, who did not realise at the time what he had found.

During the 1930s Hoffman worked for the pharmaceutical company Sandoz. He was making a variety of chemical compounds from lysergic acid, which other chemists had extracted from the ergot fungus, which grows on rye and other grasses. Outbreaks of ergot poisoning, usually from contaminated bread, had occurred throughout history, with a variety of bizarre effects. Lysergic acid therefore seemed an interesting subject of study.

With his eyes closed he saw 'an uninterrupted stream of fantastic pictures, extraordinary shapes with intense kaleidoscopic colours' and had to go home and lie down until the effect passed off.

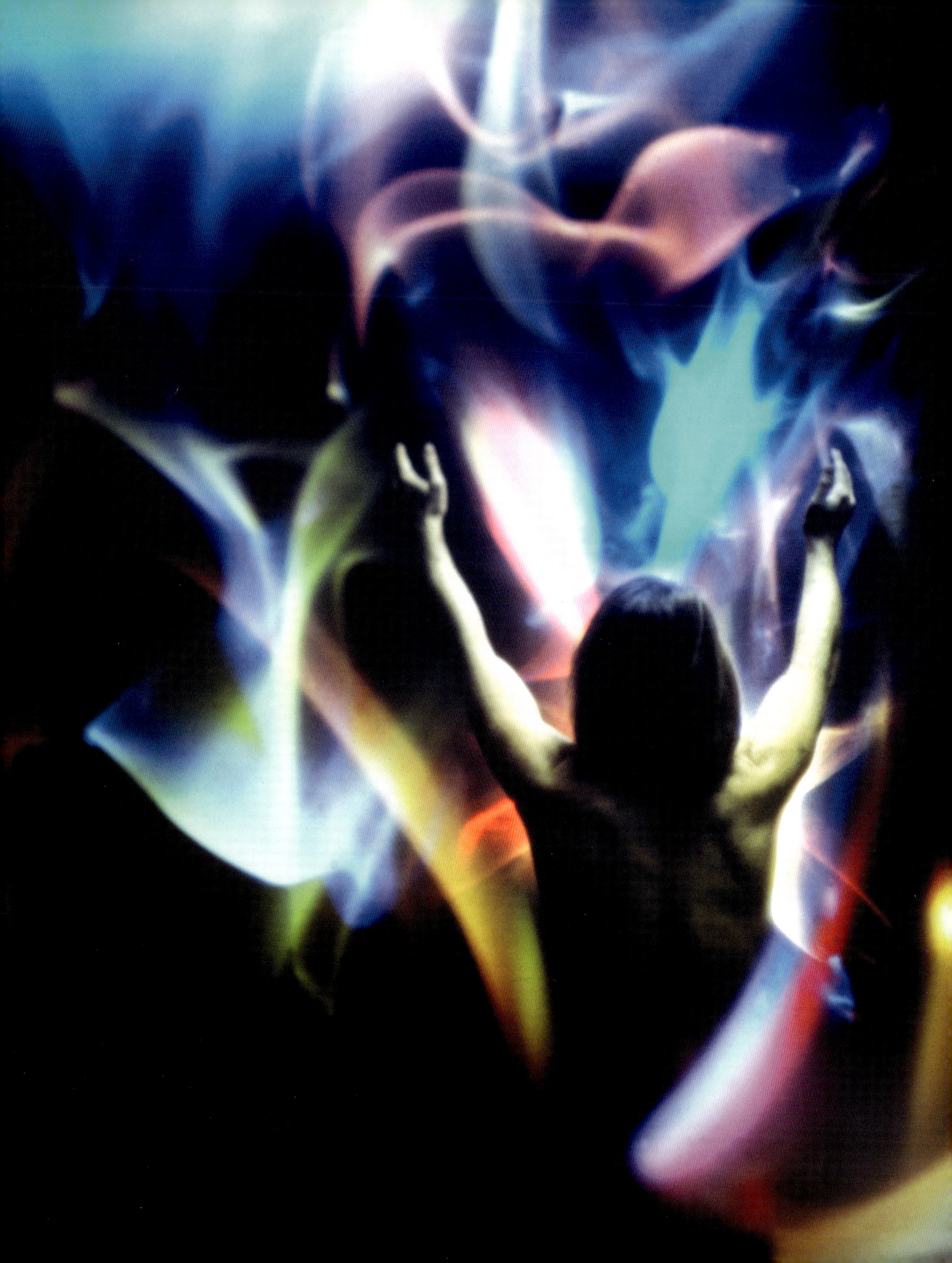

Above: Semtex plastic explosive used in anger in Afghanistan and (inset) Semtex formatted as a priming charge. The hole in the hexagonal blocks holds the detonating cord.

The 25th compound Hoffman made that year—1938—was lysergic acid diethylamide, which he dubbed LSD-25. Not finding any significant reaction to it from his laboratory animals, he put it aside. Not until 1943 did he come back to it, and apparently absorbed a tiny dose through the fingertips of his ungloved hands while making some more. He quickly became 'semi-intoxicated', in his own words, his imagination strongly stimulated. With his eyes closed he saw 'an uninterrupted stream of fantastic pictures, extraordinary shapes with intense kaleidoscopic colours' and had to go home and lie down until the effect passed off.

Wanting to know more, he deliberately swallowed 35 millionths of a gram three days later. The dose was minute, but the effects occurred again, just as vividly. LSD would prove to be one of the most powerful hallucinogenic drugs ever developed, yet it left him 'without a hangover' the next morning.

Hoffman believed that the unique mind-affecting properties of LSD would find use in neurology and psychiatry, and were worthy of intense study. In time, that gave rise to 'psychedelic therapy' as an alternative to psychotherapy, the drug successfully treated long-term alcoholics, and the CIA reportedly took an interest, on the grounds that enemy nations where ergot was common might have found strategic uses for LSD. The drug fell from official favour, however, and its sale was declared illegal in the America in 1968. It may or may not be relevant, but Albert Hoffman lived to be 100.

The explosive terrorists love
Stanislav Brebera

In December 1988 a mere 300 grams or so of the explosive Semtex, moulded into a cassette tape recorder, detonated on board Pan Am Flight 103 over Lockerbie in Scotland. The plane crashed with great loss of life. This wanton act made Semtex notorious, and it became the explosive of choice for terrorist groups around the world. Semtex, invented in 1966, takes its name from its inventor's home village in the Czech Republic. Stanislav Brebera was always saddened, he later said, when his invention fell into the 'wrong hands'.

Semtex is now the best known of the plastic explosives. Their appeal, both for legitimate uses in mining and construction, and for criminal acts, is threefold: they are very powerful; they are very stable and safe; and they can be made into different shapes like plasticine, hence their name. Much of their potency comes the explosive now called RDX— from its wartime code-name Research Development Explosive—and more ominously known as Cyclonite. Like all explosives RDX gains its destructive force from exceedingly rapid chemical reactions, breaking down to release very hot gases travelling so fast it is impossible to out-run the explosion, whatever action movies might suggest. RDX is not new, first created in the 1890s—initially as a medicine—found to be explosive in the 1920s, and much used on both sides during World War II.

Like other plastic explosives, Semtex has its RDX mixed with other compounds called plasticisers and binders to make it soft and mouldable. It can be safely shaped for maximum blast in a particular direction, say to crack open a safe, or packed into cracks in a wall and detonated to bring the wall down. Semtex, originally designed to destroy landmines, has no smell. Sniffer dogs cannot detect it, hence its appeal to users seeking to avoid detection. After the Lockerbie disaster, Brebera added metal fragments and chemicals with distinctive odours to make Semtex easier to find during security screenings.

Some energy must be supplied to get the chemical reaction going. Semtex and other plastic explosives like C-4 contain additional chemicals called stabilisers, so the energy boost that sets it off has to be very swift, from say a blasting cap; otherwise the explosive simply burns quietly.

Stranger than fiction

American soldiers in Vietnam sometimes set pieces of C-4 alight to warm their food, getting into trouble only when they stamped out the embers, the shock sometimes setting the explosive off.

CHAPTER 6:
Not found in nature

Today is often called the 'plastic age', though the term is often used critically. The term plastic highlights the capacity of these materials to be easiy moulded and shaped into useful objects. It is certainly true that 'synthetic' materials, with properties unmatched by anything we find in the natural world, have now penetrated every corner of our lives. The history of synthetics begins in the mid-19th century. They arose initially through efforts to improve upon nature using chemical manipulations. All the early synthetics had a starting point with a naturally occurring substance. The true synthetics came much later.

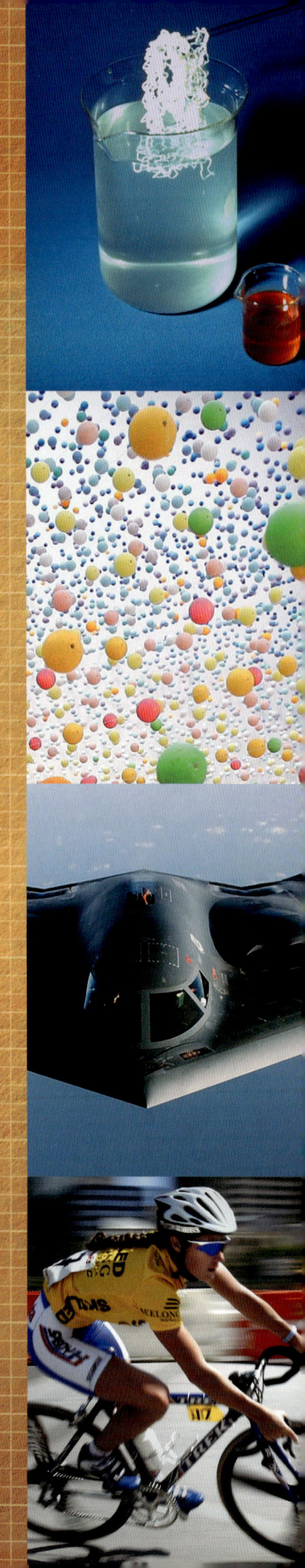

The first synthetic
Alexander Parkes and the Hyatt brothers

'Vulcanite' was first produced around 1843 by heating natural rubber with sulphur. It was hard and could be shaped but was always black, unless painted. Prolific English chemist Alexander Parkes—80 patents plus 20 children by two wives—sought to improve on vulcanite, to allow it to be coloured all through or made transparent. By 1862 he had his wonder material, soon dubbed Parkesine; it won a medal at the International Exhibition in London.

Parkes started with a known compound, cellulose nitrate, made by reacting cellulose, found in cotton, with nitric acid. Too much nitric acid produced explosive 'guncotton', a valuable gunpowder substitute. Made with lesser degrees of 'nitration', cellulose nitrate could be dissolved in a solvent like ether to become colloidin. Invented around 1848, this had a multitude of uses, in photographic plates, shatterproof glass and as a dressing for wounds.

When the solvent evaporated, a hard, waterproof and elastic residue remained. Parkes built on that. Cellulose nitrate was dissolved in a minimum of solvent, heated and rolled, coloured and pressed into shape as needed. Parkesine was the first 'thermoplastic' (one that softens when heated).

Parkes tried to market his invention, but could not balance quality against cost. He claimed it was cheaper than vulcanite, but that was possible only if the quality was poor, and it would not sell unless it was cheap. His company failed in 1868.

It fell to the American Hyatt brothers, John and Isaiah, to succeed a few years later. Their secret was to use camphor (from the laurel tree) as the 'plasticiser' in cellulose nitrate, something Parkes had not tried. This produced a 'synthetic ivory'; 'celluloid', as the material was called, was first used in billiard balls. Billiards was very popular and thousands of elephants were hunted to supply enough balls. Celluloid slowed the slaughter.

Celluloid was the first mass-produced plastic, used in toys, knife handles, collars and cuffs, photographic film and many other products through to the middle of the 20th century. At that time, its flammability and tendency to decompose spontaneously saw it replaced by safer, better materials like cellulose acetate and polythene. However, synthetics had certainly shown what they could do.

Billiards was very popular and thousands of elephants were hunted to supply enough balls. Celluloid slowed the slaughter.

Left: An early synthetic ivory brush, retrieved from the wreck of the *Titanic*. It has been well preserved at the bottom of the ocean, away from damaging ultraviolet radiation.

More synthetics: rayon, cellophane and bakelite

Jacques Brandenberger, Leo Baekeland and others

Following the invention of celluloid, which had a major impact on the advance of photography among other things, chemists continued to tinker with cellulose in search of other products. A new approach was to break it down into as tiny particles as possible by various chemical treatments and then make those same fragments 'polymerise' (link up) again. The broken-down cellulose was called 'viscose'.

There were two ways to proceed. If the viscose was forced through many small holes and into an acid bath, the fragments formed fine fibres, which could be drawn into a thread for spinning and weaving. Dubbed 'rayon' from its sheen or 'artificial silk', and first produced in 1889, it was soon very popular—and remains so.

Alternatively the viscose was sent through a narrow slit into the acid bath, forming a sheet of 'cellophane'. Swiss chemist Jacques Brandenberger first achieved this in 1912, after a long search for a flexible, water-resistant film to protect food and other perishables. The American company DuPont began large-scale production under licence after World War I. Very popular and useful in its day, cellophane is still about, but has been overtaken by newer and better materials for many purposes.

Synthetic materials more removed from the natural world began to emerge in the early 20th century. Best known of these was bakelite, invented in 1904 by Leo Baekeland, Belgian-born but living in United States. He mixed formaldehyde with phenol (found in coal tar) and made a 'resin'. When precisely heated and pressed, the resin hardened, taking up the shape of its container, and thereafter would not burn, melt or dissolve in corrosive chemicals. Nor would it crack, shatter, fade or discolour when exposed to temperature changes, sunlight, dampness or sea salt. Bakelite was unlike any material known.

This first 'thermosetting' plastic (hardening when heated) found a great many applications. Its high electrical resistance was put to use as an insulator in electrical circuits and in the new 'wireless valves'.

The military found its combined lightness and strength better than steel for many purposes, and almost all World War II weapons contained some bakelite. It is obsolete now, but what a history!

> **It would not burn, melt or dissolve in corrosive chemicals. Nor would it crack, shatter, fade or discolour ... Bakelite was unlike any material known.**

Chapter 6: Not found in nature   85

Left: Tangled threads of nylon are drawn from the interface between its two precursor chemicals, neither of them occurring naturally. Nylon was the first fully synthetic fibre.

The age of plastics
Nylon, polythene, PVC and many more

In the modern age of plastics, when synthetic materials have become almost indispensable in every aspect of our lives, nylon fills a special place. This iconic product was the first artificial fibre that was totally synthetic. It is made ultimately from minerals such as coal, water and air, unlike the earlier rayon, the starting point of which was cellulose from plants.

First produced in 1938, Fibre 66—the 'nylon' name came later—was initially used in toothbrush bristles. It was soon seen as an artificial silk and, once war came, replaced real silk in parachutes. The global conflict delayed nylon's most famous use, in women's stockings; the words 'stockings' and 'nylons' were later almost synonymous. As nylon is a thermoplastic that softens when heated, clothes made from it can have permanent creases and pleats. Its strength, lightness and toughness see nylon widely used in ropes and many engineering applications.

Nylon flowed from research at the American DuPont company, founded 150 years earlier to make gunpowder, and already the makers of cellophane. DuPont was a leader in the new field of polymer chemistry—one of the most important developments of the 20th century, if measured by the diversity of uses man-made polymers have today.

Through the 1920s and 1930s, chemists at DuPont and elsewhere sought to make artificial polymers that

Left: The amazing plastic banknote, pioneered in Australia. It is almost impossible to damage and its manufacture allows the imprinting of holograms to prevent forging.

Polymers: what are they?

The natural world is full of polymers. These long molecules are made by joining up many tiny, identical chemical units, called 'monomers'. For example, a molecule of the polymer cellulose, a constituent of cotton or linen, is built up from thousands of units of the simple sugar monomer glucose. Polymers can be formed into fibres, sheets and solid blocks. The chemical name of every synthetic material includes the word poly ... somewhere.

improved on polymers found in nature, using other simple molecules including many made from coal or crude oil. DuPont research chief Wallace Carothers, who led the nylon discovery team, was among those arguing that by carefully choosing the monomers, and even chemically altering them, the resulting polymer could have whatever properties were desired.

Everywhere we look now we find everyday materials based on this polymerisation principle. Monomers derived from acrylic acid can be joined together in one, two or three dimensions to make acrylic fibres for garments, acrylic paints that dry in minutes rather than hours or days (polymerisation here produces a thin sheet rather than a thread), and solid transparent plastic marketed as Perspex, Lucite or simply 'acrylic'—common in unbreakable drinking glasses. Early contact lenses were made from such material.

Starting with the monomer styrene, we can produce polystyrene, first manufactured in 1930. Blowing air through liquid styrene causes the polymer to form a lightweight foam (Styrofoam), used in hot drink cups, packaging and thermally insulating containers. Its tendency to crumble and blow about, however, makes polystyrene less than ideal environmentally.

Materials known collectively as 'vinyls' are now ubiquitous. The name reflects that they are polymers of the simple chemical vinyl chloride. First created in 1927, polyvinyl chloride (PVC) is inexpensive, durable, chemically inert, fire-resistant and can be moulded. No wonder it is so versatile and widely used, in floor and furniture coverings and bottles, as well as in plumbing.

The story of polyethylene—now polythene—is of particular interest, since it shows that not all these discoveries are planned. In 1933 two chemists at the Imperial Chemical Industries (ICI) laboratories in the United Kingdom were experimenting by subjecting gases such as the hydrocarbon ethylene to very high pressure. On opening the test cylinder they were amazed to find not ethylene gas but a white waxy substance like plastic. This was the first ever polythene, and by 1936 ICI could make it in bulk,

Polythene arrived in time for mass production during World War II. In this it played a vital role as a lightweight insulator for electrical equipment, such as the newly invented radar sets. Now we see polythene everywhere, especially as transparent film in plastic bags and cling wrap, and in durable containers. More polythene is manufactured worldwide each year than any other plastic.

Similar serendipity created Teflon in 1938,. Starting with the newly created freon—developed to make refrigeration safer—two DuPont chemists accidentally polymerised the gas by leaving it in a cylinder overnight. Chemically-inert Teflon is famous for its non-stick properties and valued in both the kitchen and industry.

Opposite: Technology aids art. Here lightweight polystyrene balls, painted in a dazzling array of man-made colours, create a stunning display. Many artists have taken advantage of the unique properties of plastics..

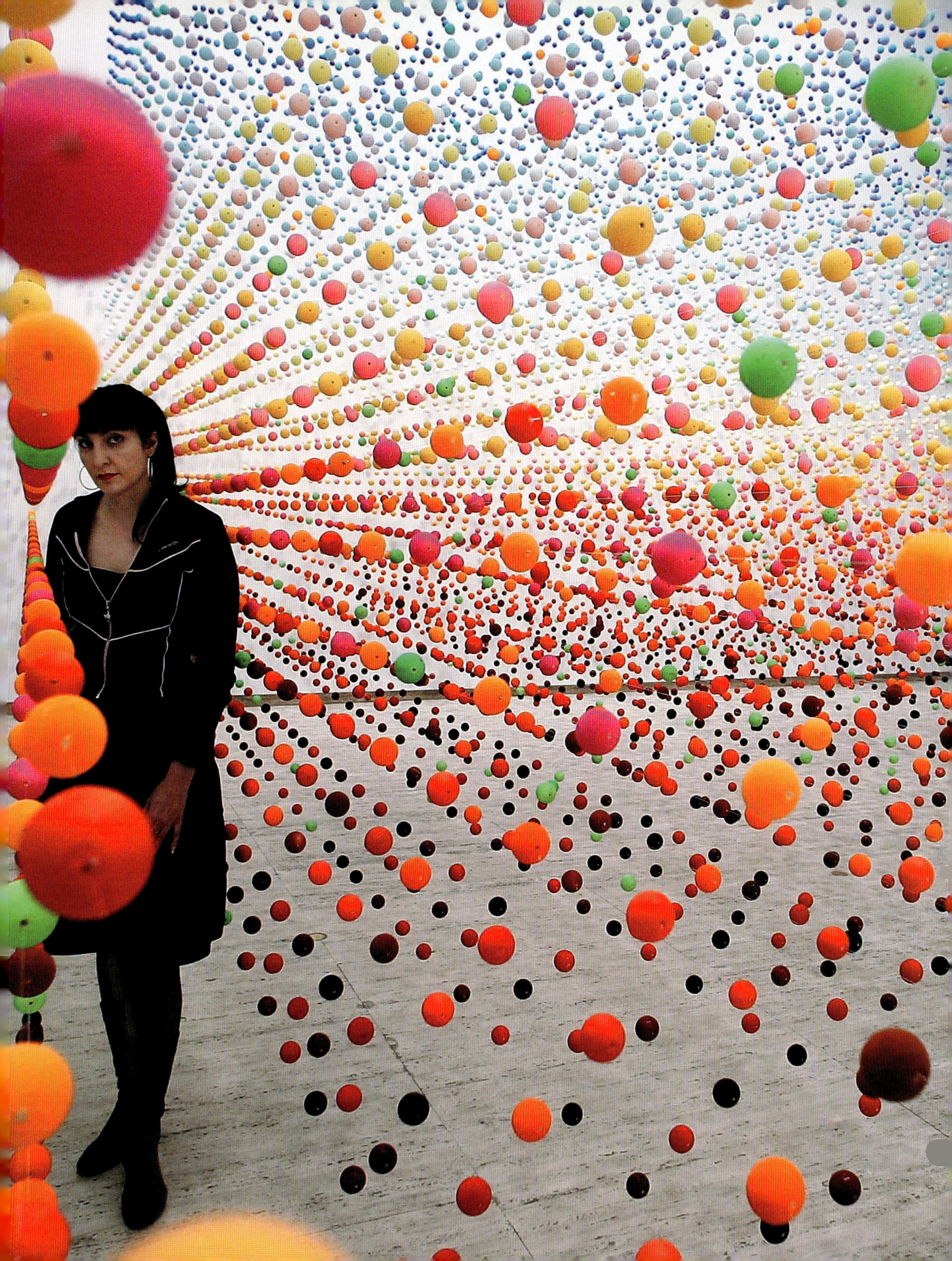

88

More polythene is manufactured worldwide each year than any other plastic.

The 'plastic age' has its doubters. Polymers are usually resistant to normal processes of decay, and so hang about in the environment for long periods, especially if carelessly discarded. For example, recent decades have seen a backlash against the ubiquitous plastic bag, with consumers encouraged to use paper or cloth bags to carry their groceries. Yet if we consider the impacts from cradle to grave, synthetic materials may prove environmentally sounder than natural ones. The production of paper for bags, for example, consumes very large amounts of water.

Spandex, Kevlar and Gore-Tex
Joseph Shivers, Stephanie Kwolek and Wilbert Gore

The golden age for inventing 'synthetics', materials that do not exist in nature, was the 1930s. This was when we were first introduced to nylon, polythene and synthetic rubber. But chemists could design other polymers with any properties they liked, so the hunt continued.

Also known as lycra or elastine, spandex was invented in 1959 by Joseph Shivers at the American chemical firm DuPont, home of nylon and many others synthetics. This remarkable fabric combines the durability of polyurethane—as in floor varnish—with the flexibility of polyethylene—as in cling wrap. It can stretch five times its normal length without breaking, always coming back to its original shape. Lightweight, soft and supple, spandex is popular in sports and exercise gear, swimsuits, wet-suits, ski-pants and the like. Comic-book super-heroes—and especially super-heroines—usually wear spandex when they appear on screen.

Also from the DuPont stable is Kevlar, which can be made into fibres and fabrics of extraordinary strength. Kilogram for kilogram, Kevlar has five times the tensile strength of steel—hence its most famous use in bulletproof vests and helmets—but it adds reinforcement in many other stressful situations, such as in car tyres. Its resistance to heat makes it valuable in fireproof garments, where once asbestos was the first choice. Invented in 1961 by Stephanie Kwolek, Kevlar is an aramid fibre (short for aromatic polyamine) and a cousin of nylon. Kevlar fibres are spun much the way a spider spins its web.

Gore-Tex is named after its inventor Wilbert Gore, who did not work for DuPont. It has been used since the 1970s in outdoor clothing and tents, as it is windproof and waterproof, but also 'breathes'. Unlike a plastic raincoat, it allows water vapour to escape, so uncomfortable humidity does not build up inside the garment. The secret is a thin layer of a polymer related to the non-stick Teflon. It has more than a billion tiny pores in each square centimetre, each much too small to let a water drop pass but bigger than water vapour molecules.

Opposite top: A B2 stealth bomber: probably the most advanced synthetic plane in the world. Created from carbon/kevlar/foam sandwich, it is not only super light, but does not reflect radar as readily as a metal plane would.

Opposite bottom: The plastic cyclist: from the carbon/epoxy frame, wheels, gears and handlebars to the PVC-coated styrofoam helmet, just about everything in this image—except the rider—is synthetic. The tyres are resin and the bright clothing is breathable lycra.

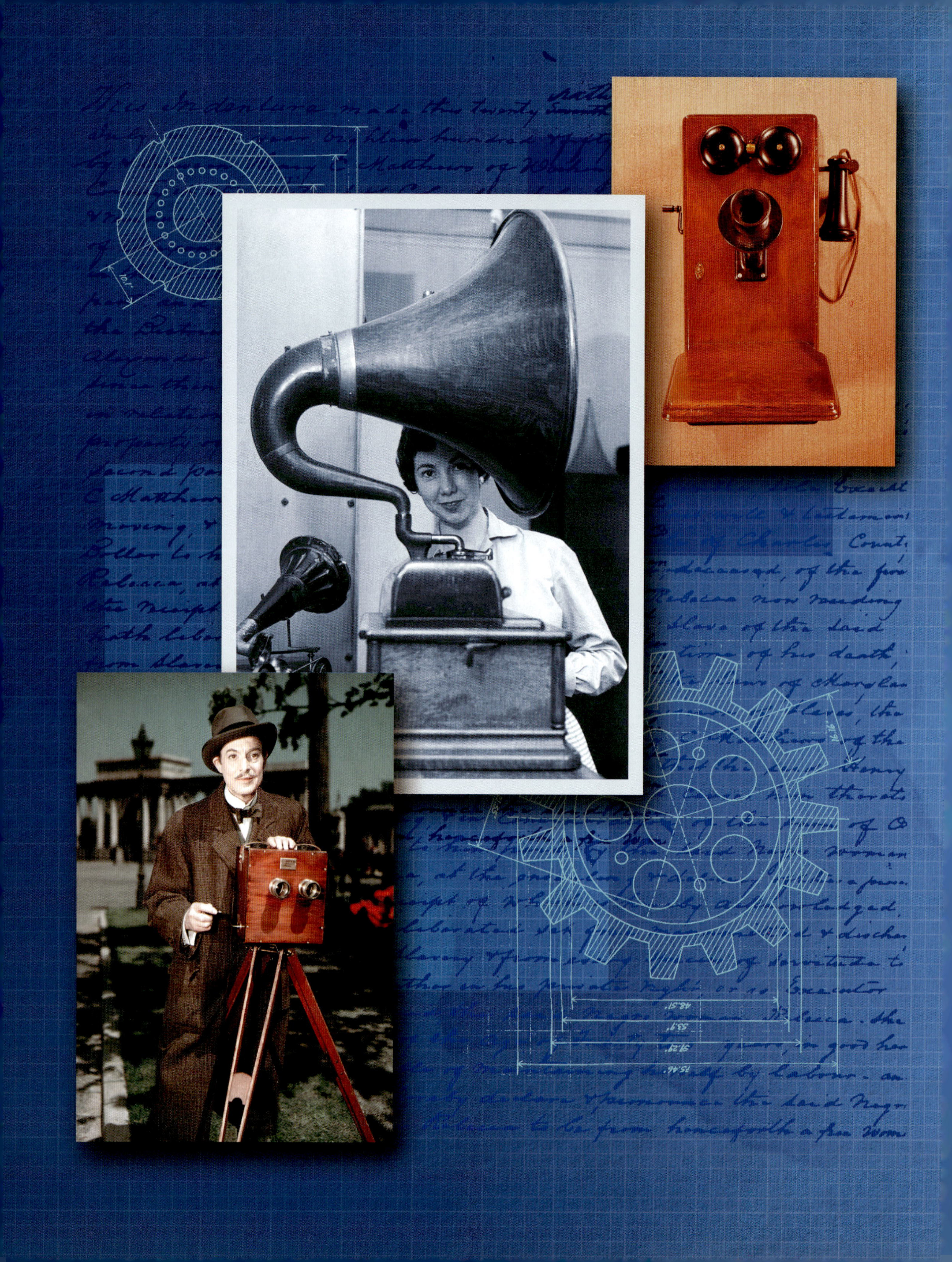

02
PART TWO:

Information impact

CHAPTER 7:
Pen, paper and printing

The ability to record information and ideas in written form was one of the first steps in civilisation. Over millennia we progressed from carving on stone and making marks on wet clay to painting symbols on papyrus or parchment. Around 700 AD we arrived at the 'quill'—a pen, made, as the name suggests, from a bird feather. The story thereafter has led to a host of inventions, all directed towards making permanent records, at least in principle: pencils, fountain pens and biros, typewriters, printing presses and photocopiers. Parallel improvements were made to the paper they made their marks on.

Something better to write with
Pencils, rubbers and fountain pens

A thousand years after its invention, the quill pen was still in use everywhere, though paper made from plant fibres had replaced parchment for everyday use. Quill pens needed careful preparation and regular sharpening, using a 'pen knife', and of course regular visits to the inkpot. However, the quill 'dip pen'—and its metal replacement—was well able to deliver the much-admired 'copperplate' script, and continued to do so through to the mid-19th century.

First to rival the quill was the pencil, invented in 1565 when a thick bed of black material called 'graphite' was found in north-west England. The stuff could make marks on paper, and the name came from the Greek 'to write'. Once it was cut into sheets, then into thin rods, and finally encased in wooden holders, we had the first pencils. The name comes from the word *pencillus*, the tiny brush Roman scribes wrote with.

If you made a mistake with the pencil, the error could be easily corrected with the aid of a new material extracted from the sap of South American trees. This could 'rub out' pencil marks, and so was called 'rubber'. For a time graphite was thought to be a compound containing lead, hence the term 'lead pencil', but before long it was correctly identified as a form of carbon.

England had a monopoly on making pencils for a long time; no other country had deposits of pure graphite. Later German and French manufacturers made pencils of different hardness by heating powdered graphite mixed with clay.

Fountain pen-making really took off around 1880, everyone wanted one, and the quill pen finally fell into the dustbin of history.

The quill pen had another emerging rival: the 'fountain' or 'reservoir' pen, which held its own ink supply. These had been around for centuries in crude form but refinement took a long time. Early fountain pens used quills as nibs, but by the 19th century manufacturers were using steel nibs, often gold-plated. Cases and reservoirs were made of rubber, hardened by a newly discovered process. Fountain pen-making really took off around 1880, everyone wanted one, and the quill pen finally fell into the dustbin of history.

Better paper, faster printing
Nicholas Robert, Friedrich Koenig and William Bullock

Paper, like many seminal inventions—including gunpowder, the compass and pasta—reached Europe from China, helped by exploration, trade, war and religious conquest. Initially all paper was made from linen and cotton rags, as are the finest quality papers today. Fabric scraps were broken down into fibres in water and treated with glue-like 'size', to hold the fibres together, and with a filler such as starch or chalk, to smooth and whiten the surface. The fibres were then spread on a sieve-like frame, which allowed water to be squeezed out and the fibres to be pressed into a sheet and dried. It was a slow process; paper remained expensive.

With the spread of literacy and the rise of magazines and newspapers—*The Times* was first published in London in 1788—the demand for paper rose rapidly. Papermakers turned to other sources of the needed fibres. In 1720 versatile French scientist René Reaumur noted the way certain wasps created paper for their nests from plant materials. This stimulated experiments, especially in Germany and France, with papers made from materials as diverse as hemp, straw, tulip leaves, potatoes, stinging nettles and cabbage stalks.

Wood was harder to turn into paper. The timber had to be ground into pulp, and it was not until 1840 that the German weaver Friedrich Keller secured a patent for a wood-grinding machine for the purpose. Early pulp paper was enriched with cotton or linen fibres for strength and quality. By the 1880s, however, all-wood pulp paper was judged an adequate substitute for cotton/linen paper. Early 'newsprint' quickly went yellow and brittle, due to the presence of lignin from the wood, giving rise to the derogatory term 'yellow press'.

The first steps towards mechanised production, to increase speed and reduce costs, were taken around 1798 by a young French student, Nicholas

The success story

Printing speeds went up fourfold, from 300 to 1200 pages per hour, and the results were impressive enough to convince the owner of *The Times* newspaper to have two such presses built. *The Times* went to press mechanically for the first time on 29 November 1814. All the other major newspapers quickly followed.

Above: Paper-making the old way. Pulped fibres of cotton or linen, mixed with size (a glue) and other chemicals are ready to be squeezed onto screens, formed into sheets and dried. It was a slow and laborious process.

Robert. His experimental machine reproduced the various steps of the hand paper-making process but made a continuous roll of paper rather than separate sheets, by placing the fibres on an endless wire mesh with rollers to squeeze it on one end.

This became known as the Fourdrinier machine, from the involvement of the Fourdrinier brothers, stationers in London, who perfected and marketed it. They had a craftsman build a working prototype based on Robert's original plans and models, and by 1803 were able to make paper of reasonable quality.

The pair spent a considerable amount of money—several million dollars in today's terms—in developing the machine, but like the original inventor Nicholas Robert they made no real profit from the invention. The rest of us have certainly benefitted, as cheaper paper permitted cheaper and more accessible printed materials. The continuous rolls of paper from a Fourdrinier machine were ideally suited for mating with the rotary printing press once that was invented.

At the time, printing too was still a one-page-at-a-time process, much as it had been since the 15th century and the days of Gutenberg and Caxton: ink the bed of type, lay on a page of paper, lower the press to make the impression, raise it again, peel off the printed page and then repeat.

The move forwards was driven by the new availability of the steam engine as a continuous source of power, and by the inventive mind of German printer Friedrich Koenig and his chief mechanic Andreas Bauer. Arriving in London early in the 19th century, they secured financial backing for their new machine, which mechanised much of the printing process. Inking the flat type-bed and pressing the paper onto it were done by rotating cylinders, but the paper was still placed and retrieved by hand.

The job of mechanised printing was only half complete. American inventor William Bullock took the final step in 1863 with the rotary printing press, which improved greatly on a machine built and patented twenty years earlier by his countryman Richard Hoe. Bullock made a cast of the type-bed and curved it to fit a cylinder. A continuous roll of paper could then be fed between one roller carrying the type and one providing pressure, with a third roller inking the typeface between impressions. Printing speeds soared—up to 12,000

Below: Paper by the kilometre. Workers in 1955 operating large-scale machinery to turn pulp into paper at a factory in Transvaal. The resulting rolls of paper would feed equally immense printing presses elsewhere.

Chapter 7: Pen, paper and printing

Stranger than fiction

In a sad and bizarre accident, Bullock died after becoming entangled in one of his printing machines and gangrene invaded his shattered leg.

pages an hour—and the machines could print both sides at once. Printing technology is fundamentally the same today, though the nature of the printing plates has changed.

A machine for writing
Christopher Scholes and Carlos Glidden

The first reference to an 'artificial machine for the impressing of letters'—today called a typewriter—is nearly 300 years old. England's Queen Anne granted a patent to one Henry Mill in 1714, the patent document claiming the results 'could not be distinguished from print'. Nor could letters 'be erased or counterfeited without manifest discovery'. Sadly, we do not know how his machine worked, or if it worked at all.

A hundred years later, the American government granted a patent for a 'typographer' to William Burt from Detroit. He did not get very far with it, especially once the only model was lost when the Patent Office burnt down in 1838. By all reports, we did not lose much either; the typographer was so clunky anyone could write faster than they could type with it.

Another quarter century passed before a practical typewriter emerged, this time from the American pair Christopher Scholes and Carlos Glidden. These two tinkerers frequented Kleinsteuber's Machine Shop in Milwaukee, Wisconsin, hoping to come up with inventions to help mankind and perhaps to enrich themselves. In the first model, around 1868, the user could press on old telegraph keys in order to bang a piece of type, attached to a swinging arm, through carbon paper to mark a page. There was no way to advance the line or provide a space, but it was a start.

Five years of work led to two patents and a machine good enough to sell. Scholes was no businessman. He sold his rights to his financial backer James Densmore, who then induced the rifle manufacturer Philo Remington to build and market the machines. Hence the Remington name is more famous in typewriter history than Scholes or Glidden.

The first model sold less than 5000 machines, but within a few years, improvements by the Remington engineers, experienced in mass-producing machinery, made the typewriter much more useful. These included the 'shift key' to type both upper and lower case. Sales soared. The lives of office workers were changed forever; new careers opened up for women, who would make the office typewriter their own for

Below: An early typewriter: heavy, clunky and totally hand-driven, but still a revolution in its day in putting words on to paper. Despite a hundred years of development, keyboards today still use the same layout of letters.

Sales soared. The lives of office workers were changed forever; new careers opened up for women, who would make the office typewriter their own for the next century.

the next century. Over that time typewriters continued to advance, especially once they became electrified. In recent years the typewriter has become the word processor and then the 'keyboard', the universal interface with computers and the Internet.

As a matter of interest, the first author to submit a novel in typescript was Mark Twain; the book was *Life on the Mississippi*, first published in 1883.

The first photocopy
Chester Carlson

Modern technology was predicted to bring the 'paperless office', yet today's workplaces use more than ever. Our paper copiers have unparalleled productivity. Yet for generations, the only copying methods were carbon paper, useful only when a document was first prepared, and the photostat, which needed chemical development and took time to dry.

Seattle-born Chester Carlson had degrees in physics and law. He headed the patent department of an electronics firm, but it was the Depression and times were tough. He set out more or less deliberately to invent what we now call the photocopier, partly to earn some more money, partly to make extra copies of patent applications. Months of library research, his background in physics and years of experimenting—in his kitchen until his wife complained—ultimately produced success. Carlson's experience symbolised Edison's famous dictum: '1 per cent inspiration, 99 per cent perspiration'.

Carlson secured a patent in 1938, but no-one else thought the invention had a future. Over the next five years, IBM, Kodak, General Electric and RCA, among others, declined to invest. Finally the non-profit Battelle Memorial Institute took an interest. Researchers there made some major improvements, using selenium rather than sulphur to coat the plate, and mixed fragments of iron, ammonium chloride and plastic as the powder.

By 1948 primitive crude photocopiers were being marketed by Haloid, a small photographic company. It took a major risk on the untried technology. Early machines needed 14 steps and 45 seconds to make a single copy; they hardly sold at all. Another ten years of work produced the first automatic photocopiers, and

Carlson's copier: how did it work?

If a light image of a document was made on an electrically-charged plate (initially zinc covered with sulphur), the plate remained charged only where the image was dark. Carlson then covered the plate with black lycopodium powder, made from club moss spores. The powder stuck where the plate was charged—on the dark parts of the image—but not elsewhere. Loose powder was blown away. Finally Carlson pressed a piece of heated waxed paper onto the plate. When he peeled it off, the adhering powder came away as a copy of the original.

Above: What we used before the ink-jet printer saved us. This popular "duplicator" forced ink through text typed on a wax stencil and onto blank paper. Messy and only one colour, but it was all we had.

these sold so well that Haloid earned $60 million in just one year, 1960. The patient Carlson got his reward, most of which he gave to charity. By then Haloid had renamed itself Xerox Corporation, and the process xerography, from a Greek word meaning 'dry'.

New ways to write (and to correct)
The Biro brothers and Bette Graham

During World War II, the Royal Air Force was looking for a new type of pen that did not leak at high altitude, unlike fountain pens. Such a pen already existed, patented in 1938 by Hungarian journalist Laszlo Biro and his brother Georg, but had not yet attracted much attention. Biro had noted that ink used to print newspapers dried more quickly than writing inks, and wondered if they could be used in pens. Ordinary nibs choked on the thicker ink, so he replaced them with tiny ball-bearings, hence the generic name ballpoint pens. The thick ink was also less likely to leak.

Sales took off once peace broke out. Competing vendors claimed the 'leak-proof' pens could write for a year, even write underwater, but poor manufacture caused many not to write at all. The ballpoint all but succumbed to consumer dissatisfaction; fountain pens returned to favour. Balls of stronger, more reliable tungsten carbide, introduced in the late 1950s, revived consumer trust. Today's major supplier, French company Bic—founded in 1950 by Baron Bich—sells 20 million 'biros' a day. Fountain pens are now more for status than everyday writing.

Meanwhile, the typewriter still dominated the office. The best typists made errors; correcting them was messy and laborious. Dallas typist Bette Graham was their saviour, inventing in 1951 what she called Mistake Out, but later universally known as Liquid Paper. She mixed up a batch of white tempura paint in her kitchen blender, used it herself to cover up and type over mistakes, and sold it to workmates. Her business boomed—even though IBM had declined to buy her patent. She sold out to the Gillette Corporation in 1979.

Initially water-based, Liquid Paper dried slowly and did not suit all papers. By the early 1980s rival products containing fast-drying organic solvents reached the market, one of the most famous being Wite-Out, which was applied with a brush and dried in seconds. In 1992 Bic bought the rights to Wite-Out, soon dominating the correction-fluid business as well as that of ballpoint pens. It also took over TippEx correcting paper, which could erase a typo by typing over it. The demand for such products was to fall before the advance of word processors, since errors could then be corrected before they ever reached the page.

CHAPTER 8:
Sending a message

The need to pass on often urgent information to someone who is not standing next to you has produced many solutions over human history: smoke signals, message sticks transported by hand, talking drums, fiery beacons, carrier pigeons. All these fell well short of the ideal. You could not be sure the message reached the intended recipient—or that it got through at all—it would often take days or weeks in transit, and the amount you could say was often limited, perhaps only 'yes' or 'no'.

Some 200 years ago, we began to find better ways to send information, initiating the growth of communication networks that now carry messages and data reliably, and all-but-instantaneously, across town or across the planet. Every aspect of our lives has been affected.

Information outruns the horse
Claude Chappe

At the end of the 18th century, the fastest and most reliable way to send a message further than you could shout was to give it to a man on a horse. Forty-year-old French clergymen Claude Chappe came up with something faster. With the aid of four unemployed brothers, Claude determined to perfect something proposed but never achieved in ancient times, a network of relay stations to transmit messages over long distances in minutes or hours, rather than days.

France being at war, the authorities soon accepted the military value of such a system. The support of a fifth brother, Ignace, a member of the revolutionary parliament, smoothed the way.

The first chain of such stations opened in 1792 from Paris to Lille—a stretch of about 150 kilometres. The stations, about 15 kilometres apart, were equipped with a tower supporting a jointed crossbar, which could be arranged at various angles and patterns to signify letters. Station operators kept watch up and down the line using telescopes, sending on messages as they arrived. At peak efficiency, Paris to Lille took as little as five minutes.

The number of 'telegraph' lines (the official term, devised by Ignace and meaning 'distant writing') soon spread across France and reached into nearby countries such as Belgium and Italy. A message could pass from Paris to Venice in six hours. Napoleon came to rely on the system to coordinate his realm in peace and war, using mobile stations on carts to bring news from the battlefront. Chappe's telegraph has a place in literature.

Above: For centuries before technology intervened, a man on a horse was the fastest and most reliable way to take a message from place to place. The famous American Pony Express lasted only a few years before the electric telegraph took over.

In Alexander Dumas's novel, the Count of Monte Cristo bribes a poorly paid operator to send a false message.

Other nations took up the idea, including Russia, Sweden and England, though in the last poor weather often reduced visibility and slowed down the traffic. Chappe's system ultimately came under challenge from the newer 'electric telegraph', though one line remained in use in Sweden until the middle of the 19th century, and officials at the British Admiralty thought the 'semaphore' they were running from London to Portsmouth worked well enough for them to turn down a proposed electric telegraph line in 1819.

However successful his invention, all did not go well for Chappe. He had founded the world's first telecommunications network, but depressed by illness, and harassed by claims from rivals, he committed suicide in 1805 by throwing himself down a well at his Paris hotel.

Chappe's telegraph has a place in literature. In Alexander Dumas's novel, the Count of Monte Cristo bribes a poorly paid operator to send a false message.

Stranger than fiction

In 1845 the new technology of telegraph helped to capture John Tawell, who had killed a woman near Slough and fled towards London, dressed in the then distinctive garb of a Quaker. This information was telegraphed ahead, but with the description spelt 'Kwaker' since the system did not code the rare letter 'Q'. Tawell was apprehended following his arrival at Paddington Station and ultimately hanged.

Messages over the wires
William Cooke and Charles Wheatstone

An electric telegraph was a possibility once Englishman Stephen Gray showed in 1729 that an electric charge could pass down a wire, but a century elapsed before it was a reality. The key problem was how to embed messages in the flow of charge. Early proposals needed 26 wires in parallel, one for each letter of the alphabet. In 1774 French inventor Georges Lesage set up such a system between two rooms of his house. Over larger distances, a practical device would demand many fewer wires.

Users also needed to know if a signal had arrived. Lesage relied on an 'electroscope'; a small ball of light wood which moved as a pulse of charge came down a wire. Some inventors proposed running the wires into phials of acidified water and watching for gas bubbles released by the inbound current. None of these schemes seemed practical.

Electromagnetism, not known until 1820, was much more promising. The pulse of current could signal its arrival by moving a small compass needle. That idea formed the basis of the first commercial electric telegraph, patented in 1837 by Britons William Cooke, a former army officer, and Charles Wheatstone, a physicist. The alphabet was arranged above and below a row of five needles; any combination of two indicated a particular letter. Only five wires were required, and the Great Western Railway Company became interested. The burgeoning railway networks needed some communication system to control the movement of trains. Cooke and Wheatstone's telegraph was installed along 20 kilometres of Great Western Railway track west from London in 1839.

The inventors later fell out over claims by Wheatstone that the telegraph was all his idea. Some dissension might have been expected, given the popularity of the telegraph and the revenues beginning to flow. By 1852, 5000 kilometres of telegraph line had been strung alongside the expanding railways and elsewhere. In other countries the electric telegraph spelt the end of the innovative semaphore systems devised by Claude Chappe and running since the 1790s. A new generation of technology was supplanting the old.

The telegraph expands
Samuel Morse

American painter and inventor Samuel Morse is often named 'creator of the electric telegraph'. He was not; others had been busy before him, but he did devise a better way to send and receive messages. Early versions required a multiplicity of wires; even Cooke and Wheatstone's successful equipment needed five. Morse knew that just two wires would suffice if letters and punctuation were represented by a code. The most common letters, such as 'E', would use the simplest combination of symbols.

Opposite: The system of signalling flags known as semaphore (Greek meaning 'carrying a sign') served the military forces well for many years. Railway signals were also called semaphores until replaced by lamps.

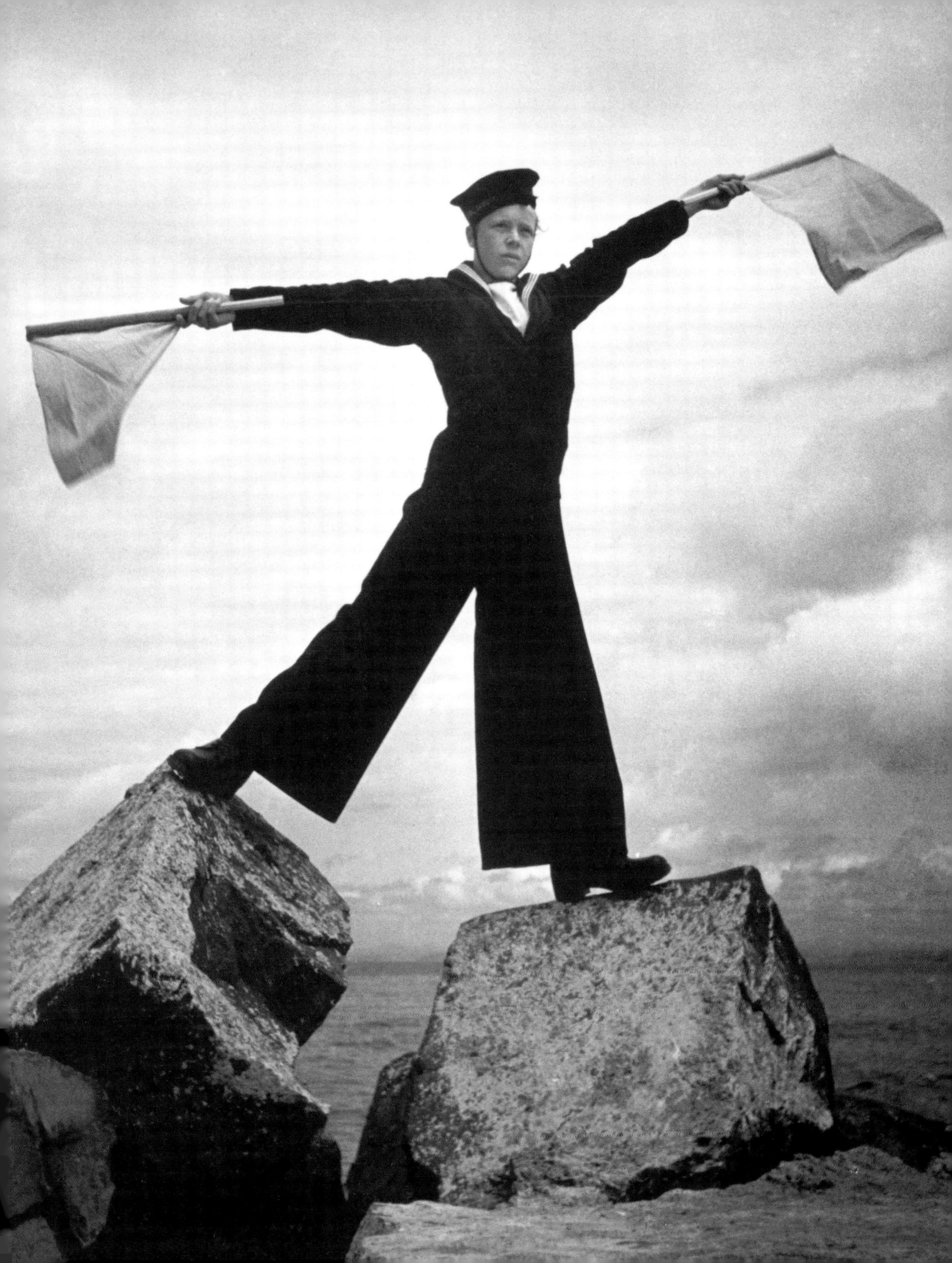

Morse and his backer Alfred Vail demonstrated his creation, with its combinations of short 'dots' and long 'dashes', before a scientific committee in 1839. 'Morse code' was an early example of 'binary code', now universal in information technology. The first telegraph line in the United States—government funded—connected Baltimore and Washington in 1844. Morse himself keyed in the first message, the biblical quotation 'What hath God wrought?'

It was suitably portentous. As in Europe, telegraph wires spread rapidly across the continent. By the 1860s a telegraph message could go coast-to-coast, ending the brief era of the 'pony express'. That expansion was paralleled by the growth of lawsuits as Morse fought to protect his patents. The various companies he and others established ultimately merged to form Western Union.

The first Morse/Vail system marked the dots and dashes on a moving paper tape. Operators soon realised they could read the message from the clicking of the electromagnetic equipment; the tape was unnecessary. Morse devised 'relays', which effectively amplified the signal and sent it over ever-increasing distances.

Below: A telegraph operator in the late 19th century marks up a Morse code message on paper tape for later transmission, increasing speed and efficiency. Note the collection of electric batteries under the desk to power the system.

Morse himself keyed in the first message, the biblical quotation 'What hath God wrought?'

As the decades passed, Morse code was supplemented by one using dots and spaces in groups of five. Such 'letters' could be punched into paper tape, automating transmission and allowing for messages to be prepared ahead of time. Still other technologies, devised by Thomas Edison among others, allowed 'multiplexing', sending more than one message at a time. The productivity and convenience of the telegraph multiplied, costs plummeted, messages sent 'by wire' became commonplace.

Though revised in 1848 to become 'International Morse', Morse code is mostly now obsolete, a historical curiosity, like the electric telegraph itself. Yet it began the shrinking of the world into a 'global village', where information and personal contact is only a few seconds away. At the end of the century, wires were no longer essential; the 'wireless telegraph' was the first step towards today's radio.

Going global
The telegraph spreads

With swarms of telegraph wires beginning to draw together cities and towns across countries and continents, the next challenge was linking the continents by well-armoured and waterproofed cables laid on the beds of the seas. Britain and France were so joined in 1851 by a cable between Dover and Calais. Britain became a part of the information network spreading across Europe.

As always, the Atlantic was the great barrier. A cable from Ireland to Newfoundland, laid in 1858 by Brunel's massive ship *Great Eastern*, soon failed from corrosion and penetration of sea water. A second endeavour in 1866 proved successful, though it required greater understanding of electric currents in wires and new technology able to make sense of the very faint signals arriving at the far end. Much of this was provided by the great British physicist Lord Kelvin.

By the turn of the century, a total of fourteen undersea cables sewed the hemispheres together. Messages could pass between major cities across the planet in minutes and hours rather than weeks and months. A single wire strung across Australia in 1876 connected the cities of the south and east to Darwin and thence to Europe by cables through southern Asia and the Middle East. Messages from the Old Country, which had previously taken half a year to arrive, were now received as fast as they could be written down.

Early on, the main users of the telegraph and the new cables were not ordinary citizens in their everyday lives. Traffic was mostly official—government business, defence matters, police reports, railway operations—as well as news reports wired in by correspondents, commercial matters such as orders for goods, currency and share transactions, share prices at the various exchanges. With the invention of the telegraphic money order, you could 'wire' somebody some money, across the world if need be.

The advent of the telegraph had many spin-offs. For example, it gave a boost to the weather forecasters by making possible the drawing of synoptic charts, the 'weather maps' on which the meteorologists base their prognostications. To draw such a chart, a large number of measurements, particularly of air pressure, had to be taken over a wide areas at the same time and communicated to a central point within a few hours, otherwise the chart would be out of date and useless. The telegraph made that possible.

Messages could pass between major cities across the planet in minutes and hours rather than weeks and months.

The birth of the telephone
Alexander Graham Bell and Elisha Gray

On 14 February 1876, lawyers for Alexander Graham Bell and Elisha Gray contacted the US Patents Office within a few hours of each other. Bell's lawyer lodged a patent; Gray's gave notice that one would be lodged within three months. Their clients had invented, simultaneously and independently, devices to send sounds, including speech, as an electric current on a wire. For reasons of priority, Bell got the patent; Gray did not, despite the backing of the telegraph company Western Union. So Bell is the name we remember.

The two men had very different backgrounds. Bell was the Scots-born son of a university professor, trained in medicine, with expertise in working with deaf children. Gray was a Quaker from rural Ohio, a decade older than Bell, who had worked as a blacksmith, boat builder and carpenter, later paying his way through college to a two-year science degree.

Both had worked to improve the operation of the 30-year-old electric telegraph, designing, for example, ways to send more than one message at a time. Both had devised systems that made an electric current vary in time with the minute changes in air pressure that constitutes sound. In Bell's design a metal diaphragm (a vibrating membrane or disc like the skin of a drum, but made of thin metal) responded to the pressure changes and moved a coil of wire near a magnet. The movements 'induced' a varying electric current. At the receiver, these changing currents powered an electromagnet, which pulled a second diaphragm to and fro, simulating the original sound. That is how Bell's assistant, the electrician Thomas Watson, heard the boss calling from several rooms away, 'Mr Watson, come here. I want to see you,' not through the air but over the wire.

Bell and Gray were not alone in this field. The German Philip Reis, among others, had already built primitive devices, and still others would make improvements on Bell's design. In 1877 Thomas Edison replaced the coil-and-magnet receiver with one containing a button of carbon, improving fidelity (sound quality). Even so, the telephone was not quite ready for mass use. Over the next few years, Bell, Watson and others added a hand crank to signal down the line that a call was being made, and put the receiver and transmitter on a common handle. This hand-piece hung on a hook; lifting it opened the line to the 'exchange'. There 'operators', later mostly women, put plugs into sockets to connect callers to their desired destinations. 'Dial phones', which instructed exchanges operating automatically, were still 40 years away.

The appeal of the telephone over the telegraph was its immediacy, convenience and privacy. You heard the voice of the person you are talking to, rather than just reading their words. Messages could be sent from your own home, rather than needing a trip to the telegraph office. Once the call was established between the two parties, no-one else need interfere.

Bell's telephone caused a sensation at the 1876 Centennial Exposition in Philadelphia; the Emperor of Brazil reputedly put a receiver to his ear and dropped it with a start, crying 'It talks!' The phone was not, by

Opposite: A classic wall-mounted telephone, still used in rural areas well into the 20th century. It had no dial and the user had to 'ring' the exchange (by turning the handle on the left) to have their call 'connected'.

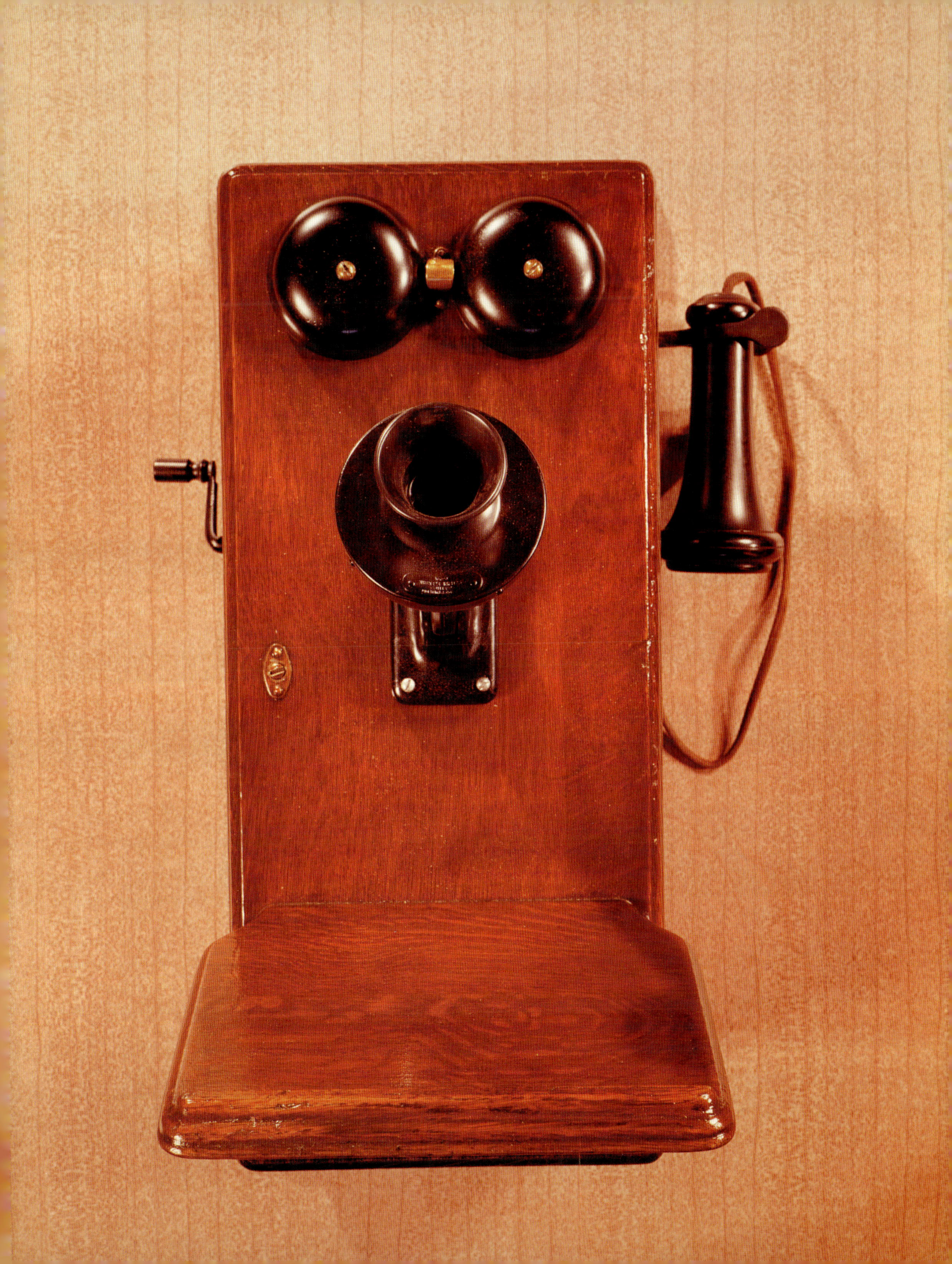

Left: Telephone pioneer Alexander Bell listens to the earpiece of one of his early creations, while his indispensable assistant Thomas Watson looks on. From this picture, it is hard to guess how convenient the telephone would later become.

industrialised nations had one, though a threat to the dominance of the 'landline' would emerge a few decades later from the 'cell-phone'.

Despite losing the patent battle over the telephone, Elisha Gray did not fade into obscurity. He became a well-regarded university professor, and held 70 patents. His company, Western Electric Manufacturing, became today's Lucent Technologies. A deeply religious man, he wrote extensively on the nexus between science and faith, and advocated the concept of 'intelligent design', much discussed today. He died in 1901.

today's standards, an overnight commercial success. In 1895, twenty years after its invention, less than 300,000 were in use in the United States, serving about one in 50 of the population. Long-distance calls, possible with the development of 'electronics', made the telephone more popular; the number of subscribers trebled over twenty years. By the 1950s just about every home in the

A foretaste of the Internet
ARPANET and Ray Tomlinson

One of the enduring myths about ARPANET, which was like a trial run of the Internet ten years earlier, claims it was devised so the American military command and control system could survive a nuclear attack. If the

The success story

Aged only 30, Bell was rich and famous, married to the daughter of his financial backer, a Boston lawyer keen to break the Western Union monopoly. With his Bell Telephone Companies rolling out wires on poles throughout cities and towns to link his telephones, Bell continued to innovate. Among his many creations, he thought most highly of his 'photophone', which sent sounds on a modulated beam of light, anticipating today's 'optic fibres'. The death of his newborn son from respiratory problems led him to construct a metal jacket to aid breathing. This presaged the 'iron lung', used by polio victims many decades later. Water desalination, hydrofoils, sheep breeding to increase multiple birth, all these caught his attention and he was an early and passionate pioneer of powered flight. Bell's mind remained active until his death at 75.

Packet switching: what's that?

A piece of information, such as a document, image or sound passing along the network, is initially broken up into many small messages or packets, all the same size, like electronic envelopes. Each packet is given an identification number, showing where it fits into the document, and a destination and is then sent on its way. The network is studded with special computers known as switches; these read the destination of each packet and point it down the right wire. Whatever path the individual packets might take across the network—the Internet now has thousands of possible routes—all the packets ultimately arrive at their destination within a few seconds and can be re-assembled to present the complete original information.

large computers that had become so important for the national defence could be linked together to exchange information, the loss of one or more from a sudden nuclear strike would not mean the information was lost, and the whole system would not go down. While this is a reasonable motivation, those who were there at that time generally agree that the defence justification came later. There was a more immediate need.

ARPA stands for Advanced Research Projects Agency. This organisation encompassed—and paid for—researchers across the United States involved in projects of military significance. They could all benefit from access to the large computers of the day, but the computers were relatively few in number and widely separated geographically. Putting these computers 'online', as we would say today, so that anyone could access them from anywhere, seemed a good idea, so it was done. The first computers, four of them in all and all involved in research rather than operations, were hooked up to ARPANET in 1969. They connected Los Angeles, Santa Barbara, Stanford and the University of Utah through 56 kilobit-per-second links—top speed for the time. Through the 1970s, ARPANET connected computers in research labs supported by the Department of Defense.

To make the network operate, and to allow different sorts of computers to talk to one another and exchange information, the engineers devised some vital new technology, including the rules we now know as the Internet Protocol (IP) and the File Transfer Protocol (FTP). Central to these was 'packet switching', a vital part of the Internet's operations today. This was the beginning of something really big.

ARPANET pioneered another major application, one about which there are some mixed feelings. In 1971 Ray Tomlinson pioneered email. A form of email already existed; people using the same computer could leave messages for one another. Each had an electronic mailbox, which only they could open but in which others could leave messages. A number of computers were already talking to one another through ARPANET. Why couldn't we send a message to someone on another computer?

Once Tomlinson had thought of it, making it happen, by devising a new protocol, was easy. He simply piggybacked on an existing program that allowed computers on the network to exchange files. Tomlinson initially sent messages from one computer to another standing right beside it; these were connected only through the network. Legend has it that his first messages were something like QWERTYUIOP, generated by simply dragging a finger over the top line of the keyboard. This is not of the

A terminology so widely used today was, he later admitted, the result of only a half minute's thought.

same impressive tone as Alexander Bell's, 'Mr Watson, come here. I want to see you,' or Samuel Morse's 'What hath God wrought?' but the impact on our daily lives has been at least as great.

Tomlinson's contribution to computer speak was the @ symbol. He needed a way to distinguish between messages going to people using the same computer and those going out onto the network, to someone somewhere else. By his own account he settled on the @ (meaning 'at') since it was not used in anybody's name and could be taken to designate a place. It fitted comfortably between the user's log-on name and the name of their home computer. A terminology so widely used today was, he later admitted, the result of only a half minute's thought.

Tomlinson would also later admit that no-one was clamouring for email at the time he invented it. It just seemed a 'neat idea', something that could be done. Yet few computer applications have had the impact of email, especially once it was married to the Internet. Some people hate it, but others would confess they could not run their businesses, or even their personal lives, any other way.

The telephone goes mobile
Martin Cooper

On 3 April 1973, Dr Martin Cooper, employed by electronic firm Motorola, made a phone call, not from his desk but while walking in a New York street. He was using a prototype of the 'mobile' or 'cellular phone', which Cooper, as much as anyone, could claim to have invented. That telephone was the size of a brick and weighed a kilogram, far chunkier than today's mobiles—100 grams or less and able to slip into a pocket—but it proved the concept.

Person-to-person communication using individual networks of mobile radios had been used for decades by emergency services, like police and ambulances, as well as by taxis and long-distance truck drivers. Motorola supplied much of the equipment they used. There was also the now barely-remembered portable pager, invented in 1949 and widespread in the 1980s and 1990s, which could receive a small message by radio—such as 'ring your office'—but not send one. In time, the mobile phone would let the pager talk back.

The concept of the cellular phone went back at least to 1947, the year of invention of the transistor, which would make so much difference to the size, cost and power of mobiles. The company AT&T (aka Bell Telephone) was an early mover—Cooper's call in 1973 was to an executive of AT&T, then their rivals in the business. Unfortunately, the Government

The cell-phone system: how does it work?

The system operates by dividing an area into small zones called cells, each with a radio base station that connects to mobiles in its cell and is in touch with other cells. As a user moves from one cell to another, the system adjusts automatically. Transmitting and receiving power can be kept low and the available radio frequency channels used more efficiently.

Chapter 8: Sending a message

Above: Instant connectivity as we enjoy it today. Three telephone conversations going on simultaneously without a wire in sight. The mobile phone has become one of the most rapidly-adopted new technologies ever.

regulator, the Federal Communication Commission (FCC), allocated so few radio channels to the initiative that hardly anyone thought it had a future; there was little incentive for research and progress very slow.

Not until 1968 did the FCC reconsider its position; if the technology could be proven, more frequencies would be allocated. The chase was now on, with AT&T and Motorola in the lead. By the late 1970s trials of the new technology were running in cities like Chicago, with a few thousand handsets. A decade later the mobile was fast becoming mainstream.

The United States did not have one million mobile phone subscribers until 1990, but by then growth was becoming exponential, as costs fell and the early systems were creaking under the load. Today mobiles number in the billions globally, with more mobiles than landlines now connected.

Consumer demand has melded with technological

Right: Connectivity 2.0. Nowadays, small handheld devices, such as this iPhone can link us to many data sources, act as a camera, a note book, diary and address book as well as a simple telephone. Taking and transmitting photos is now almost as easy as talking on the phone.

progress to make the mobile phone one of the most convenient and widely-used inventions ever. Handsets have dwindled in size and weight through more powerful computer chips and smaller batteries More phones have permitted smaller cells and driven down the required power. We moved from analog operation, in which an electric current varies in time with speech sounds, to digital, in which sounds are transmitted in binary code, with a major improvement in reception. Successive generations of technology, such as the 'third generation' (3G), have used higher operating frequencies, carrying much more information. 3G phones can handle moving images as well as high-quality speech. 4G phones, not far away, will do even better. It has been a remarkable twenty years.

The other trend is convergence, with one device performing functions that previously needed separate machines. Most new phones include a digital camera. Many are pocket computers, storing data, sending and receiving e-mails, playing video games, accessing the Internet. Phones now often contain an MP3 player.

However, the cost of buying phones remains low and stable, even as functions are added. Some are even given away by the phone networks. The declining costs reflect better technology and rising demand, with economies of scale. Access to data networks like the phone system and the Internet is what gives real value to the user, and how the supplier secures their profit.

Right: The increasingly ubiquitous 'Internet cafe' lets us surf the net or send e-mails over a cup of coffee, no matter how far from home we are. Internet cafes have become the global traveller's lifeline, allowing financial transactions, travel and accommodation bookings and contact with friends and family, all for a very reasonable fee.

Consumer demand has melded with technological progress to make the mobile phone one of the most convenient and widely-used inventions ever.

Commentators may argue about the balance of impacts of the 'anywhere, anytime' capacity for communication that mobile phones bring. The mobile phone's impact has been profound, perhaps greater than that of any other invention since the car and the personal freedom of movement that provided.

The ultimate data network
Tim Berners-Lee

With hundreds of millions of people sitting at their computers and accessing the Internet every day, connecting to each other almost instantaneously around the world, few would doubt its power and influence. But the Internet is just a computer network, or a network of computer networks. Its value comes from the way we use it, to send emails, for example, or to access the World Wide Web (the 'www' in Internet addresses).

The Web is a remarkable system, enabling anyone to access a publicly-available document or other information source, wherever it might be, all but instantaneously, to read it, to print it out, to store a copy on their own computer—and for no cost other than the Internet connection. It is the most far-reaching and extensive means for exchanging information that has ever existed on this planet.

For all its reputation, the Web started small, initially involving just the scientists at one European research centre, the CERN particle physics laboratory near Geneva. One of these was Briton Tim Berners-Lee. Many of the issues about sending documents through the Internet had already been solved through the workings of ARPANET in the United States. The question now was how to find what you wanted to have sent to you.

In 1989 Berners-Lee proposed 'a global read-write information space'. Each discrete information source, such as a scientific paper or a press release, would have its own unique ID, a 'uniform source identifier'. Programs on your computer called browsers or search engines could find the source for you. Within two years, the World Wide Web—his term—was in use across the particle physics community; within another two years, other academics and businesses were taking notice. The latter soon realised that enabling customers to find you on the Web ('having a Web presence') through their own webpage or website was more that useful; it was essential.

In 1994 CERN announced that anybody who could connect to the Internet could use the Web, and it would be free. Not surprisingly, the years following were a period of extraordinary growth in Web use and in Web presence, the latter including the relatively new phenomenon of blogging.

The success story

Berners-Lee's honours include a knighthood in 2004, the Fellowship of the Royal Society in 2001 and being named by *Time* in 1999 as one of the 100 most influential people in the 20th century. Given the pervasive influence of the World Wide Web, the last may understate the case.

CHAPTER 9:
Making an image last

We collect images of the world around us all the time through our eyes, but these soon lose detail once stored in our brains, and most disappear completely. Furthermore, we cannot share these images with others; we generally rely on words to describe a scene. Artists have long tried to capture their impressions of the world, but the results are often subjective, or deliberately not true to life. Hence the immense and ongoing impact of images preserved through technology.

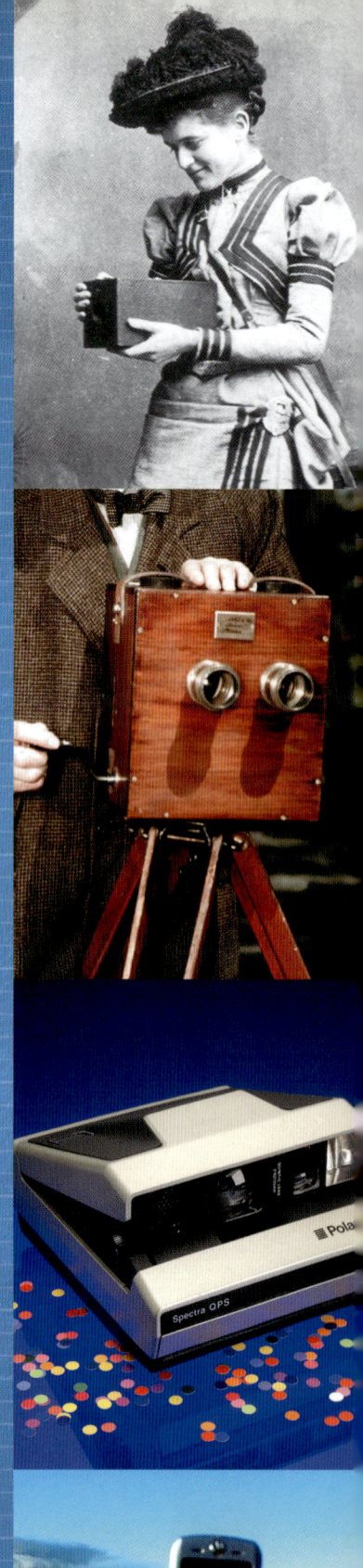

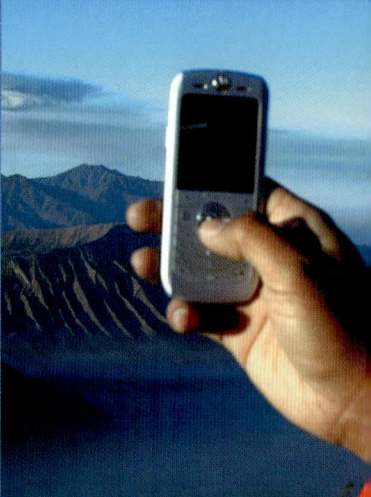

Beginnings of photography
Joseph Niepce and Louis Daguerre

Photography ('writing/drawing with light') is one of the great 19th-century inventions, but its origins lie much further back. As early as 1727 German doctor Johann Schulze found that certain 'salts' (compounds) of silver, like silver nitrate, which are normally pale in colour, went dark when light fell on them. Fifty years later Swedish chemist Karl Scheele found that silver chloride darkened under the influence of light, because it decomposed into silver and free chlorine.

Thomas Wedgwood, of the famous pottery family, and leading scientist Humphry Davy experimented with these chemicals, and found they could make silhouettes of leaves and other objects by placing them on paper covered with silver salts and exposing them to sunlight. But they could not make the image permanent; undarkened areas soon went black, and the image vanished.

Meanwhile something apparently unrelated had been going on. The familiar term 'camera' (Latin for 'room') goes back to the phenomenon of the 'camera obscura' ('hidden room'). Light from a brightly lit scene was let through a small hole in one wall of a darkened room. It threw a faint but distinct image of the scene upside down on the opposite wall. Since the 17th century artists had used miniature versions of this—a small box on legs with a lens in front and an opaque glass screen at the back—to view the image. It was a useful tool to frame a scene for painting.

Someone needed to make the leap of imagination that would bring together the silver salts and the camera obscura, to capture, and maybe preserve, the image on the opaque glass by substituting it with a piece of glass covered in light-sensitive silver chemicals. That someone proved to be Joseph Niepce, a French amateur scientist. He made images in this way, and even partially 'fixed' them (prevented fading), but they were 'negatives', with black where white should be. He did not know how to make 'positives'. The images also had poor definition and contrast, but it was a start.

More than a decade later, Niepce was still struggling to find the right combination of chemicals to develop and fix his images. Around 1829 he linked up with artist Louis Daguerre, but died a few years later. Daguerre continued with some new ideas of his own, involving silver plates, iodine, mercury and common salt. Making his 'daguerreotypes' required much chemical manipulation, but the images were bright and clear. After some improvements in technique, an image

Below: Taking your own picture. Late 19th Century photographer Hippolyte Bayard with his camera in one of the first self-portrait photographs ever taken.

Eastman once said the name Kodak had no meaning; its appeal lay in its sound, like a camera shutter opening and closing.

could be made in a short enough time for photographic portraits to be feasible. Alert entrepreneurs were soon setting up studios where people could go to 'have their picture taken'.

Progress through many minds
John Herschel, William Henry Fox Talbot, Frederick Archer and George Eastman

Lots of people had ideas on how best to proceed, especially in terms of the chemistry. Astronomers were among the very first to make use of the new technology, and English astronomer John Herschel proposed new chemicals for fixing the image. He coined the terms 'positive', 'negative' and 'photography'. English archaeologist William Henry Fox Talbot found out how to make many prints from one negative, impossible to do with daguerreotypes.

Around 1850 Fredrick Archer of England developed a process to produce sensitive 'wet plates'. These gave good results but needed to be prepared immediately before exposure and developed straight after. This could be done in the increasing numbers of photographic studios but was difficult outdoors. By the 1870s photographers could buy 'dry plates', glass rectangles covered with gelatin to hold the light-sensitive chemicals. These needed much longer exposures than the wet plates but worked with landscapes and were much more convenient to use. Photographers no longer had to lug a trolley laden with chemicals.

Photography as we know it today, done by amateurs without complex equipment or poisonous chemicals, dates from the 1880s, and the key name is American George Eastman. The newly invented material 'celluloid' was now available in strong, flexible sheets, and Eastman began to market such sheets coated with light-sensitive 'emulsion' for imprinting photographs.

Once made into rolls, these 'films' fitted into something very new, an easily-carried camera called the Kodak. Eastman once said the name Kodak had

Right: When photography was hard work. A little boy watches a 'wet plate' photographer at work at a fete. At the time, glass plates covered in wet chemicals had to be prepared just before use and developed immediately after.

no meaning; its appeal lay in its sound, like a camera shutter opening and closing. First marketed in 1888, the Kodak was sold already loaded with enough film for 100 pictures. When these had been exposed, the camera was sent back to the factory, where the pictures were developed and printed, and the camera reloaded.

As popularity grew, other developments inevitably followed: films able to be loaded into a camera in daylight so the user could do it herself; films producing coloured images; safety films that did not burn; ever simpler and more powerful cameras, which were cheaper in real terms. This outpouring of innovation continued until the arrival of digital technology and electronics more than 100 years after Eastman's revolution. Even Kodak stopped making film cameras in 2004.

Pictures that move
Thomas Edison and the Lumière Brothers

Motion pictures (the 'movies') were arguably the dominant cultural influence in the West in the 20th century, at least until television arrived, but their origins, a century earlier, lay in toys.

These toys had a variety of names. In the thaumatrope ('wonder turner') of 1826, each side of a card carried a different image, such as a bird and an empty cage. Spinning the card rapidly on a twisted string seemed to put the bird in the cage. A few years later, the phenakistiscope ('deceptive view') contained a number of images set around a drum. The viewer saw them reflected in a mirror as the drum was turned. Abstract patterns moved; acrobats tumbled. The simpler zoetrope ('wheel of life'), initially ignored but revived in the 1860s, allowed more than one person to view at a time, the later praxinoscope ('seeing something happening') set the moving images against a background.

None of these shows could last very long, a few seconds at most, before repeating. Longer

> **Moving image toys: how do they work?**
>
> These made use of the slow rate at which the human brain processes information. A series of still images, each slightly different from the next, appear blended into a moving image if they are presented to the eye at least at 16 images per second and with a blank space before each image.

performances needed many more images. This was made possible when George Eastman began marketing photographic 'film' made from celluloid. 'Magic lantern' shows, which projected photographic images onto a screen, were already popular for entertainment and education. Combining that with the idea behind the persistence-of-vision toys ultimately produced the first movies, but it took several decades, and many active minds including Thomas Edison's.

A kick-along came from the famous series of images taken by English-born Californian photographer Eadweard Muybridge in 1872 to prove that a galloping horse has all its hooves off the ground together at one point during each stride. He arranged dozens of cameras in a line beside the horse's path, exposing them one after the other by tripwires. Edison knew of this work, but thought that a single camera taking multiple images on a long strip of film would be much more practical. By 1891 he had developed the kinetoscope ('seeing motion'), though his assistant William Dickson did most of the work. This was a 'peepshow' for one viewer at a time, marketed like his coin-in-the-slot phonographs. These became very popular in amusement parlours, showing a minute of moving image for 50 cents.

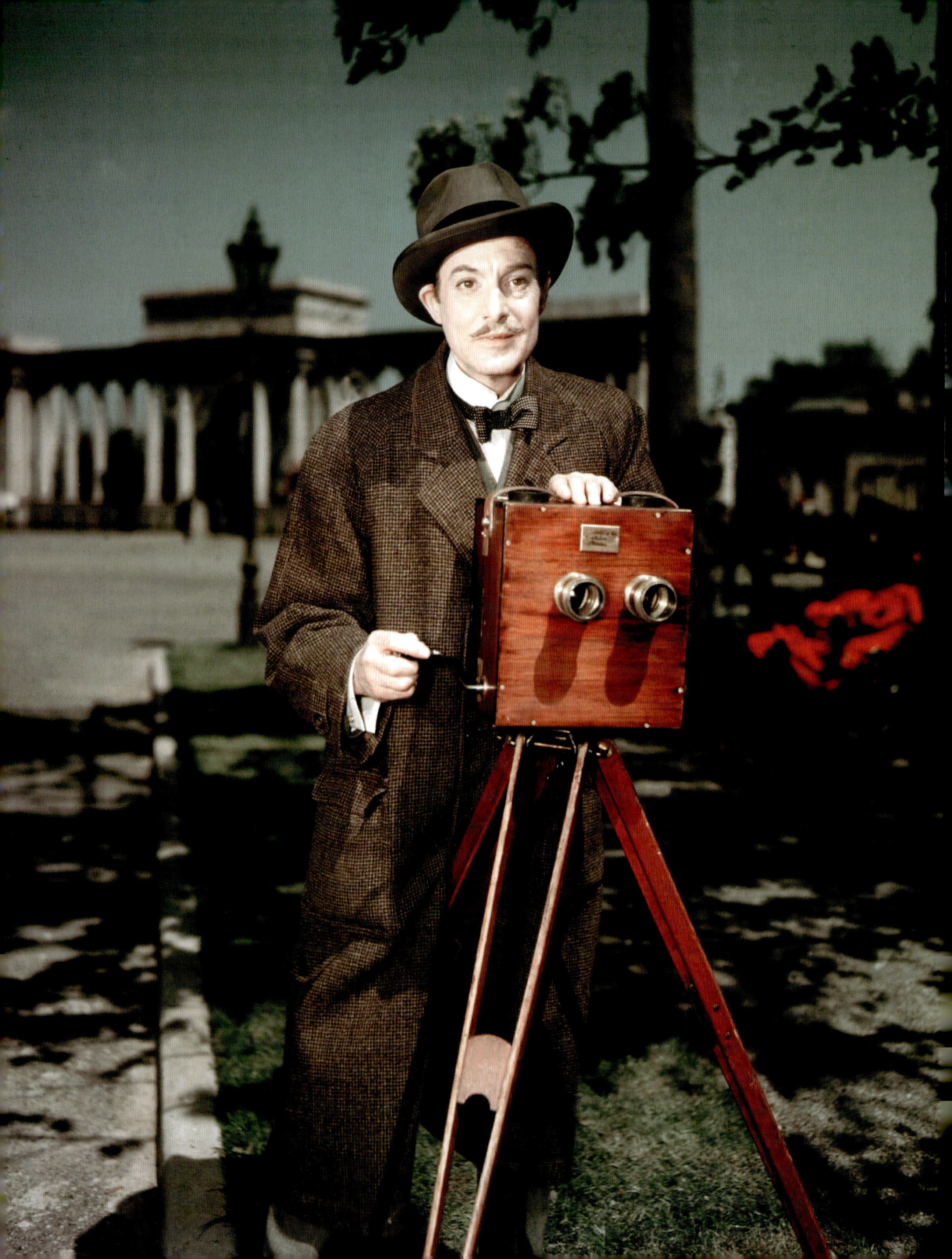

A train pulling into a station—so realistic that people ducked for cover, fearing they were about to be run over.

For others, the future lay in entertaining or amazing a whole audience at once. The leading players in America included young collaborators Charles Jenkins and Thomas Armat. They had devised a motion picture projector, the Phantoscope, by 1895, demonstrating it at an exposition in Atlanta. The pair split up soon after, each claiming to have invented the machine, which indicates the interest it had aroused. Armat sought help from Edison, who agreed to build the machine provided it was marketed as Edison's Vitascope. The first movie screening in the United States took place in a theatre off Herald Square in April 1896. Images of rough seas off the Dover coast were the most popular.

Edison left his imprint on the new technology. His film stock was 35 millimetres wide, still the standard size for professional moviemaking today. He placed holes down the side of the film to pull it through the projector. He also developed a revolving shutter to flash the images and a 'Maltese cross' device to stop and start the film many times a second. Emerging moviemakers everywhere used very similar ideas.

Nonetheless, Edison had been beaten to a key milestone by August and Louis Lumière, the 30-something sons of a Paris photographer. The brothers used equipment very similar to Edison's, but combined the camera and projector into one relatively lightweight machine called a cinematograph. They staged the world's first public screening in the basement lounge of a Paris café on 28 November 1895. The program of ten films lasted only 20 minutes, but the patrons marvelled at moving images of everyday scenes, a rowboat in a harbour, workers leaving a factory, a train pulling into a station—so realistic that people ducked for cover, fearing they were about to be run over.

Though pleased with their achievement, Louis Lumière saw no future in the 'cinema'. Why would people pay to see things they could observe just by stepping outside? Yet the patrons came in increasing numbers. Into the new century, the technology found its role as entertainment, filming vaudeville acts and scenes from successful plays, and ultimately creating new stories on film, presenting action not possible on the stage. A technological curiosity was about to become a major art form.

Adding colour
Herbert Kalmus

Despite the early triumphs of photography, capturing realistic colour proved difficult. For many decades both still pictures and movies were 'monochrome' (usually black and white). We knew the science behind colour perception well enough. In the 1860s, Scottish genius James Maxwell had created coloured images by taking three separate pictures through red, green and blue filters and then projecting them top of each other through the same filters. Though complex, this additive technology delivered the earliest colour prints.

It was used in the first successful process for colour movies. Technicolour was invented by Herbert Kalmus and his colleagues early in the 20th century—the 'Tech' in Technicolour recalling their alma mater, the Massachusetts Institute of Technology. Light entering the camera was split into separate beams, passing through

Opposite: A movie about movies. Actor Robert Donat plays pioneering photographer Willie Friese-Green taking his first moving pictures in Hyde Park, London, as part of a film called (appropriately) *The Magic Box*.

Left: An early advertisement for the new Kodak camera, with 'the most advanced ideas in camera construction'. The prices quoted would be hundreds or thousands of dollars in today's money.

Opposite: An elegantly-dressed woman around the turn of the 20th century, with the new wonder. The image emphasises that anyone could use one, although the camera looks clunky by today's standards.

coloured filters: initially two, red and green (a third, blue, filter was added later) then onto two, later three, strips of monochrome film joined side-by-side.

In projection, the developed strips of film ran through the projector side-by-side. Separate light beams passed through filters and blended the film into the coloured moving image on the screen. Much depended on the skill of the projectionist. Kalmus was on safer ground with a subtractive process, using the black and white originals to print three coloured images one above the other in a single film and projecting one beam of light through the lot. The original films being black and white, the product was very resistant to fading. Properly cared for, Technicolour movies should retain their colours for centuries. The effect is brilliant. Walt Disney adopted Technicolour very early and that helped explain the popularity of his animated films in the 1930s.

The system was costly and cumbersome, however, and the hunt was on for a comparable single-film process. This involved passing the light through three transparent light-sensitive layers one after the other, each recording a primary colour. When the three developed layers were viewed together, the result was very effective. Eastman Kodak put the first still film using this process onto the market in 1935—the same year *Becky Sharp*, the first movie using three-colour Technicolour, came out—and marketed a movie version called Eastmancolour in 1952.

The technology continued to advance, and manufacturing colour films became a remarkable achievement, requiring a dozen separate chemical-packed layers, precisely laid to make a strip less than a millimetre thick. Yet, however clever, it is all old-hat, now we have digital cameras.

Pictures in an instant
Edwin Land

The Polaroid 'instant' camera is now obsolete, overtaken by the advance of digital or electronic cameras. Yet for 30 or more years around the middle of the 20th century, it offered something no other camera could offer,

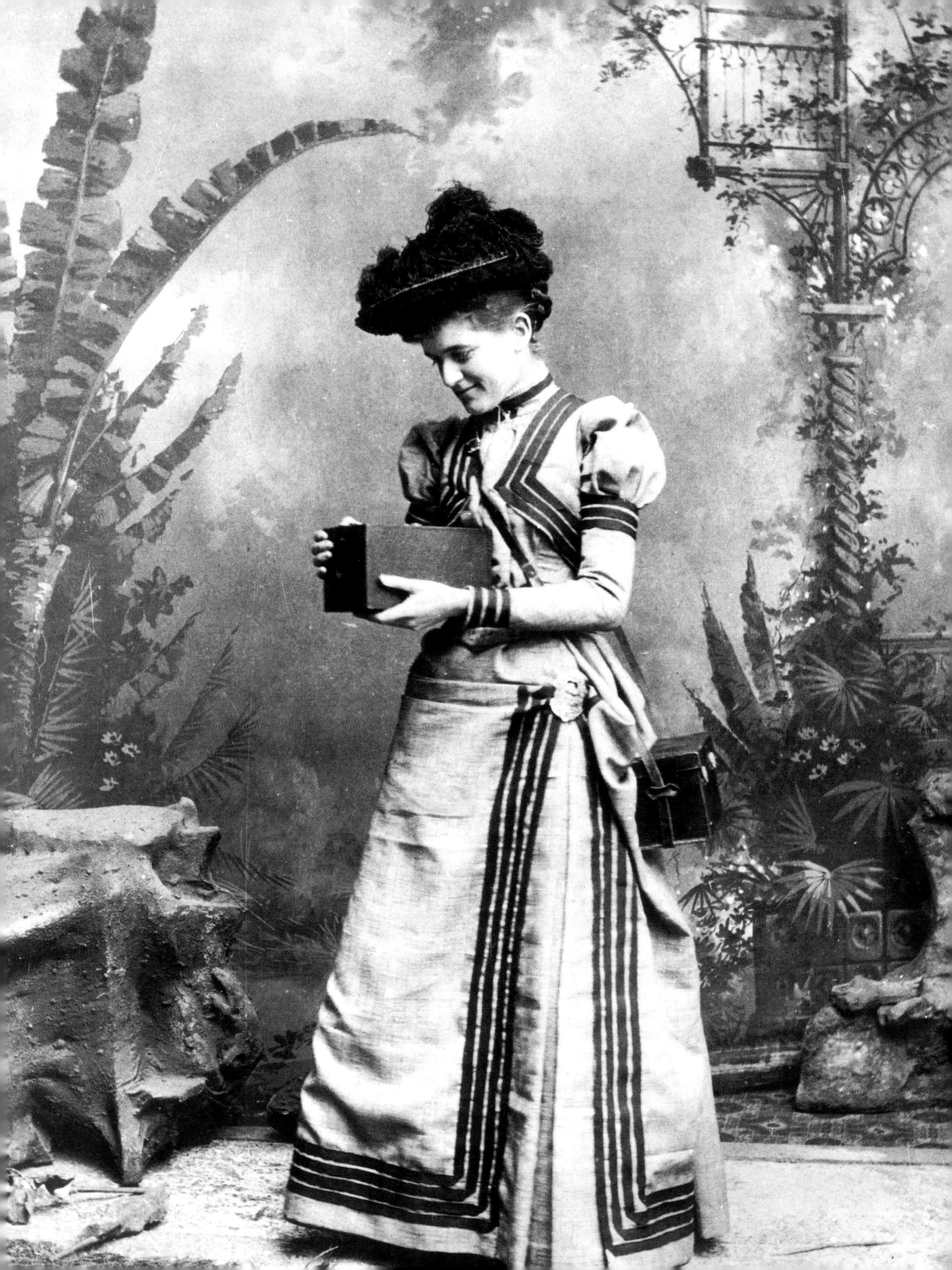

pictures able to be admired—or discarded—within a few minutes of being taken.

Edwin Land was the innovative mind and driving force behind both the Polaroid camera and the widely-used Polaroid filters. The phenomenon of light polarisation caught Land's attention while a physics student at Harvard. In a light wave, the energy fields vibrate from side to side at different angles. In polarised light the vibrations all line up. A century earlier, French researchers had found that Iceland Spar, a naturally-occurring crystal, filters light according to the angle the vibrating fields take. While still a student, Land put that property into plastic sheeting. Everyday unpolarised light battles to get through a polarising filter, so its intensity is greatly reduced. Land's 'Polaroid' had many applications.

A Polaroid camera does not use Polaroid, however; the name is simply that of its manufacturer, the Polaroid Corporation. The inspiration for an instant camera was, we are told, Land's three-year-old daughter. While on a holiday in Santa Fe, she asked why she could not see the picture of her he had taken a minute earlier. Why not, indeed? Land at once sketched out the many processes needed to develop exposed film automatically inside the camera, so it could eject a finished print within a few minutes.

Below: Although now come and gone, the Polaroid instant camera was a marvel in its day, delivering a freshly-taken image 'hardcopy' in a few seconds. Its demise underlines the dynamic nature of modern technology.

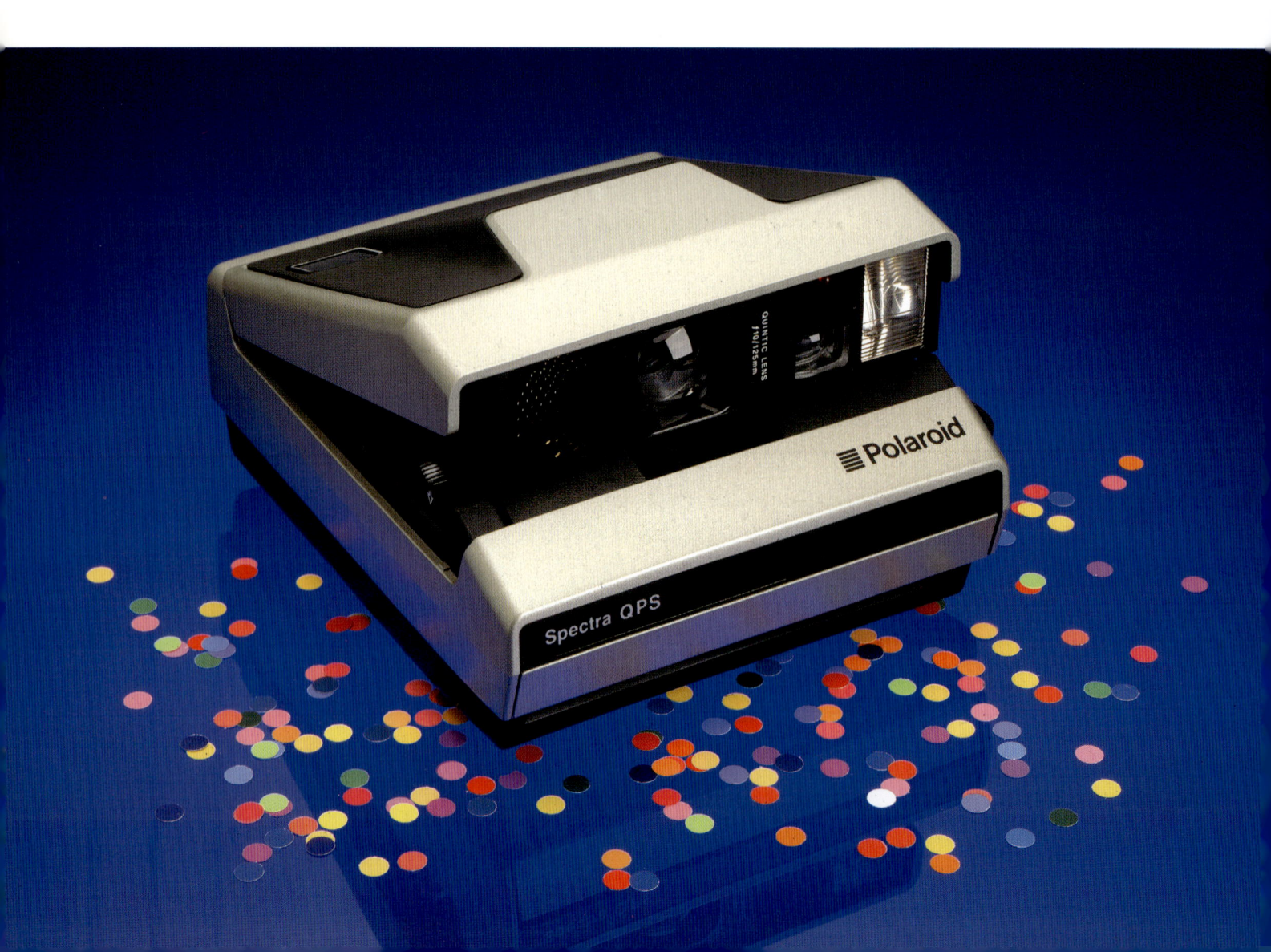

A multitude of uses

Land's Polaroid technology was ideal for sunglasses, reducing glare without affecting colour. It could also separate the two images presented simultaneously in 3D movies so that each eye sees only one image. There were other uses in photography and science. Major firms like Kodak and General Motors bought Land's unique product. It was a goldmine.

The first version, making sepia prints, took three years to perfect, reaching the market in 1947. Instant black-and-white pictures took another two years, instant colour pictures a further decade. Given the complexity of the processes, the technology was remarkable; through the 1970s sales soared. Film for instant movies, however, failed to sell; strong competition came from the equally new, and equally 'instant', camcorders which saved moving images on magnetic tape. Land dismissed the threat, believing that photographic film would always outperform electronics. The ultimate demise of the Polaroid camera settles the argument.

Capturing the third dimension
Denis Gabor

Our world is three-dimensional, with depth as well as height and width. Yet an ordinary photograph loses the third dimension. For example, it does not show parallax, the relative movement between parts of the scene at different distances as the eye moves. We have to infer that quality from our previous experience.

A solution to this challenge was created by the Hungarian-born British scientist Denis Gabor; the invention won him the 1961 Nobel Prize for Physics.

While working at Imperial College in London in 1947, he invented 'holography' (from Greek meaning 'the whole message'). At the time, Gabor was trying to improve the amount of detail visible on an electron microscope picture. He had experienced such serendipity previously, inventing the mercury vapour lamp, while trying to make one using cadmium.

Gabor quickly realised why photographs have no third dimension. The film records only the intensity of light that strikes it, discarding another vital piece of information, the phase. The phase tells us how long the light has taken to reach the film from various parts of the object, whether the arriving light wave is at the peak of its cycle, at the trough or somewhere else. Capturing the phase required the original scene to be illuminated with coherent light, which ensures that the peaks and troughs of the light beams are initially in step. The only coherent light source then available, Gabor's own mercury vapour lamp, fell short of what was needed. In that sense, holography was ahead of its time. Only once the laser was invented could it flourish.

A hologram does not look like a picture, but its swirling patterns can recreate a three-dimensional image once suitable light is shone through it. Holograms have become very useful in many areas, for example, measuring very small distortions of objects due to stress. Holograms imprinted on banknotes and credit cards make them much harder to forge. Moving picture holograms are now available, and many artists have explored holography's potential.

Pictures on tape
From the VTR to the camcorder

Except for at the high end—Hollywood et al—movies on traditional film are obsolete. Decades ago, bright minds realised that if sound could be recorded for playback on magnetic tape, the same could surely be done with moving images, since those too could be reduced to varying electrical currents. There was certainly a need. Without any means of recording, all early television programs had to be live, and broadcasts were crudely archived by filming the transmission as it appeared on a screen.

Moving pictures carry much more information than sound does, so the technical challenge was daunting. The early mover was the American firm AMPEX. In 1956 it demonstrated a practical machine that used tape 5 centimetres wide racing through the recording and playback heads at 40 centimetres a second. AMPEX soon sold a hundred of their massive videotape recorders (VTRs) to enthusiastic television studios and production houses. Despite new technology, which provided features like slow-motion and stop-action, the open-reel VTRs remained too big to sell into homes, despite the efforts of Japanese firm Sony, which was soon to be a major player.

Video continued down the path already pioneered by audio recording. In 1971 Sony introduced the first videocassette, the widely used U-matic, followed in 1975 by the ill-fated Betamax. Compact enough for recording and playback at home, with the tape neatly packaged out of sight, the Betamax was still expensive. The next year Beta had a rival from the firm JVC, the VHS cassette, which was not only cheaper but ran for two hours rather than one.

Battle commenced in the first 'format war'. The two protagonists pushed their technologies, particularly seeking the ability to record more onto each cassette, ultimately reaching nine hours. They spent lots on advertising. Movies had to be put out in both formats, adding to costs. Beta probably produced better pictures, but VHS had more industry supporters, especially in the rental market, and ultimately won out. Beta mostly disappeared from view. The dominance of victorious VHS machines was relatively brief. By 2000 VHS was yielding ground to DVDs, and by 2005 most manufacturers no longer sold them.

Meanwhile camcorders (light-weight video cameras using small cassettes) were transforming the home-movie market. Cameras using film were soon on the way out. New formats, recording on tape only 8 millimetres wide, took over, soon joined and surpassed by digital recording on mini-discs. Every advance cut size and cost. Costly two-package 'luggables' when first introduced, camcorders were becoming truly portable, fitting into the palm of the hand.

Pictures with pixels
Steven Sasson

Photography has a long history, approaching 200 years. For most of that time, cameras all had one thing in common: images were recorded on glass plates or plastic film using light-sensitive chemical emulsions. They were developed and fixed (made permanent) by chemical processing. That, of course, took time. Other than the briefly successful Polaroids, no camera took a picture you could look at immediately. Furthermore, the only way you could send a good quality photo to someone was printed out on paper.

Digital imaging is a recent arrival, for the everyday photographer at least. The first affordable electronic

The success story

The comparison between the first cameras and 'electronic' or 'digital' versions is startling. These let you view the image as soon as you have snapped it; you can send it down a phone line or cable, confident it will arrive looking good. And processing is easy. You can play with images, enlarge them, change the framing or colour balance—things that previously took hours in the 'dark room'.

Chapter 9: **Making an image last**

The CCD: what's that?

A CCD is an array of 'pixels' (light sensitive cells), which accumulate electric charge in proportion to the brightness of the light that falls on them, much as solar cells do. The levels of charge on the pixels can be 'read' one by one. That data can be stored and processed to recreate the scene. Any picture can be reduced to a string of numbers.

cameras reached the market only in the early 1990s. Their origins lie a few decades further back, and derive much from the development of television. That too required technology to convert images into electric currents.

In most digital cameras, the key component is the charged coupled device (CCD), based, like the computer chip, on semiconductors such as silicon. The first CCDs, built in the 1970, were crude, with only 10,000 pixels; modern cameras boast five million or more. The images were correspondingly fuzzy. The first effort to build one of these into a camera, undertaken by Steven Sasson at Eastman Kodak in 1973, was equally clunky. The camera weighed 4 kilograms and took 20 seconds to create an image in black and white only, but it was only a technical exercise to see if it could be done, and not intended for the market. Kodak would not produce a consumer-friendly version for another 20 years.

However, Sasson's exercise began an inevitable trend, one that will ultimately kill off the business that Kodak pioneered, everyday cameras using film. Such cameras are not yet dead, but they live on mostly in the niche market of cheap, disposable cameras. These cameras trade off cost and disposability, losing image quality and flexibility in the process. It is a rather sad farewell.

The evolution of the digital camera has been swift, in line with the rest of information technology. We see the same trends; ever-increasing capabilities at ever-decreasing cost. Early digital cameras cost $10,000 or more; that price has shrunk twentyfold. Progress has been aided by the development of international standards for storing images in digital code, such as the JPEG format devised by the Joint Photography Expert Group. Among other things, JPEGs allow images to be compressed. Unnecessary information in the image can be discarded without an unacceptable loss of quality, as with DVDs and MP3 players. As a result, images take up much less space in computer memories and can be quickly transmitted by email or over the Internet.

New sorts of data storage, such as flash memory, let a camera store more pictures. Inkjet printers generate quality copies on demand. Cameras continue to get smaller; they are now often barely larger than a credit card, or able to be secreted in a mobile phone.

Below: The modern descendants of the Kodak (and the Polaroid) vastly surpass them in performance, with colour pictures, built-in zoom, instant storage and recall, yet small enough to slip into your pocket.

Early digital cameras cost $10,000 or more, that price has shrunk twentyfold.

Most electronic cameras will now take both still and moving pictures. We have come a very long way in little more than a decade.

The digital camera transformation has raised a host of issues. Their simplicity and ubiquity have encouraged some inappropriate use, with invasions of privacy that can verge on assault. The capacity to manipulate images by digitally altering the location and content of individual pixels has damaged our confidence that 'the camera never lies'. Surveillance in public places is easier, cheaper and therefore more common, which has benefits for law and order but asks a question about our right to privacy. Digital cameras enable us to create and access information freely, but are open to both use and abuse.

Below: Today you hardly need a separate camera to take pictures. The camera in your mobile phone can do the job, though not yet with the same quality. As a bonus you can instantly send the picture across the world or to your home computer for later printing.

More movies in less space
The rise of the DVD

From the 1960s, with CDs replacing vinyl discs for music recording, a push began for an optical disc to record movies. For home viewing, movies had already been transferred to video cassettes. Industry leaders remembered well the costly battle between the Betamax and VHS formats, so IBM president Lou Gerstner acted as mediator between rival videodisc formats, one pushed by Phillips and Sony, the other by Toshiba. Phillips/Sony ultimately withdrew, but some special features they wanted were included in the victorious—and more heavily-backed—Toshiba technology.

The compromise produced the Digital Video Disc (DVD)—or Digital Versatile Disc—since they can do more than record movies. DVD players and discs first went on sale in Japan in 1996, with other countries quickly following. VHS, dominant until then for home video entertainment, was doomed. In the United States DVD rentals surpassed VHS hire in 2003; in a few more years VHS will disappear altogether, except for nostalgia.

Like a CD, a DVD records information in digital code as pits in the bottom of a spiral groove, burnt by a laser and read out by another. However, DVDs can hold six times as much data, 4.5 gigabytes (GB) rather than the 700 megabytes of the CD. The grooves are twice as fine, the pits half as long; more space is saved by efficient error correction procedures. Despite the fact that DVDs could hold more data than CDs, fitting a two-hour movie onto a 12-centimetre disc required innovation.

The now-familiar DVD is not for the longterm, however. Already rivals and successors are mustering. High-density DVDs use blue lasers rather than red ones to record and play back; the pits can be smaller and more data—up to 15 gigabytes—recorded. Those could still be outclassed by the rapid growth in data capacity of hard drives and even microchips. Time will tell.

The DVD: how does it work?

Moving images contain immense amounts of information, but much of it is redundant. If a large area of blue sky on one frame does not vary significantly from place to place, and is the same in the next frame, it is enough to record that it has not changed. Other details can remain unrecorded because the eye cannot perceive them. Using the data compression technology MPEG-2, the DVD can store 40 times less data than the printed film without the viewer telling the difference.

CHAPTER 10:
Saving sounds

Sounds of all kinds, be they bird song or orchestral music or human speech or traffic noise, are merely transient fluctuations in air pressure impinging on the ear. Once those have died away, the sounds exist only in the memory. At least, that used to be the case.

What happened when?

1877	Thomas Edison invents the phonograph
1887	Emile Berliner records sound on a flat disc containing a spiral groove
1898	Valdemar Poulsen imprints sound on a magnetic tape
1927	The first talking movie, *The Jazz Singer*, released
1929	The first sound-on-film movie, *Broadway Melody*, released
1930s	Magnetic recording machines reach the market
1948	Peter Goldmark introduces the long-playing record (LP)
1965	Phillips markets the first compact audio cassette
1979	The Sony Walkman appears on the market
1983	The compact disc is introduced
1997	Korean firm SaeHan Information Systems creates a digital audio player

The first recordings
Thomas Edison and Emile Berliner

In 1877 the prolific American Thomas Edison was trying to improve telegraph operations. He built a machine that marked the dots and dashes of Morse code as indentations on paper tape, ready to be run through a reading device to send the message. As the paper passed through the machine at high speed, the spring on the stylus reading the impressions made a musical sound. Edison grasped that such indentations could record and play back sounds by connecting the stylus to a diaphragm (vibrating membrane or disc), as in the newly-invented telephone.

The prototype phonograph (literally 'writing in sound'), made to Edison's specifications by one of his craftsmen, used metal foil around a cylinder that carried the indentations. The first sounds ever recorded were of Edison himself reciting 'Mary had a little lamb'. Soon after, Edison demonstrated his new machine to the staff of *Scientific American*, having it wish them 'goodnight'.

He started selling the machines the next year, but foil cylinders were awkward and fragile. Interest in the novelty soon waned. Before moving to other pressing matters such as the light bulb, the visionary Edison wrote down a list of potential uses for the phonograph: 'talking books' for the blind, reproducing music, recording telephone messages, taking dictation, 'talking clocks', preserving the last words of the dying.

Coming back to the phonograph ten years later, he found that others, like telephone pioneer Alexander Bell, had made improvements, using wax instead of foil for the recording. Machines became 'spring-powered', driven by clockwork rather than turned by hand. Edison soon had a range of pre-recorded cylinders for sale—sentimental ballads, marches, comic monologues, opera excerpts. Most people heard these through coin-operated machines in amusement arcades, but recordings were limited to two minutes in most cases, and there was difficulty in reproducing cylinders in large

> **The first sounds ever recorded were of Edison himself reciting 'Mary had a little lamb'.**

Left: The first-ever sound recording machine, made to Edison's design by one of his craftsmen. Hand-cranked and crude, it was not yet a serious technology, but Edison had many ideas how it might one day be used.

Sound from a tape
Valdemar Poulsen, Phillips and Sony

While Thomas Edison and Emile Berliner were recording sound on cylinders and discs, other experimenters were using the electric currents that represented sounds—as in the telephone—to alter the magnetic field imprinted on a steel wire or tape. The Dane Valdemar Poulsen succeeded first in 1898; his magnetic recording of Austrian Emperor Franz Joseph is the oldest known. Poulsen, however, soon turned his attention to radio; dictating machines using wire or tape could not compete with cheaper, more reliable cylinders from Edison and Dictaphone, by now a major company in this field.

Not until the 1930s did practical magnetic recording machines reach the market, especially in Germany. These took advantage of advances in electronics and recorded on wire or tape on reels and in cassettes, and later on plastic tape coated with iron oxide, as today. Other innovations improved sound quality.

Tape had advantages over discs for some purposes. It could be physically cut and rejoined to allow content to be edited, which was impossible on discs.

Hollywood studios quickly seized on the new technology to record speech, music and sound effects for film soundtracks.

Cracking the home market was much harder. Tape recorders were big and heavy, as well as expensive; only music buffs and 'hi-fi' (hi-fidelity) enthusiasts were willing to wrestle with the technology. 1965 was the

numbers. By the time of World War I, the Edison cylinder had given way to disc recordings.

German-born Emile Berliner had emigrated to Washington DC aged 19. His technique, devised in 1887, recorded sound in a spiral groove on a flat disc. The impressions coding the sound went side-to-side rather than up-and-down as in the Edison machines, making for better reproduction. The recordings could run for four minutes and, most importantly, could be mass-produced by stamping an etched metal 'master disc' onto blank discs of shellac (a purified resinous material) softened by heating. His name 'gramophone' reversed Edison's title. His system—with modifications—underlay all recorded sound for more than half a century, until challenged by sound on film, tape recordings and compact discs.

Stranger than fiction

Much of the post-war demand for 'tape recorders' resulted from publicity generated by entertainer Bing Crosby, who needed to record his popular radio shows for later broadcast in other time zones across the United States.

Opposite: Some early 'record players', with the big horns needed to amplify sound in the pre-electronic age. The Berliner disc machine on the left is hand-cranked. Sound quality was mediocre by today's standards.

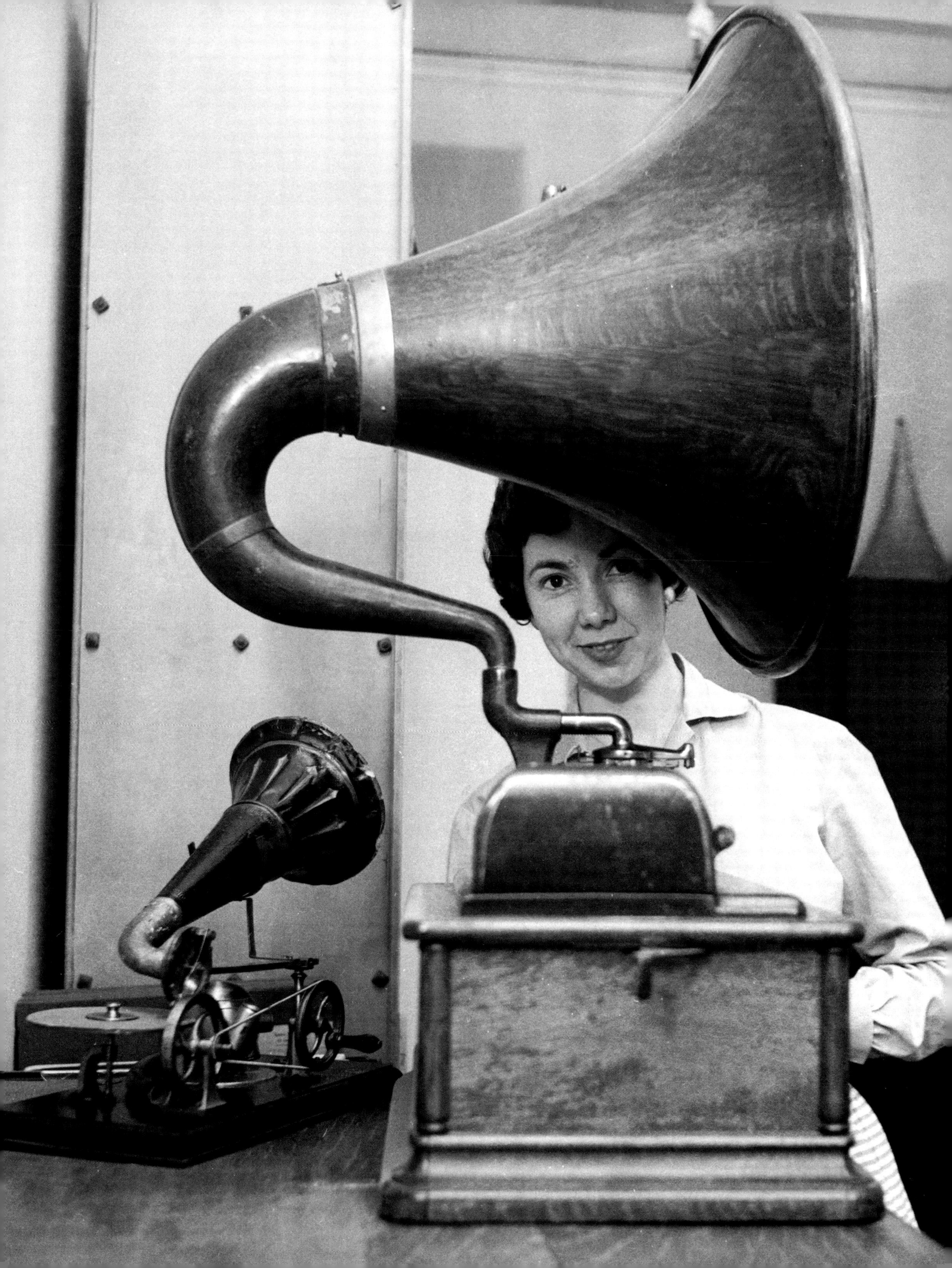

Above: Though we would think it cumbersome and inconvenient today, the reel-to-reel tape recorder could record sound instantly, impossible with the existing vinyl discs, and sound quality was good. Nonetheless the technology was doomed.

key year, with Dutch electronics firm Phillips first marketing the compact audio cassette. A quarter the size of other cassettes, this recorded 30 minutes of music on each side. Machines using it were equally compact. Phillips originally intended the cassette for business dictation, but the market was soon dominated by consumers who wanted a cheap, simple way to record and replay music.

The peak of the audio cassette era was 1979, 80 years after Poulsen's first recording, when the Japanese company Sony released the Walkman, barely larger than a cassette, and able to be carried—or rather worn—by users as they walked. That changed forever the way people listened to music, preparing customers to accept—indeed, to desire or demand—the portable CD player and later the MP3 player. Of course, these two innovations have made the cassette player mostly obsolete.

The movies learn to talk
Sound-on-disc versus sound-on-film

In the early 1920s, Hollywood feigned indifference when the technology for 'talkies' (movies with speech synchronised to the actors' lips), was convincingly demonstrated. Studios had huge inventories of silent films and actors on long-term contracts, many with heavy foreign accents. The cost of converting theatres to sound would be colossal.

In their hearts, of course, the studios knew that making movies talk would greatly increase their appeal, making them mainstream entertainment. Progress might have been delayed, but it could not

'Talking audiences for silent films were replaced by silent audiences for talking films.'

be denied. In October 1927 Warner Brothers released *The Jazz Singer* with Al Jolson, the first feature-length talking picture, though not the first movie with sound. There was no going back.

The move to talkies saw a battle between competing technologies, sound-on-disc and sound-on-film. The first was pioneered by Thomas Edison, inventor of the phonograph, but he later decided it had no future. The big electronics firms Western Electric and Bell Telephone used the new electronic amplifiers to create the Vitaphone ('living sound') process, used in *The Jazz Singer*.

Vitaphone—a sound-on-disc process—had problems; it was notoriously difficult to keep the recorded sound exactly in time with the film. The record could jump a groove; the film could be damaged, requiring some of it to be cut out. Either event could destroy synchronisation, needing swift action by the projectionist. This was hilariously recalled, decades later, in the classic comedy movie *Singing in the Rain*. Frequent playing wore the discs; they had to be replaced every twenty showings.

Initially, discs gave better sound quality, but sound-on-film quickly improved, and its other advantages soon won out. Soundtracks started to be recorded directly onto the film beside the picture, so synchronisation was automatic. Victory for the new technology was swift, beginning with the 'all talking, all singing, all dancing' *Broadway Melody* from MGM in 1929, which won Best Film at the Oscars. Warner Brothers stopped making new Vitaphone movies in 1930. Alternative sound-on-disc copies were still needed until 1937, because of the cost of converting theatres to the new technology, but sound-on-disc had no future.

Something else was already doomed, the habit of talking while the film was screening. This was acceptable when films were silent but was frowned upon in the new talkies era. As one commentator put it, 'talking audiences for silent films were replaced by silent audiences for talking films'.

Playing a soundtrack: how does it work?

To play sound recorded on film, a light was shone through that part of the film, the 'soundtrack', which varied in opaqueness according to the sound recorded on it. On the other side of the film, the newly invented photocell registered the amount of light shining through the film. This was converted into a varying current, which went through amplifiers to speakers.

The music goes on and on
Peter Goldmark

By the early 20th century, recorded sound on cylinders and discs, pioneered by Thomas Edison and Emile Berliner, allowed anyone to hear great music, performed by great artists, in their own homes. The first radio stations had plenty of material for their programs. Early recording systems were mechanical, but electrical recording, using electronic amplifiers, arrived in the 1920s. Music sounded more and more like it did in the concert hall or ballroom.

Above: The 'jukebox', with its large repertoire of music on records, continued to enliven parties and dances for many decades after the first machines to play music on demand were pioneered by Thomas Edison.

Limitations remained, particularly the length of the recording. The standard 'record' was 10 or 12 inches (25 or 30 cm) in diameter, turning 78 times a minute and holding only three to four minutes of music on each side. For that reason popular songs were commonly written to last three minutes. Replaying a complete symphony, let alone an opera, involved a great pile of discs, and much getting up and down to change from one to the next.

A major step forward came in 1948. Working for Columbia records, Hungarian-born American Peter Goldmark perfected the 'long-playing record' (LP) after four years of work. On older records, 30 grooves fitted into each centimetre; the new 'microgroove' technology fitted 100 grooves into that space. The discs also turned half as fast, 33⅓ times a minute, so that each side held six times as much music—twenty minutes, enough for two movements of a symphony. The new format saw a vast expansion in the repertoire of music on record.

Playback was now electrical too; sound quality improved still more. The old steel needle was replaced by a 'stylus', which had a synthetic gemstone at its tip to reduce wear and prolong record life. Devotees now spoke of 'high fidelity' music ('hi-fi'). Unbreakable vinyl replaced shellac in the discs themselves.

Goldmark and Columbia planned to share their new format with their competitors, but major rival RCA responded a year later with a third format, the '45', smaller and lighter than the old 78 but holding as much music. These could be played on the same equipment as the new 33s, and were especially in vogue for popular music and jazz.

The 78 was now obsolescent technology, even obsolete. Few would have foreseen that 30 years on, the 45s and 33s would themselves be threatened with extinction.

The compact disc takes over
David Gregg

The arrival of the compact disc (CD) in the marketplace in 1983 was both a beginning and an end. It brought a profound change for music lovers. Audiophiles and classical music buffs embraced it at once for its increased fidelity and convenience; the rock and pop communities were somewhat slower to enthuse but ultimately did.

At the same time it signalled the close of almost a century of sound recording on vinyl discs, since the days of Edison and Berliner. It was to bring the end of analogue recording, with CDs using digital techniques, and of the seemingly inevitable clicks and pops that plagued the old 78s and LPs. CDs seemed almost impervious to dust and damage.

The idea of a disc holding information that could be read by a beam of light rather than a needle seems to have come first to inventor David Gregg in the late 1950s. He wanted to record movies; his early optical disc was soon called a 'video disc', then a 'laser disc' when the laser became the ideal tool for both recording and playback. Gregg's patents—and his company— were acquired by the firm MCA, which, in partnership with Dutch electronics giant Phillips, commercialised the LaserDisc for movies, reaching the market in 1978. This was 15 years ahead of the DVD, which would ultimately do the same job much better. LaserDiscs never really caught on: they were inconveniently large—30 centimetres; played only for an hour so—half a typical movie; and were not digital, with the limitations that brought.

Phillips was also experimenting with using laser-disc technology to record sound only. Others were trying to do the same, including Japanese electronics leader Sony. It seemed sensible for Philips and Sony to cooperate. Combining ideas and agreeing on standards, they created what we would now recognise as compact discs, made of high quality, polycarbonate-plastic discs, backed by reflective aluminium and protected by a lacquer coating, on which labels could be printed.

The key development was 'going digital' or, more precisely, 'going binary'. The processed music was laid down—and still is—in a spiral groove cut into the polycarbonate, as on an ordinary record but only about a micron (a thousandth of millimetre) wide. The digitised data was encoded as changes in the depth of the groove, much as in the original Edison phonographs, rather than the side-to-side wobbles popularised by Berliner. Just two distinguishable depths ('pits' and 'lands') accommodate the two-symbol digital code. On playback, a laser beam reads the pattern of highs and lows; decoding technology turns the signal back into music.

Below: Any music shop has CDs by the thousand, many of them at bargain prices. Low cost is the result of improving technology and mass production, which will see the inevitable demise of the CD as newer technologies emerge.

Storing music as strings of 0s and 1s, MP3 is yet another manifestation of the digital revolution.

CDs had great appeal from the start. Smaller and more convenient than the 30-centimetre LP, they can hold four times as much music. The target playing time of around 70 minutes was set, so the story goes, by Beethoven's *Ninth Symphony*. They are much less susceptible to damage, and some players can be carried about and even bumped without interrupting the music. In the judgement of most people, the sound quality is at least as good, though some traditional audiophiles will never accept that. Some claim, for example, that the dynamic range (from loudest sound to the softest) is compressed compared to that of an LP.

Although CDs arose originally from a search for a way to record movies, and were initially used exclusively for music, we soon saw them as a convenient and powerful way to store any information that could be put into binary code. This includes just about everything; text, numbers and still pictures were soon being recorded for playback from CDs, and we saw the rise of the 'compact-disc read-only memory' (CD-ROM), supplementing and later replacing the floppy disc in personal computers from the 1990s.

Given its power and convenience, one would have thought the CD was here to stay, but it is already under threat, not least from the increasing capacity of microchips with no moving parts that store large amounts of information, hence the proliferation of iPods and other MP3 players. In today's technology, nothing is forever.

More music in less space
Music compression becomes mainstream

Just about everyone knows the term MP3. They know it lets them replay dozens of popular songs from a machine the size of a deck or cards or smaller. MP3 players are transforming the convenient enjoyment of music. But the name has a deeper meaning. MP3 is shorthand for MPEG-1 Audio Layer 3. MPEG stands for Moving Picture Expert Group, a group of boffins who have been meeting since 1988 to set internationally-accepted standards covering the encoding of video and audio signals (pictures and music) into forms that can be moved quickly and electronically, say over the Internet.

The MP3 players many people have today draw together several technologies. Storing music as strings of 0s and 1s, MP3 is yet another manifestation of the digital revolution. The player holds these bits of information for quick retrieval. Unlike a CD player, an MP3 player has no moving parts; the data is imbedded in a microchip. The operation of Moore's Law over

CD recording: how does it work?

Rather than continuously recording every minute fluctuation in the sound waves conveying the music, engineers devised ways to take 'snapshots' of the sound wave, relying on the human ear and brain to fill in the gaps. It parallels the way a series of still images projected quickly enough can give the illusion of movement. A measurement of the strength of sound wave, taken 44,000 times a second, is converted into a binary number, usually one with 16 'bits' allowing for thousands of distinguishable levels of sound. For most of us, the ear itself cannot do better.

decades has produced chips able to hold immense amounts of information.

How to store the music wasn't the only problem. The sampling methods used in compact discs mean that a typical three-minute song needs 32 megabytes (32MB) of data to capture all its nuances. Downloading that from the Internet took an inconvenient twenty minutes through a 56 kilobit per second dial-up modem. To make music-sharing through the Internet practical, we had to reduce this download time by compressing the data needed to encode the song without making it sound noticeably different.

Hence MP3, the current answer to the problem as devised by the bright people working under MPEG—mostly European engineers funded through the European Union. Applying a few rules to remove redundancies—such as not coding sounds the ear cannot hear, and storing only the louder of two sounds when they sound together—we can store the three-minute song in only 3 megabytes, 90 per cent less space. This cuts the download time to two minutes—less time than the song takes—but

Below: A collection of MP3 players, which allow you to hold hours of high fidelity music in the palm of your hand and play it where ever and whenever you want, such as when you are out jogging. They will only get smaller.

Moore's Law: what's that?

Attributed to Intel founder, Gordon Moore, and more an observation than a 'law', this famous pronouncement states that the number of transistors which may be squeezed onto a computer chip doubles every 18 months. This means more than a thousand-fold improvement every decade in computer memory and processing power, driving information technology inevitably forward. This vision has held true for nearly four decades and has not yet run out of steam.

maintains, just about, 'CD quality'. With broadband, it's much faster still.

MP3, which followed industry standards MP1 and MP2, entered our lives in 1994, when the first programs to convert music to the MP3 format went on the market, and in 1995, when MP3 files began to flourish on the Internet, ready for download to portable players. Once more, life would never be the same again.

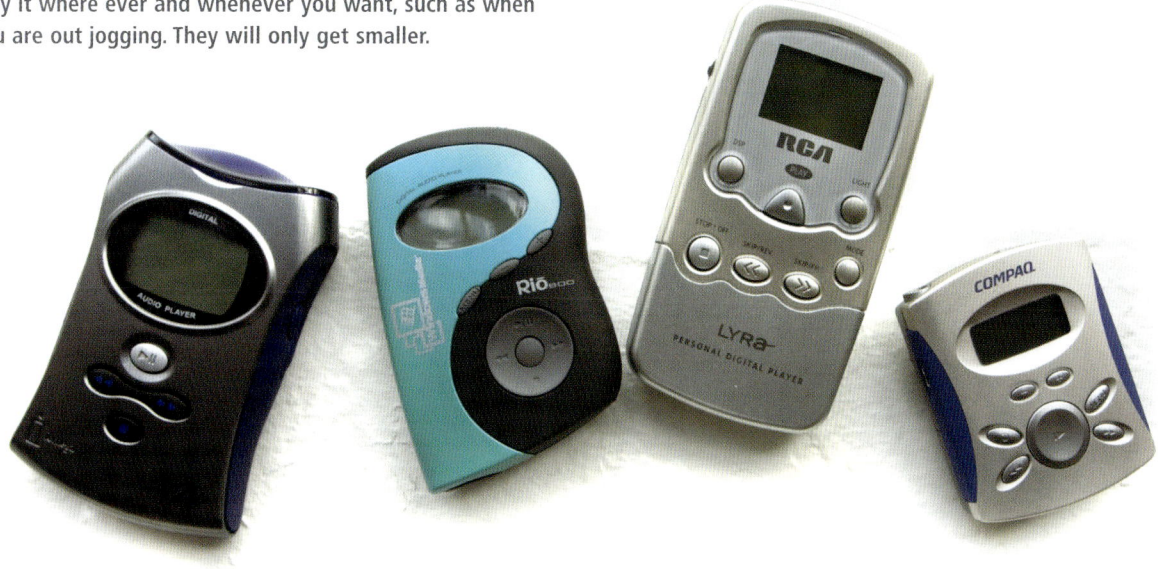

CHAPTER 11:
The wireless world

By the dawn of the 20th century, a vast and growing network of copper wires and cables carried telegraph messages and telephone calls between cities, across continents, beneath the oceans. Messages that once took weeks or months to reach their destinations now took seconds. But they could not go anywhere without the wires going there first.

Another approach was stirring, one that would do away with the wires and unite our planet as never before. But its history would be intertwined with the growth of another technology, electronics.

Chapter 11: The wireless world

The beginnings of the 'wireless'
Heinrich Hertz, Eduard Branley and Guglielmo Marconi

The saga of radio, aka 'wireless'—an early name, now coming back into favour in the digital age—begins in 1863. The Scots mathematical genius James Maxwell predicted the existence of a then unknown form of radiation, similar to light, generated by rapidly moving electrical charges such as in a spark. The story progressed in 1887, with confirmation that such radiation existed by the physicist Heinrich Hertz, whose name is still used today in the unit of frequency. He found that sparks did indeed make such radiation, which behaved as predicted, particularly by travelling at the speed of light.

Hertz was delighted to have confirmed Maxwell's prediction, at least inside his laboratory, but he was pessimistic that any use would be found for the 'Hertzian waves'. Many agreed, but some did not, wondering if the waves might be used to transmit messages, say in Morse code. There were practical problems to be solved. One was detecting when the rapidly oscillating radiation reached the receiver.

To meet the need, Frenchman Eduard Branley invented the 'coherer', a glass tube filled with iron filings. Normally the coherer was an insulator, but it would allow electricity to pass when it was struck by the radiation. This was much better than watching sparks jumping across a gap as Hertz had done, since it effectively amplified the incoming signal. With such

Below: The youthful Italian-British Guglielmo Marconi, with some of his early 'wireless' equipment. Driven forward by Marconi and others, the technology revolutionised communications and became big business.

He initiated the age of 'wireless telegraphy'. A revolution in communications was stirring.

primitive technology, Branley succeeded in detecting radiation 150 metres from the sparks that generated it.

Other bright minds such as the Englishman Oliver Lodge and the young New Zealander Ernest Rutherford were also at work. The most famous of the early experimenters was Guglielmo Marconi. Starting in 1895 he tried to pick up Hertzian waves at greater distances—up to a few kilometres. A gun was fired to signal from receiver to transmitter that the signal had come through.

Failing to get any official support, Marconi moved to England where his English-born mother had relatives in the Post Office. With their aid he sent signals from Portsmouth to the Isle of Wight, then across the English Channel and, ultimately in 1901, across the Atlantic from Newfoundland to Cornwall. Stopping and starting the waves in accordance with the Morse code—the transatlantic message was the letter 's'—he initiated the age of 'wireless telegraphy'. A revolution in communications was stirring.

Below: An X-ray picture of early wireless equipment, notable for the presence—on the right—of one of Lee de Forest's 'valves', the offspring of the light bulb. This vastly improved performance, and initiated the era of 'electronics'.

The Edison effect: what's that?

Now called thermionic emission, the Edison effect refers to the flow of charged particles from a charged metal surface. It is caused by thermal vibrational energy overcoming the electrostatic forces that normally keep electrons on the surface of the metal. When Edison introduced an extra metal plate into his light bulb and gave it a more positive charge than the filament, the electrons emitted from the hot filament were attracted to the plate, causing current to flow.

Electronics stirs
Thomas Edison and John Ambrose Fleming

The American inventor Thomas Edison has been called the 'man who created the future', but among his many discoveries was one he did not know he was making. He found the 'Edison effect' accidentally in 1882, when he was trying to stop carbon particles coming off the glowing filament of his light bulb and blackening the inside of the glass.

Edison's lamps used direct current; one end of the filament was therefore positively charged, the other negatively. He noted that the carbon particles seemed to come mostly from the negative end. He wondered if there was a way to repel them before they hit the glass, and had light bulbs built with an extra metal plate inside. When the plate was positively charged, he found a tiny but steady current flowing out of it, apparently coming from the filament through the near-vacuum inside the bulb. When the plate was negative, no current flowed. Though he patented this effect and discussed it with colleagues, he did nothing more, busy with more pressing issues.

The current was due to a flow of minute particles that soon came to be called electrons; the technology that grew from the Edison effect would be dubbed electronics. Pioneering English 'electrical engineer' —a new term in his day—John Ambrose Fleming kick-started that in 1904.

The new technology of 'wireless' had posed some challenges, particularly how best to 'rectify' or 'detect' the incoming signal, that is, make the alternating current flow only one way so that the receiver could make sense of it. Many ideas had been tried, including the 'cat's whisker', a piece of thin wire resting on a crystal of the mineral galena. It was crude but it worked.

Fleming, a consultant to the Marconi company, saw a possible answer in the Edison effect. A glass bulb containing a glowing filament and a metal plate would allow current to flow in only one direction (from the filament to the plate), so it was like one of the valves in a blood vessel, The device was soon called a 'wireless valve'; it was also called a 'diode', from its two electrodes. Such diodes, enhanced in performance, would remain in use for half a century—until the arrival of the transistor, and made possible the growth of radio and television.

Music in the air
Reginald Fessenden

In the century-long history of radio, the name of Reginald Fessenden does not come quickly to mind, unlike Marconi or perhaps Edwin Armstrong, the inventor of FM. Yet the Canadian-born American was a real pioneer, broadcasting music and speech by wireless when everyone else was still tapping out Morse code. The head of the General Electric laboratories later called him 'the greatest wireless inventor of his age, even greater than Marconi', which is quite a call.

Unlike Marconi and most others, Fessenden believed that wireless waves could be imprinted with speech and music. On Christmas Eve 1906, he proved it was so. Radio operators on ships off the coast near his laboratory on Brant Rock, Massachusetts, were amazed when their headphones delivered not the familiar dots and ashes of Morse code, but Handel's *Largo*, reproduced from an Edison cylinder, and the man himself playing 'Silent Night' on the violin and reading from the *Bible*, before wishing his listeners 'Merry Christmas'.

Fessenden's feat was possible through use of a very high-speed electrical generator to produce the high-frequency oscillating currents he fed into his antenna, rather than the commonly-used, but crude, spark gaps. Nonetheless, most people, even Fessenden's own backers, believed Marconi; such a broadcast, even if feasible, was of no value or interest. More than a decade went by before 'radio broadcasting', rather than mere wireless telegraphy, began its phenomenal rise in popularity.

Fessenden was very early in proposing what was later called 'radar'. He argued that a system that detected objects like icebergs by reflecting radio waves off them could prevent disasters like that of the *Titanic*. Again he was ahead of his time; 20 years passed before radar was ready for use. Fessenden did not live to see it, he died from heart disease in 1932, aged 62.

Electronics stand up
Lee de Forest

Electronics, with all its immense potential, had begun with Englishman John Ambrose Fleming's radio valve, patented in 1902. This solved one of the big problems with the still embryonic technology of wireless, how to 'rectify' (change from alternating to direct current) the oscillating currents from the receiving antenna so that their arrival could be detected.

The success story

Fessenden was prolific. His 500 patents included signalling systems for submarines; a smoke generator to hide tanks on the battlefield; and a depth-sounder for shipping, which gave him a comfortable income. Early in his career, he had worked as a chemist for Thomas Edison and later for George Westinghouse, helping him to light the 1893 Columbian Exhibition with electricity. His greatest triumph was to outdo Marconi, who in 1901 had sent the first Morse code signal across the Atlantic.

The technology was moved forward by American electrical engineer Lee de Forest, son of a white Congregational pastor who had championed the education of Afro-Americans. He had hoped his son would follow him, but young de Forest had been more interested in electricity and wireless. In a crucial development, de Forest added a third electrode to Fleming's diode. This was in the form of a wire mesh between the cathode (heated filament), which gave off electric current as electrons, and the anode (cold plate), which received them. By varying the current in this 'grid', he could control the much bigger current flowing through from cathode to anode, in other words amplify weak signals into much stronger ones.

De Forest's creation, the 'audion' or 'triode', patented in 1906, could do something else even more startling. Connecting the grid and plate together with some external circuitry made the valve oscillate,

Opposite: His Majesty's voice. King George V of Britain makes a broadcast to his people. Wireless (soon to be called 'radio') enabled national and political leaders to be heard around the country, and later across the world.

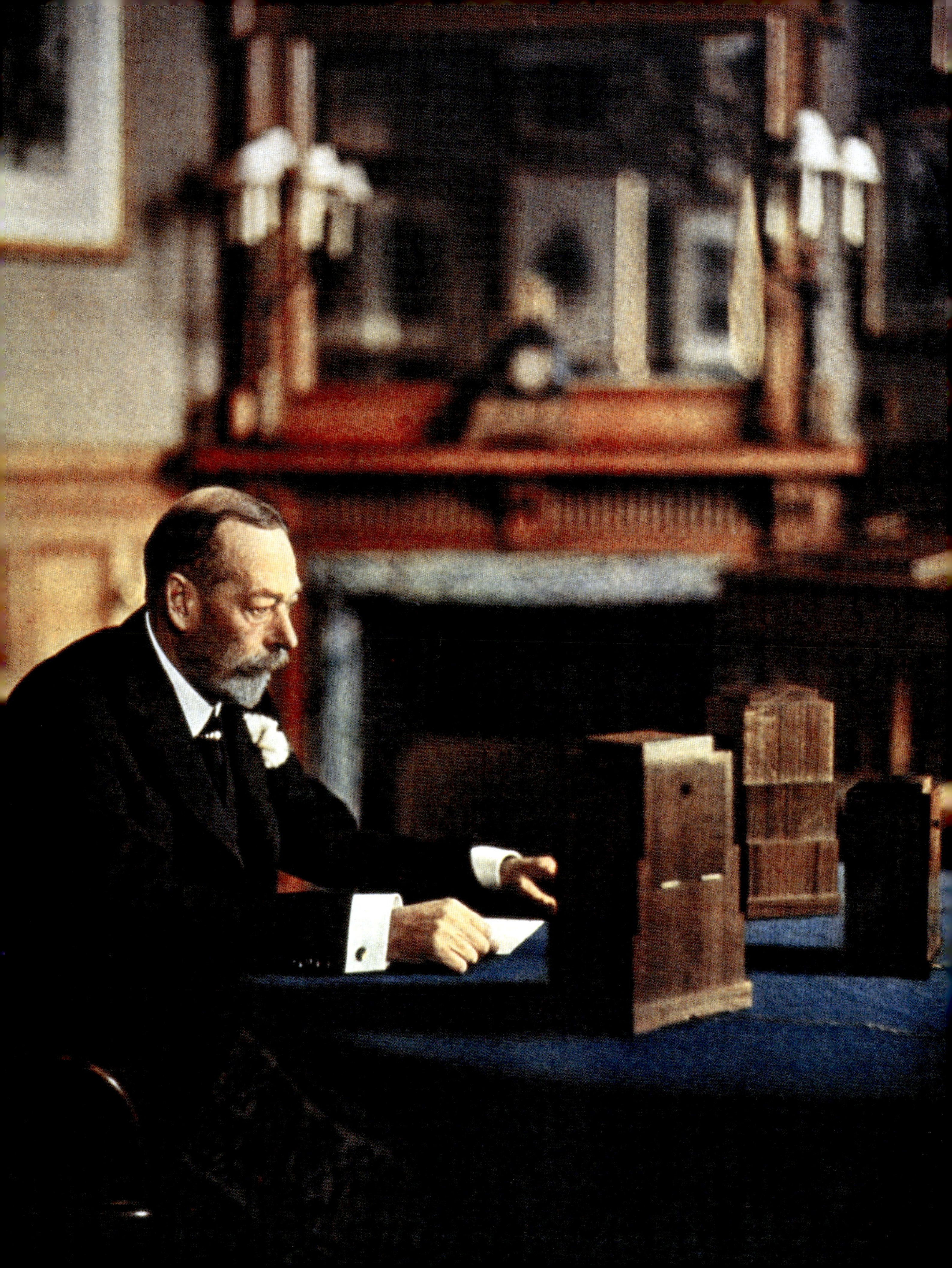

Right: One of the many impacts of radio: German dictator Adolf Hitler was able to pound out his message to a large and widely dispersed audience. For the first time in history leaders could send their voices, and their ideas, into people's homes and workplaces.

generating the high frequency alternating current needed to transmit wireless signals but which until now had been made much more crudely by sparks jumping across a gap. It was Western Electric, part of the Bell Telephone (AT&T) empire, that turned de Forest's bulky prototypes into cheap, reliable and long-lived devices. Bell needed the audion to amplify telephone signals, which inevitably weakened as they travelled, and so make possible long-distance telephone calls.

The triode, able both to amplify and to oscillate, was arguably the greatest advance in communications between the pioneering efforts of Tesla and Marconi and the invention of the transistor. For more than 50 years, the burgeoning power of radio—and later television—would depend on de Forest's invention and its descendants.

The achievement was not de Forest's alone. Austrian Robert von Lieben had come up with the idea of a three-electrode valve at the same time; such an occurrence is not uncommon in the history of invention. De Forest lived to be nearly 90, but von Lieben died in 1914 aged only 36, too young to develop his technology and play his proper role in the new age of radio.

The birth of radio broadcasting
David Sarnoff

Following Marconi's first signal across the Atlantic in 1901 the new technology of wireless slowly made its mark. Early use concentrated on messages between

Right: American actor (and later President) Ronald Reagan plays the role of an on-the-spot reporter who lets his listeners know what is happening, even as it happens, thanks to the new 'radio'.

For more than 50 years, the burgeoning power of radio—and later television—would depend on de Forest's invention and its descendants.

ships, or from ship to shore, replacing the technologies of the day—flags, flashing lights, carrier pigeons. It soon proved its worth. When two ships collided in fog off New York in 1900, a message from the Marconi operator on one ensured the passengers were quickly rescued.

Longer-range wireless telegraph services soon followed. In 1903 the US president and the British king exchanged telegram greetings across the Atlantic. Reports of the attack on Port Arthur during the Russo-Japanese war in 1905 came in by wireless. Explorer Robert Perry sent a telegraph message from the North Pole in 1909. A year later radio aided the capture of escaped British murder Robert Crippen while he was still at sea. This reprised an early triumph by the electric telegraph.

The eruption of war in Europe in 1914 provided an immense stimulus. Ground controllers could direct aeroplanes and airships—more new technology—to their targets using radio; and quick warning could be given of gas attacks. Intercepting radio signals enabled one side to know the other's plans in advance; the 1917 Zimmerman Telegram that brought America into the Great War is the most notorious example. Growing use was matched by increasing power and sophistication—better valves derived from de Forest's triode, new radio circuits that improved reception.

When peace came in 1918, we could contemplate civilian uses for radio, as a source for entertainment and information. David Sarnoff, working for the Marconi company, had outlined his vision in 1916: a 'radio music box' in every home, educational lectures, news reports and sports scores broadcast across the country. People in remote areas would particularly benefit. Sarnoff was later to head up the Radio Corporation of America (RCA), a leading player in radio and later founder of the NBC network.

By 1920 Sarnoff's vision was taking shape. Radio broadcasting as he had envisaged, it began in November that year, when radio station KDKA went to air from the Pittsburgh garage of an employee of the Westinghouse Corporation, bringing regular broadcasts of music and news. One of its first reports told of the election of Warren Harding as US President. Many others followed. We had entered the age of radio.

Seeing by radio
Robert Watson-Watt

An early discovery about radio waves was that things reflect them, just as they do visible light. Sharp minds, such as radio pioneer Reginald Fessenden, argued that such reflections could reveal the presence of objects rendered invisible in darkness, fog or by distance. This might, for example, stop ships colliding with each other or things like icebergs.

Stranger than fiction

A beam of radio waves was initially planned as a potential 'death ray', disabling the pilot of attacking aircraft by heating his blood. That quickly proved impractical, but the radio reflections from his plane could certainly be detected.

Lack of real interest, and the limitations of the technology of the day, saw the idea languish until the 1920s, when its potential military value emerged. Could attacking aircraft or warships be detected when still a long way off, giving time to organise defence? Research began in several countries, but again there was no sense of urgency.

In the mid-1930s British authorities suddenly realised the potential risk of devastating air attacks by the formidable German airforce, against which there seemed to be no real defence. 'The bomber will always get through,' cried the pessimists. Interest in radio-location revived in 1935, led by Scottish-born engineer Robert Watson-Watt, who can thereby claim a prominent role in the invention of 'radar', as it was later called.

It was discovered that if radio waves were directed at aircraft, it was possible to detect their reflections. Systems were soon developed to measure the distance, height, direction and numbers of attacking forces. The first 'radio direction finding' (RDF) stations were operational in 1936, and a major push had a defensive chain in place around the south and east coasts of Britain when war began in 1939.

FM threatened the profitability of the hundreds of AM radio stations then on air.

Was radar 'the weapon that won the war', rather than, say, the atomic bomb? Certainly radar made a crucial difference in many significant theatres of the conflict, helping the Royal Air Force triumph in the Battle of Britain in 1940, blunting the impact of German night-bombing during the Blitz, winning the Battle of the Atlantic against the U-boats, guiding bombing raids into occupied Europe. Only the technology of 'code breaking' had anything like the same impact.

As with radio in 1914–1918, military demands drove the technology of radar rapidly forward, incidentally readying it for civilian use when peace came. One of the many outcomes was the mastery of microwaves, which would find a multitude of uses.

A life for radio
Edwin H Armstrong

When American radio pioneer Edwin Armstrong jumped from his 13th storey New York apartment in 1954, penniless and distraught, he ended a brilliant career of innovation that still brings everyday benefits. In particular, his promotion of frequency modulation (FM) broadcasting ultimately transformed the radio industry. His genius had been recognised early. In 1917 he had been the first recipient of the Medal of Honour of the Institution of Radio Engineers.

Opposite: The pivotal struggle in 1940, known as the Battle of Britain, depended heavily on the new technology of radar to detect the German bombing attacks and direct defending fighters like this Spitfire and Hurricane. The British were outnumbered but ultimately victorious.

Ironically, FM drove him to suicide. Armstrong had wanted to improve the quality of radio reception, particularly for music. Since the days of Lee de Forest, radio used amplitude modulation (AM), the amplitude (wave height) of the radio signal varying in line with the sound signal being impressed upon it. Unfortunately, electrical sparks, from such things as lightning, passing tramcars or faulty motors, generated similar waves, causing troublesome static or interference. Armstrong had devised another approach, slightly varying the frequency (the number of cycles per second) of the radio signal to code the words or music. The result was clearer, static-free sound.

Armstrong, then working for the Radio Corporation of America (RCA), patented FM in 1933. By the early 1940s, a number of FM stations—which became known as the Yankee Network—were operating in the New England region, using a radio band allocated by government regulation. FM threatened the profitability of the hundreds of AM radio stations then on air. RCA owned many of these and fought back by getting the Government to shift the FM band, ostensibly to make room for the coming of television, but really forcing FM into a less suitable frequency range and undermining the stations.

RCA also claimed patent rights over FM, since Armstrong had been working for RCA when he invented it. Armstrong lost the subsequent court battle. Unable to claim royalties from the FM sets then in use, he was soon ruined; suicide quickly followed. His widow continued the fight, finally prevailing in 1967. After decades of technical development, FM also triumphed.

Armstrong was no stranger to patent battles. In the 1920s, he had struggled against de Forest and his backers over patent rights to a new sort of radio receiver using the 'regenerative principle'. Most of its operations took place at one frequency, so it was simpler to tune from one station to another. The decade-long lawsuit was ultimately settled in de Forest's favour, though some observers thought the Supreme Court erred in its judgement.

Mastering microwaves
Edward Bowen, Mark Oliphant and Percy Spencer

The development of radar, particularly in Britain in the 1930s, had a profound influence on the course of World War II. Early radar sets used radio waves 10 or 20 metres in length, but those had limitations, especially when the military wanted to install radar sets in fighter aircraft so they could find and destroy attacking bombers at night. Both the size of the gear and the precision with which a radar beam could be directed onto its target depended on wavelength. The new need called for radio waves of about 10 centimetres—'microwaves'. The crucial calculations were made by Welshman Edward 'Taffy' Bowen, one of the very first to join the British radar effort in 1935.

At the time, microwaves could be produced at a power of only a few watts, nowhere near the kilowatts needed to get reflections from planes and ships kilometres away. The physics laboratory in Birmingham University, headed by Australian Mark Oliphant, soon had the answer. His colleagues John Randall and Harry Boot invented a unique device; it made electrons spiral around magnetic fields in holes bored in a solid block of copper. This converted their energy into electromagnetic waves.

Well connected in defence matters, Oliphant—who also played an important role in the development of the atomic bomb—brought the invention to the attention of military authorities, who seized upon its potential. At

Above: Microwaves in your kitchen. The wartime technology of radar turned to peaceful applications, such as this early model microwave oven at the 1947 Ideal Home Exhibition in Olympia, London.

its first firing up in early 1940, the 'cavity magnetron' produced over 400 watts of power at 10 centimetres. That grew to kilowatts once engineers at General Electric got to work on it, turning the temperamental laboratory prototypes into equipment robust and reliable enough to go to war.

Microwaves and the magnetron transformed the use of radar. Night fighters could now find their targets easily in the dark, attacking bombers could navigate and strike their targets using a radar set that generated an image of the ground over which they were flying. The Allied victory against German submarines in the vital 1943 Battle of the Atlantic was largely due to microwave radar. The Americans, for all their expertise, had developed nothing like the cavity magnetron. When alerted to its existence by Bowen and others, so they could undertake further development, American experts called it 'the most valuable cargo ever to cross the Atlantic.' German engineers soon came up with something similar, and technically at least as good, but they never integrated the equipment into their overall scheme of defence as well as the British.

The mastering of microwaves, initially a necessity of war, continued over coming decades. Radar soon found all sorts of civilian applications, including air traffic control and weather forecasting. In the field of communications, microwaves, able to be beamed very precisely and to encode large amounts of information, began to replace buried copper cables in carrying large numbers of telephone calls. Microwave relay chains now cross the continents and send messages to and from orbiting satellites, though on high volume routes, the even more capacious optic fibres have begun to take their place.

Microwaves oscillate billions of times a second and have now found a vital role in the particle accelerators physicists use to seek the ultimate structure of matter. Generated by high-power machines called klystrons,

Microwave ovens: how do they work?

When microwaves are passed through food, the molecules that make up the food, particularly water, fat and sugar molecules, absorb energy. These molecules have a positive electric charge at one end and a negative charge at the other. As they try to align themselves with the electric field generated by the microwaves, the molecules rotate, so hitting neighbouring molecules and causing them to move also. All the movement generates heat.

'The most valuable cargo ever to cross the Atlantic.'

microwaves are ideal for rhythmically pumping in energy to speed up subatomic particles as they orbit the accelerators or sprint down a straight track kilometres in length.

There is, of course, a more homely use for microwaves. Their heating effect was obvious from the start. Scientists in Birmingham in 1940 could light cigarettes in the beam crackling from the first magnetron. With a focus on, as one wag said, 'cooking the Nazi goose', domestic applications had to await the end of hostilities.

The concept was vigorously taken up by Percy Spencer, an engineer at the American electronics firm Raytheon, who found a chocolate bar in his pocket melted when he walked through a microwave beam. The first production microwave oven, known as the RadarRange and marketed in the 1950s, used a magnetron similar to those in a radar set, but it was massive, expensive and didn't sell well. Both size and price came down through the 1970s, and 'microwaves' took hold, especially when controlled by microchips so becoming very simple to operate. Nineteen out of every twenty households in the United States now has a microwave, with the popping of popcorn one of the most popular uses. The global population of microwave ovens is estimated at 200 million.

Wireless by satellite
Arthur C Clarke

The first communication satellites—'comsats'—were not lifted into orbit until the 1960s, but the idea was at least 20 years older. In a 1945 issue of *Wireless World*, science fiction author Arthur C Clarke proposed a

For the first time, the phrase 'live via satellite' appeared on television screens.

system of three satellites, equipped with radio receivers and transmitters and each in geosynchronous orbit. Circling the earth at an altitude of 36,000 kilometres, such satellites would pass once around the earth from west to east each 24 hours, keeping step with the rotation of the planet beneath and appearing from the ground below to hang stationary in the sky.

Clarke's vision needed three satellites so that each was simultaneously in view of the other two, as well as in sight of ground stations. The ground stations would transmit messages upwards, these would be relayed around the system, then transmitted down again to destinations way out of sight of the origin of the message. Clarke drew his inspiration from the experiments with rockets carried out in Germany and the United States in the 1920s and 1930s, culminating with the use of such machines as weapons of mass destruction in the closing months of World War II. A decade after he wrote, the Soviet Union placed the first artificial satellite, Sputnik I, into orbit in October 1957, opening the way for space communications. The start was modest. In 1958 the US Army Score satellite had only one voice channel but could transmit messages directly or store them for a later playback.

Clarke's idea certainly had appeal. A network of comsats would be an advance on the practice of sending messages across land areas by microwave relay. Tall towers with transmitting and receiving antennas are built in sight of one another. Clearly we cannot place such towers in the middle of the ocean. A satellite is therefore like a microwave tower tall enough to see both sides of, say, the Pacific Ocean and relay messages between stations on either side. Furthermore, a satellite can send a message onward to any receiver within its field of view, not merely to a single station as in a conventional microwave relay. The cost of international phone calls would fall sharply as capacity grew.

In 1960 the United States launched Echo, an aluminium-coated balloon 30 metres in diameter, able to reflect radio signals beamed up to it from Earth. Soon after experiments were underway with 'active' comsats such at Telstar and Relay, equipped with low-noise radio receivers and transmitters to amplify the oncoming signals and send them onward. These

The success story

The power of a network of comsats to draw the global community together was first demonstrated in 1967 with the program Our World. Short segments showing facets of everyday life in countries around the planet were carried simultaneously by satellite and relay to hundreds of millions of viewers, stitched into a two-hour celebration of diversity. We saw babies newly-born on four continents. In a London recording studio, the Beatles sang *All You Need Is Love*, a few minutes later the first tram of the day rattled out of a shed in Melbourne. The response was much like that of the first viewers of the moving picture show staged by the Lumiere Brothers in 1895. The subject matter was mostly familiar; the means that delivered to our eyes and ears was extraordinary and unprecedented.

experimental rigs could carry hundreds of voice calls simultaneously, and that capacity could be blended to deliver a complete television signal, video and audio. However, they were in low orbit, only a few hundred kilometres above the ground and therefore appeared to move swiftly across the sky. Ground stations had to track them mechanically in order to remain in contact, and they did not stay in sight for long.

Much closer to Clarke's vision was Syncom II, the first geosynchronous communication satellite, launched by the United States in 1963 in time to relay live coverage hour by hour of the 1964 Olympic Games in Tokyo. For the first time, the phrase 'live via satellite' appeared on television screens. Nowadays, of course, we are so used to seeing things as they happen on the far side of the world, from wars to royal weddings to tennis matches, that we no longer remark upon it—and the banner has long disappeared. A global network of satellites, in contact with thousands of ground stations in many nations, sustains an intimate and immediate worldwide exchange of experience through words and pictures that led one early commentator to suggest that we were already in a 'global village'. Consider the reach and impact of a global television news network like CNN, only possible in the satellite age.

Below: Dozens of communication satellites, such as this one, Syncom II—lifted into orbit by powerful rockets and powered by solar cells—now girdle the globe in a network of instant connectivity.

CHAPTER 12:
The box in the corner

The impact of television today is immense. Most people get their up-to-the-minute news, their knowledge of the world, their entertainment from television. Everywhere you go there is likely to be a television screen within eyeshot, and the picture you see may just have arrived from the other side of the planet. You can watch 'the box' in your car—though preferably not while driving—even on your mobile phone. For so powerful an influence, the technology had an obscure beginning in the phenomenon of 'cathode rays', which fascinated physicists a century and a half ago.

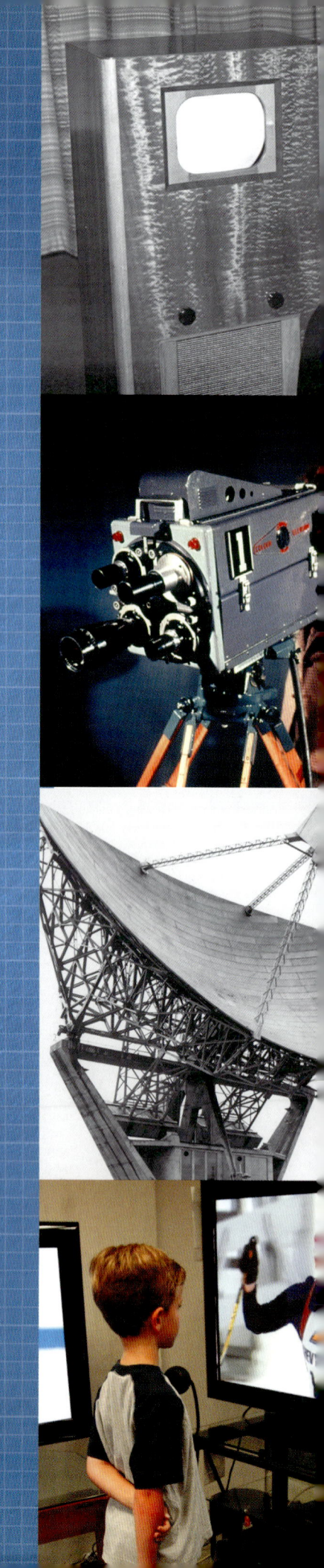

The Father of the CRT
Karl Braun

Italian Guglielmo Marconi shared the Nobel Prize for Physics in 1909 for pioneering work in wireless telegraphy with German Karl Braun. During a long career as a university professor, Braun increased the range of Marconi's transmitter and invented the 'crystal rectifier', which improved reception. Today we mostly—and gratefully—recall Braun as the inventor of the cathode ray tube (CRT).

As Marconi had refined Hertz's primitive radio equipment, Braun had taken on the discharge tube, a laboratory tool used by physicists for decades, particularly in Germany and England, leading to the discovery of X-rays and the electron. In discharge tubes high-voltage electricity was made to pass between two electrodes (metal plates) in a sealed glass tube almost empty of air.

All sorts of fascinating phenomena were seen. With the right shape and arrangement of electrodes, one end of the glass tube could be made to glow, especially if coated on the inside with a chemical like zinc sulphide. The effect was traced ultimately to the impact of some otherwise invisible 'rays' that came from the negatively charged electrode, which is known as the cathode; hence, the term 'cathode rays'.

In 1897 Braun honed the old discharge tubes into something much more versatile. Controlling the cathode ray beam with electric and magnetic fields focused the glow into a small spot and allowed it to be scanned from side to side across the flattened end of the tube and in rows from top to bottom. This produced the first 'oscilloscope' (sometimes abbreviated as CRO, for 'cathode ray oscilloscope'), which displays variations in electric currents over time. Much improved in operation over the years but essentially unchanged in principle, the CRO remains vital equipment in every electronics workshop and laboratory.

Stimulated by Scots-born Alan Campbell-Swindon, cathode ray tubes morphed into the first television cameras and monitors, and later still into monitors for

Above: The Scottish engineer John Logie Baird, regarded by some as the real inventor of television. He is holding a cathode ray tube, but long promoted mechanical scanning which was abandoned in the mid 1930s.

All sorts of fascinating phenomena were seen. With the right shape and arrangement of electrodes, one end of the glass tube could be made to glow.

Left: A television set from the early days: a tiny screen with rounded edges that showed only fuzzy black-and-white pictures, prone to interference, and cost half a year's salary. But there were plenty of buyers.

And few would have expected otherwise. The first 'moving pictures' were still 30 years away.

There were always two challenges with 'television': how to break a picture up into elements that could be transmitted, and how to convert those fragments of image into an electric current which could pass down a wire. 1873 saw an answer to the second. French physicist Henri Becquerel—who was later to discover radioactivity—found that the newly discovered element selenium (technically a semiconductor) could be used to make a 'photo-electric cell' to turn light into electricity. The brighter the light falling on the cell, the stronger the current it produced.

As for analysing the picture, the English inventor Shelford Bidwell tried moving a selenium cell across the face of a picture using cams, cogs and levers. That was in 1881. Three years later, Germany's Paul Nipkow had a much better idea, a concept used in some of the first television transmissions 50 years on. Nipkow punched a series of holes arranged in a spiral pattern into a metal disc. He placed his 'Nipkow disc' between the picture and the photo-electric cell and spun it rapidly. This broke up the picture, as seen by the selenium cell, into many dots; the brightness of each dot controlled the amount of electric current leaving the cell.

In 1908 Bidwell suggested to the science journal *Nature* that 'electric vision' was possible if just less than 90,000 selenium cells in a grid—each looking at one segment of an image and connected by its own wire to a similar point in the receiver. The breathtaking impracticality of this scheme made Scots-born electrical engineer Alan Campbell-Swinton wonder if the same result could somehow be achieved with just a single wire between transmitter and receiver.

Campbell-Swinton was highly talented. He had taken some of the earliest X-ray photographs and helped the

radar sets and computers. Only now are CRTs giving way to liquid-crystal displays and plasma screens, but Dr Braun's creation has had many offspring.

Pictures from afar
Paul Nipkow, Shelford Bidwell and Alan Campbell-Swinton

If we look go looking for the pioneers of the technology of television, we can start with the Italian-born priest Abbé Caselli. By 1862 he had a system to send pictures by wire. The results were crude, the images little more than shadows. Nonetheless, Caselli quickly gained the useful patronage of Napoleon III and set up a number of stations across France for sending hand-written messages and drawings over telegraph wires. The venture never made money. Interference was one problem. Dots and dashes from Morse code messages sharing the same line marred the images. Of course the good Abbé's pictures did not move. His technology was more like a facsimile than television as we know it now.

struggling Marconi by writing a letter of introduction to the head of the British Post Office, Sir William Preece. Most importantly, he knew that the German scientist Karl Braun had invented the cathode ray tube, in which electric and magnetic fields could move an electron beam across a surface inside a vacuum.

Campbell-Swinton reacted to Bidwell's 90,000 cell proposal in his own letter to *Nature*. He argued that the Braun tube could make electric vision possible by serving as both transmitter and receiver, as both camera and screen. At one end of the system, light from the image or object would fall on a selenium plate segmented into many pieces, with each cell of the plate accumulating electric charge in proportion to the intensity of the light falling on it. Scanning an electron beam across the face of the plate would pick up the accumulated charge cell by cell, resulting in an electric current that varied in time with the changing brightness of the scanned image.

The current, passed down a wire to a similar tube some distance away, would cause changes in the strength of a second electron beam, causing it to paint in the picture on the face of the CRT tube as a pattern of dots of fluctuating brightness. The slowness of the human eye would do the rest, as indeed it did with the illusion of movement created by the relatively new 'moving pictures'. If the whole process could be

Below: Much of early TV production was 'live', captured by cameras like this—with its turntable of lenses—and transmitted immediately, mistakes and all, for lack of any recording medium. Then, around 1956, came videotape.

He had anticipated the form and function of the cameras and display systems that formed the heart of video technology almost to the present day.

repeated quickly enough, the images provided by 'electric vision' would also move.

In 1911 Campbell-Swinton patented this idea. 'It is just an idea', he wrote in his application, 'and the apparatus has never been built. Furthermore, I do not for one moment suppose that it could be made to work without a great deal of experimentation and probably much modification.' It did take a quarter of a century to produce a really workable system based on his idea, but Campbell-Swinton's vision was brilliant. He had anticipated the form and function of the cameras and display systems that formed the heart of video technology almost to the present day.

Television goes on air
John Logie Baird, Vladimir Zworykin and Philo Farnsworth

Science and technology often take wrong turnings, and end up exploring blind alleys. So it was with embryonic television. The notion of using Karl Braun's cathode ray tube to capture and display images, proposed in 1908 by the Scots visionary Alan Campbell-Swindon, was sidelined for several decades by the mechanical system of scanning first proposed by the German Paul Nipkow. In the 1920s, two inventors took this up, American Charles Jenkins and Scotland's John Logie Baird, the man often—and incorrectly—credited in the public mind with inventing television single-handed.

Both of these men developed television systems based on scanning discs and both had many followers. In the receiver, the image was reconstructed by using the incoming signal to change the brightness of a lamp, which shone through a second scanning disc onto a screen. Dot by dot the image would appear, with the slow workings of the eye blending the trail of bright spots into a complete picture. Baird's system was publicly trialled in Britain in the 1930s, but ultimately the crudeness and lack of definition of the images—Baird's system, for example, had only 30 scanning lines to the frame—saw them abandoned in favour of the electronic systems prophesied by Campbell-Swinton.

Below: With the advent of satellite communications in the 1960s and using antenna dishes like this, television went global. The pictures came 'live via satellite', we were told. In time, the novelty faded: now we expect it.

Above: The lettering on the arch looks a bit dated, maybe from the 1940s when the American Broadcasting Company first transmitted television pictures (it already ran radio stations). The images below the arch show how far television has come since, particularly moving into cable.

Others had tried using Braun's invention to make television work, such as Russian Boris Rosing in 1907, though the scanning in his transmitter was done with a Nipkow disc. Rosing did manage to transmit some crude geometrical shapes, but half-tone pictures were beyond him and without the aid of electronic amplifiers—then only just being invented—the images were very faint.

It was Rosing's student, Vladimir Zworykin, who had migrated to the United States in 1919 and joined the staff of Westinghouse, who did most to realise Campbell-Swinton's vision. In 1923 he patented an electronic camera based on the cathode ray tube. Initially it was no match for the ubiquitous scanning discs, but by 1928 the 'iconoscope' was ready for use and for a decade of legal battles over patent rights to the invention.

At least one other name must be mentioned, that of Philo Farnsworth, who made so many contributions many regard him as the real inventor of television. Professor Farnsworth in the television series *Futurama* carries his name. He devised his first system at the age of 15 as a Utah farm-boy, inspired by taking part in his first telephone conversation. Through the 1920s, he pioneered electronic systems in parallel with Zworykin, in particular building his 'image dissector'. Modern video cameras are based on a combination of Zworykin's and Farnsworth's ideas.

It was reputedly Farnsworth who convinced Baird that his mechanical systems could not compete with electronic ones. By 1935 Farnsworth had built a

The success story

Sales picked up after the war, as public acceptance grew, along with the quality of equipment and programs. In 1945 there were 7000 sets in the United States, with only five stations broadcasting; by 1950 this had grown to 10 million sets and 100 stations. Serious drama, news and public affairs filled the schedules until the mid-1950s and the rise of the 'game shows' and the 'sitcoms'. Television spread quickly throughout the industrialised world, reaching Australia for instance in 1956, in time to broadcast the Olympic Games from Melbourne.

complete television system, used a year later to transmit entertainment programmes. Farnsworth struggled with depression late in life and, until his death in relative obscurity in 1971, was preoccupied with trying to control nuclear fusion as a power source.

The first regular television broadcasts went to air in 1936, less than two decades behind radio. Transmissions by the British Broadcasting Corporation (BBC) originated from Alexandra Palace in London, using both electronic and mechanical scanning for a time, and televising the coronation of George VI. The outbreak of war in 1939 closed television stations down, though tremendous technical advances made through the war, such as those of radar, made television a far more powerful medium when large-scale transmission resumed.

In the United States television quickly came under the control of commercial radio interests, led by David Sarnoff's Columbia Broadcasting System (CBS) who had paid for much of Zworykin's research, and the National Broadcasting Company (NBC). They saw television as a way of supplementing the mix of entertainment and information that was working so well on radio. First to air was NBC in 1939, telecasting President Roosevelt opening the New York World's Fair. Public acceptance of the new wonder was surprisingly slow. After two years of BBC telecasts, only 3000 sets were in use in Britain, and only 500 were bought in the first six months after NBC went to air. The high cost of the equipment and the crudeness of the images were major disincentives and programming was rudimentary compared with the sophistication of radio.

Today's television
Colour, cable and digital

Television today is a world away from the flickering, blurry, monochrome, interference-prone images that people peered at with such wonder in Britain in the 1930s, and in greater numbers just after World War II. Many fast-evolving technologies have combined to enhance, even transform, the look of television pictures and the way they are made and transmitted. These included the use of videotape, microwave relays, satellite communications, transistors and screens using liquid crystals.

The 1930s had seen the invention of co-axial cable, designed to carry rapidly-varying electric currents like those needed by television. The first cable networks, some connecting television stations, others bringing television to regions with poor broadcast reception, were in place in the late 1940s. Similar cables are one way of connecting to the Internet 'broadband'.

Many today may not remember or even believe that there were dull, grey days in the years BC (Before Colour).

Chapter 12: The box in the corner 159

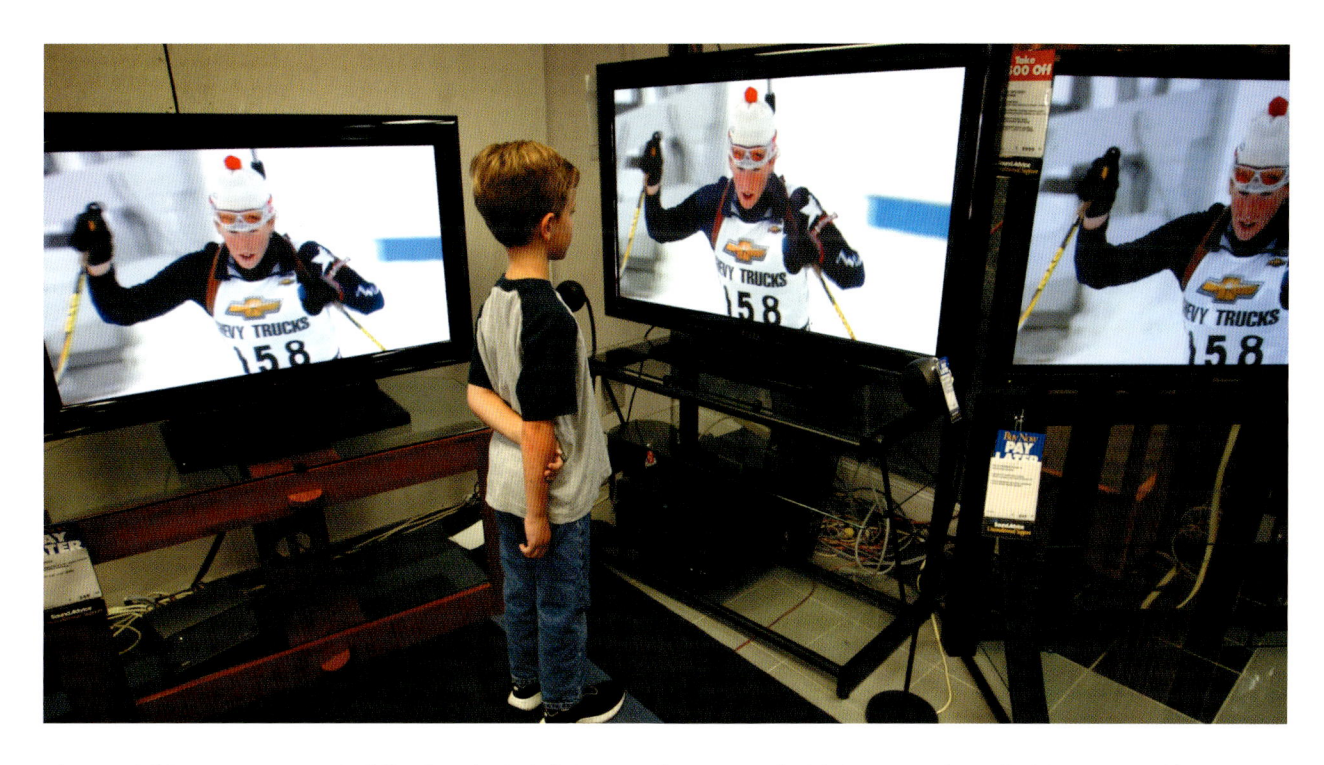

Above: Visible progress. Now in full colour, high-definition, widescreen and with instant replay, television is a world away from the unsteady monochrome images this boy's grandfather might have watched with equal fascination 50 years earlier.

Many today may not remember or even believe that there were dull, grey days in the years BC (Before Colour). Television went to colour first in America in 1951, though the transmissions were shut down a few months later when war broke out in Korea. As always, early sets were expensive and penetration slow, but everyone expects to see colour on television now. Even old black-and-white movies are often 'colourised'.

Today the momentum is towards High-Definition television (HDTV), though it will be a decade or more before that is well-entrenched. The move is not very old, starting in 1989 with a working party of the International Telecommunications Union, with standard-setting, trials and equipment development in the 1990s, and the first transmissions (by Home Box Office in America) in 1999.

HDTV uses digital (binary) data processing, which we now have in music with CDs, in place of the old analogue (continuously varying currents) process with which television began. Digital images can be compressed, as in DVDs. With redundant information removed, more programs or data streams can be fitted into the same transmission space. This allows refinements such as 'chose your own camera angle' when watching sport. What would Logie Baird and Philo Farnsworth have thought of it all?

HDTV: what's that?

An image in HDTV contains two to five times as much information as a regular television image. This comes from more scanning lines per screen, which sharpens the picture; a wider range of colours, which provides greater realism; and a widescreen format—which has a width by height ratio of 16:9 instead of the old 4:3, which came from the movies.

CHAPTER 13:
The computer story

Today 'computers' are everywhere, though such is their versatility that they would better be called 'information processors'. They are, at heart, calculating machines—and unsophisticated ones at that, counting merely from 0 to 1. Their power comes from the fact that all sorts of information, text, images, sounds, can be represented by strings of numbers and so can be processed in a multitude of ways simply by adding and subtracting. This they can do at dazzling speed, empowering the many forms of 'information technology'.

The other virtue of today's computers is that they can be 'programmed', told what to do and left to get on with it without continuous intervention from a human operator. That was certainly not the case with the first mechanical calculators.

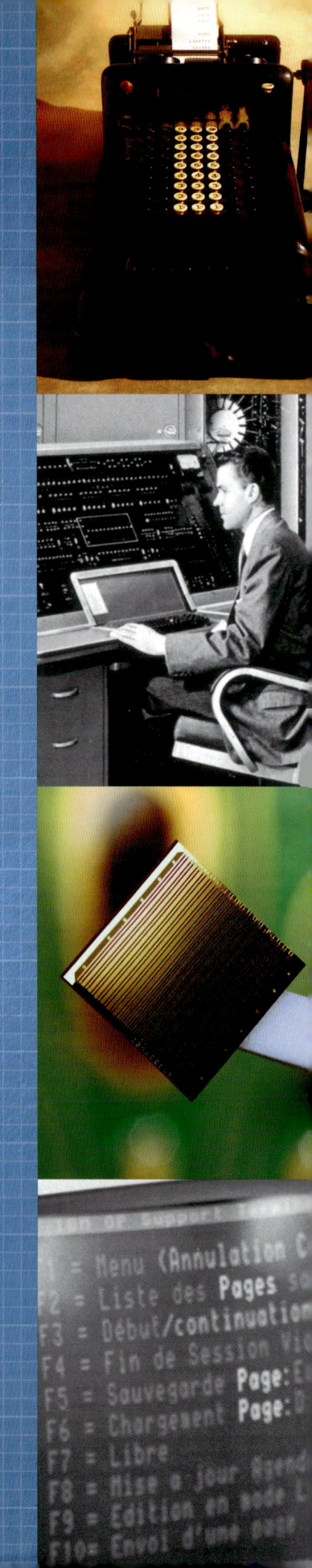

Mathematics made easier
William Oughtred, Blaise Pascal, Wilhelm Schickard and Gottfried Leibniz

As the deeply-religious Scottish genius John Napier put it: 'seeing there is nothing that is so troublesome to mathematical practice, or doth more molest and hinder calculators, than the multiplications, divisions, square and cubic extractions in great numbers, I began therefore to consider in my mind what by certain and ready art I might remove those hindrances.' Napier's response to the challenge in 1614 was the system known as logarithms.

Around 1622 the English clergyman and mathematician William Oughtred inscribed logarithms onto two metal discs on one axle, so they could be lined up for the purposes of calculation. It probably justified being called a machine. From this he developed the 'slide rule' around 1633, a device found in the breast pocket of most scientists for the next 300 years, until the electronic calculator took over in the 1970s.

Long before electronics, other clever minds were devising 'calculating machines', in which gear wheels and cranks would replace mental effort in doing sums. The German Wilhelm Schickard was the first with his 'calculating clock' in 1623. This could add and subtract six-digit numbers and devise tables giving the positions of the planets and other astronomical objects.

Others to go down the same path included the pensive French philosopher Blaise Pascal. His 'Pascaline', invented in 1642 when he was only 19, was designed to relieve his father, a clerk, of much tedious calculation. It worked tolerably well but only Pascal himself could repair it, and the need to cut the gear wheels by hand meant that it cost more than the people it replaced. Still, mathematicians began to fear for their jobs.

The German Gottfried Leibniz, the great rival of Isaac Newton, worked on the problem later in the century and found a way to speed up multiplication and division by treating them as a series of additions and subtractions. Any mechanical calculator still working today uses this principle.

For all their ingenious complexity, the machines were crude and unreliable, prone to breakage and hard to repair, a consequence of the ill-formed state of manufacturing at the time. Furthermore, they could not be 'programmed'; they needed constant intervention by the operator. Not until the 19th century could a mechanical calculator work unsupervised.

Without human control
Joseph Jacquard and Charles Babbage

Devices like 'computers', able to carry out complex tasks without continual human guidance, are not merely 20th-century marvels. In the previous century,

Logarithms: what are they?

Also known simply as 'logs', logarithms make every number a 'power of 10', reducing multiplying to adding, dividing to subtracting. To find a square root, simply divide the logarithm by two. It was a revelation.

Blaise Pascal invented his 'Pascaline' in 1642, when he was only 19. It was designed to relieve his father, a clerk, of much tedious calculation.

Right: A masterpiece of engineering, this two-century-old Jacquard loom wove intricate patterns into cloth and carpets following programmed instructions, stored on 'punch cards'. It was also a glimpse of the automation to come.

the Jacquard loom could do its job unsupervised, and Charles Babbage's 'analytical engines' were an early vision of calculations controlled by a 'program'.

Through hard times and the Revolution, French silk-weaver Joseph Jacquard kept trying to automate weaving. An endless loop of punched cards controlled a loom he built in 1801. The holes in the cards determined which longitudinal (warp) threads were depressed during each pass of the shuttle; complex patterns in cloth and carpets could be woven without the weaver intervening.

At first Jacquard's looms attracted the ire of weavers who feared unemployment, but with government support the number of looms soared. Eleven thousand such looms were in use by 1812. A threshold had been crossed.

Charles Babbage, son of a banker, was a brilliant mathematician, holder of the Lucasian Chair at Cambridge previously occupied by Isaac Newton. His lifetime achievements include the establishment of the British Association for the Advancement of Science and advances in cryptography. He is mostly remembered today for proposing—and failing to complete—the 'difference engine' and the 'analytical engine', the first attempts at calculations without continual human involvement.

To reduce the error rate in the laborious calculations of mathematical tables, including those needed for navigation by ships at sea, the steps in the calculations would be controlled by punched cards, as in Jacquard looms, rather than by a human operator prone to fatigue. In the days before electronics or even electricity, the 'engines' would be purely mechanical, with thousands of gear wheels and axles and likely to weigh tonnes when completed. Such an enterprise was unachievable without government support. This Babbage initially secured but later lost, as the practical usefulness of his project came under fire, and as he encountered the limitations imposed by the technology of the day. He was also constantly at odds with the craftsman he employed from the 1830s to build the engines. Only fragments were ever completed and after a lifetime of labour, Babbage died embittered by failure.

Babbage's vision was not achieved for a century, until electronics emerged as the technology for the task. In honour of Babbage, some computer programs, such as those used to search the Internet, are called 'engines'.

The census on cards
Herman Hollerith

Now little remembered, Herman Hollerith has several claims as one of the progenitors of today's information technology. For one thing, a company that he founded late in the 19th century became, after several mergers and changes of name, International Business Machines (IBM), highly visible in the computer market a century later. And he automated the counting of the 1890 United States national census, the biggest data-processing endeavour ever undertaken.

Hollerith was only in his twenties, and on the staff at the Massachusetts Institute of Technology, when he devised a process to encode data, such as the census collectors gathered, as patterns of holes on cards. These were not unlike the cards used to control a Jacquard loom, but Hollerith was reputedly more influenced by the way tram and bus conductors punched holes in tickets to indicate the fare paid.

In trials Hollerith's method easily outpaced manual counting, and the government bought his machines in time for the 1890 census. It did the whole job in seven years instead of nine, even though the census was far more comprehensive than previous head counts and called for more complex analysis. It did, however, cost twice as much as the census a decade earlier, largely due to the immense amounts of power consumed by the electrically-driven card readers and sorters.

Hollerith's triumph was widely noted. Scientific American devoted its August 1890 edition cover to the innovative technology. Hollerith's Tabulating Machine Company struggled to keep up with demand. Orders for the machines—rented rather than sold—came from other census authorities around the world and from government departments and private firms swamped with paperwork and mountains of figures. Some companies such as Prudential Insurance devised their own punch-card systems.

Above: Logarithms on a stick. For 300 years, until the invention of the pocket calculator in the 1970s, a slide rule was the fastest way to complete a calculation quickly. Who would know how to use one now?

Hollerith's tabulator: how did it work?

Stacks of cards were sent through reading devices with spring-loaded pins that completed electrical circuits when they passed through the holes. Numbers corresponding to particular categories were automatically totalled on dials. The cards could also be sorted, to select and count people who corresponded to a particular grouping of characteristics, such as white women living in the Bronx, aged between 20 and 30 and employed as shop assistants.

Above: Hand-punched and hand-cranked, this 20th-century calculating machine was still far better for toting up lists of numbers than a pencil and paper or fingers and toes. Many an accounting department would have ground to a halt without one (or many).

None of these were computers as we know them today, but they did foreshadow the powerful impact automatic data processing would make. Punching and sorting cards remained a key activity in offices—with key-punch operators allocated to the task—for most of the next century.

The first modern computer?
John Mauchly and J Presper Eckert

You can have a good argument over who invented the modern computer. In reality, the features we associate with computers today—that they are electronic, use binary code, can be programmed—did not emerge all at once. Some early computers had a few of these properties but not others. Some used electromechanical technologies rather that electronics. Some stuck with ten-digit numbers rather than binary code. Some were designed for one given task, while others were more versatile.

Several nations were involved, including America, Britain and Germany, inventing machines in the late 1930s or early 40s, but all their first—or intended—uses related to defence or national security, such as cracking codes or calculating tables to improve the accuracy of artillery fire. Peacetime uses came much later.

Combining these factors suggests that priority should go to the Electronic Numerical Integrator and Calculator (ENIAC). Construction of this began at the University of Pennsylvania in 1943 under the leadership of John Mauchly and J Presper Eckert, and it went live three years later. This was certainly electronic and could be reprogrammed—though with some difficulty—but it was not binary, continuing to use decimal numbers. ENIAC was colossal in every dimension, weighing 30 tonnes, consuming as much electricity as 50 homes, with 17,000 vacuum tubes and 5 million hand-soldered joints.

It was ponderous by modern standards, with a clock speed of 5 kiloHertz—today's laptops are a million times faster, supercomputers a billion times. Data was fed into the machine by punch cards and came out the same way. Originally intended to figure firing tables, its first calculations supported the development of the hydrogen bomb. With upgrades, it kept running for a decade.

ENIAC was a one-off. Nothing quite like it was built again; the technology was already picking up speed. It did have offspring in the form of UNIVAC, the first electronic computer to reach the market in America—although comparable machines were already on sale in Germany and Britain. The name stood for Universal Automatic Computer, and 'universal' was the right term.

Chapter 13: The computer story

The features we associate with computers today—that they are electronic, use binary code, can be programmed—did not emerge all at once.

From the start the builders had business needs in mind; customers would be able to assemble an information processing system to meet their precise needs.

UNIVAC was conceived in 1946 by Eckert and Mauchly, who went soon after into the computer-building business, but its godfather was the statistics-obsessed US Bureau of the Census, which put up the first funding—just as they had helped Herman Hollerith half a century before. The money wasn't enough, however, and a project for the Northrop Aircraft Company, a lightweight binary computer to test if ballistic missiles could be controlled in flight, was signed up to make ends meet. Even so, Mauchly and Eckert were short of cash and, early in the 1950s, their UNIVAC company passed into the control of Remington Rand, the leading maker of office equipment and adding machines.

Although much smaller than ENIAC, UNIVAC was still a monster. Around 5000 vacuum tubes filled a

Below: Sleek, shiny and above all, massive, this UNIVAC 'mainframe' computer was considered a wonder in the 1960s. Now your average laptop can calculate much faster than this air-conditioned room-full of electronics.

space 5 metres long, 3 metres high and 2 metres deep. It was fast, twice as quick as ENIAC at adding numbers—two 12 digit numbers in one ten-thousandth of a second—and had for the first time a significant internal memory, about 12 kilobytes in today's terminology. It was still not binary, working with ordinary decimal numbers, but it was showing the way, particularly by having a stored program, a set of instructions that told the central processing unit (CPU) the steps to take to complete the task. This vital feature was the creation of the brilliant Hungarian-American John von Neumann. The task could be changed just by substituting one program with another, rather than rewiring the machine as ENIAC had required.

Already optimists were predicting that the next generation of computers would 'weigh less than a tonne', though few could have conceived of a computer that could sit on a desk, let alone on a lap—but that was coming.

The father of hypertext
Vannevar Bush

Hypertext is not new. In defining a word, a dictionary will often include another word in small capitals, indicating that there is a separate entry for it. Cross-referencing by footnotes adds to the information stock. An early worker on computer-based hypertext was the creator of the mouse, prolific American inventor Douglas Engelbart. In 1968, he introduced his NLS (oN-Line-System), with a large number of documents held in an accessible memory, all cross-referenced and 'hot linked'.

But a founding visionary of hypertext, an inspiration for Engelbart and others, was American engineer, scientific statesman and World War II presidential adviser Vannevar Bush. He oversaw most of the technical developments affecting the war effort, including the Manhattan Project. In July 1945, around the time of the first atom bomb tests, Bush wrote about the Memex, an imagined machine able to hold an immense amount of information in words and images, but with a mechanism which could link any particular item in the data-store to any other item, as seemed appropriate ...

A user moving from one document to another would create a trail of links, which could be retrieved so that he or she, or others, could follow the same path. Any one document might be reached by a multitude of paths, whereas documents stored alphabetically or numerically or by subject can be reached by only one set of rules. These latter storage systems do not take advantage of the powerful capacity of our minds to make associations, as Bush claimed Memex would do.

The success story

The first UNIVAC machine went to the Census Office in March 1951 after six years of development. Over time nearly 50 were sold at a cost of $500,000 each in 1950s money. The demand came from private sector firms and government departments preoccupied with numbers: the Prudential Insurance Company, market researcher AC Neilson, big companies needing help with their payroll, as well as the military, universities and research laboratories. At its first public demonstration, a UNIVAC successfully called the outcome of the 1952 American presidential election. Basing its judgement on previous voting patterns, the UNIVAC—hired from the Atomic Energy Commission by the CBS television network—predicted victory for Eisenhower over Stevenson with only one per cent of the vote counted. No wonder the media were beginning to talk about 'electronic brains'.

Memex was never built, perhaps could not be built, even with the rapid improvement of microfilm, which Bush envisaged as the way to store documents. He described the Memex as the size and shape of a desk, with images projected onto sloping screens for easy reading. Does it sound familiar?

Getting data in a second
Reynold Johnson

An early impediment to the growth of the computer industry in the 1950s was getting rapid access to the data needed for the calculations. Early machines used punched cards, pioneered half a century earlier; newer computers were storing data on magnetic tape, like that used to record music. Both methods were slow: cards took time to prepare and to run for the reader; tape had to be spooled through from beginning to end. Neither could give quick access to any particular piece of information. As a result, calculations were slower and costlier than they should have been.

Hypertext: what's that?

Open a document on any website, and you will find some words underlined, bolded or in a different colour. Point your mouse at one of those and click, and a 'hot link' opens, taking you instantly into a new document relevant to the keyword clicked. The new document might be a definition of the word itself, the biography of a person named, the full text of a document referred to and so on. That is an example of hypertext: 'non-sequential writing; text that branches and allows choice to the reader, best read on an interactive screen'.

IBM, the big player in large computers, realised a new approach was needed. It set one of their leading engineers, 20-year-old Reynold Johnson, a task: devise a memory to hold as much data as 50,000 punched cards, as well as a system to access any particular piece of data in a second or less. With a hand-picked team to work with and a new laboratory near what would become Silicon Valley, Johnson came up with a solution in two years. It was simple but high-risk: disks of magnetic material were spun rapidly; sensors floating a few millimeters above their surface recorded and accessed the data on the disks.

His machine was also massive, weighing a tonne, occupying several cubic metres of space, with 50 large aluminium disks spinning twenty times a second. By modern standards, it did not hold much data, only 5 megabytes, much less than a single CD, but that first hard drive, launched in 1958, made the modern multi-billion-dollar computer industry possible. As with every area of computer hardware, progress since has been measured mostly by miniaturisation. A typical desktop computer now has a hard drive holding 200,000 times as much data as Reynold's monster, but operating on essentially the same principles.

Reynolds had a prolific mind, but a somewhat retiring disposition. His portfolio of 90 patents included the video cassette, devised while he was on loan to Sony, and a way of automatically reading cards that had been marked with a pencil rather than punched with a hole. The latter was based on an idea Johnson had first had as a high school science teacher. IBM had been so impressed that they had bought the system and offered him a job. It was a good move.

A computer on a chip
Ted Hoff

In 1969 the newly created microelectronics firm Intel, holder of the patent on the equally new integrated

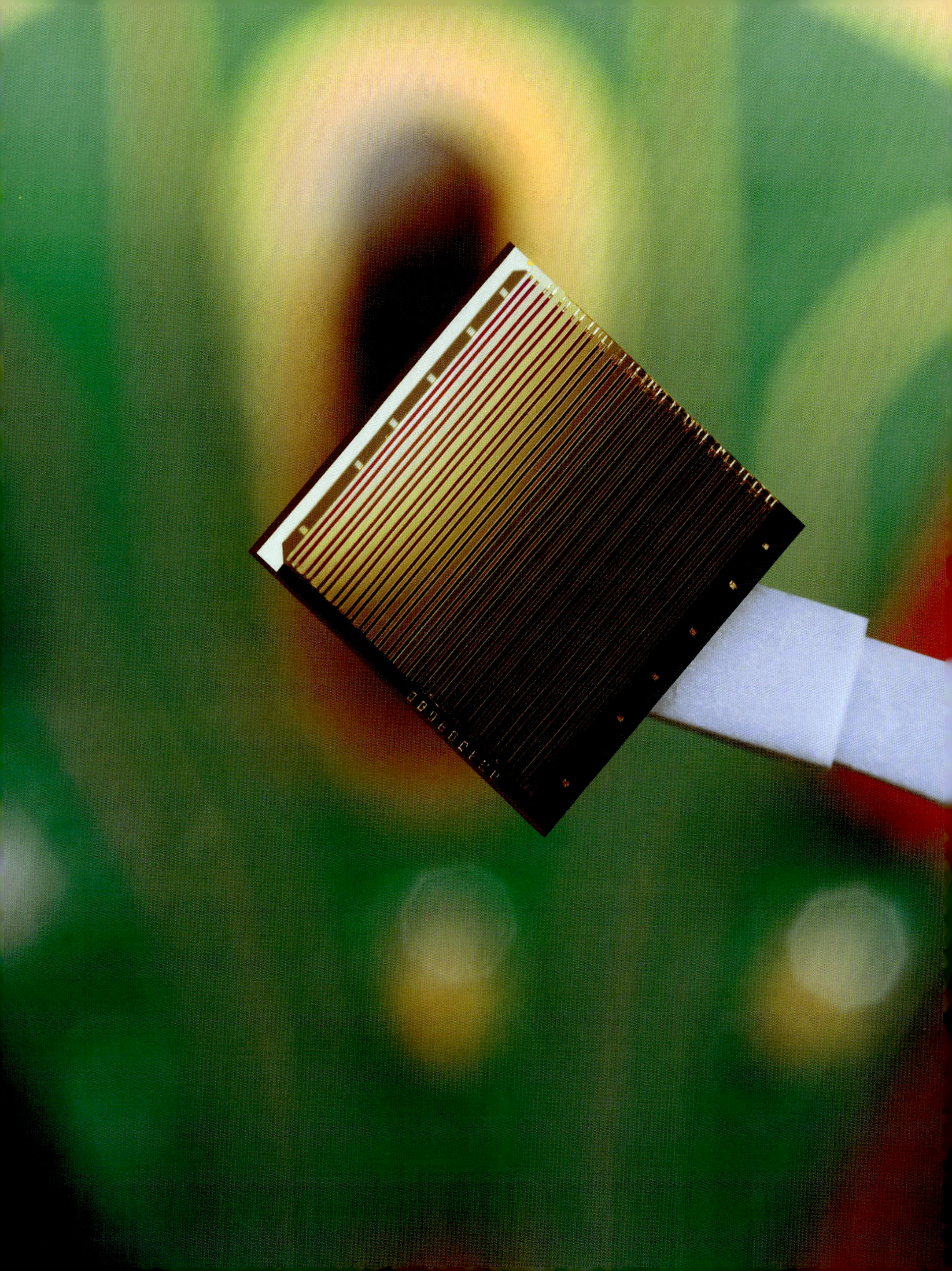

The invention changed not only computing—still in its infancy—but the whole industrial world.

circuits, was asked by Japanese manufacturer Busicom to make a set of integrated circuits (ICs) to power a proposed line of programmable hand-held calculators. The job was given to employee number twelve, bright Stanford graduate Marcian Hoff, known as Ted. Twelve was also the number of ICs to be created, each with a different job—to receive information, to do calculations, to control the display screen and so on. Hoff argued that designing so many different chips would push the cost of the device up and decided to combine some of the functions, reducing the 'chip-set' to just four.

The most important of these would be a 'general processing' or 'logic' chip, able to do a variety of calculations and control tasks as requested ('programmed'). With the help of Stan Mazor, Hoff designed what would soon be dubbed the 'microprocessor' or 'central processing unit' (CPU). Others at Intel converted the design into transparent sheets, which enabled them to print the many components on silicon. Intel launched the now legendary 4004 chip in 1971. It held over 2000 transistors, and by itself was as powerful as the entire ENIAC computer of only two decades earlier. It was literally a computer on a chip. The invention changed not only computing—still in its infancy—but the whole industrial world. Intel co-founder Gordon Moore—of Moore's Law—called it 'one of the most revolutionary products in history of mankind'.

Intel was not alone in this field. Texas Instruments (TI) was also active. Indeed, Gary Bourne of TI secured the first patent for the single-chip microprocessor architecture in 1973. By 1981 Intel, barely ten years old, had 20,000 employees, making nearly $200 million a year from microchips. Its success had spurred competition, often from firms like itself, set up by ambitious young engineers leaving more established companies. These were the boom times in Silicon Valley. By the mid-1970s, 50 or more types of CPU were on sale and the cost of any of them was down to a few dollars, as the saturated market led to price wars. Microchips were now finding homes not only in calculators but in the earliest personal computers, as well as arcade games and home video games.

Within a few years Intel replaced the 4004 with the 8008, which could handle data 8 bits at a time rather than 4, and so was markedly faster in its operations. The 8008 was the ancestor of a dynasty of CPUs made by Intel through to the Pentiums and beyond. Since that time the progress of microprocessors has paralleled the other type of integrated circuit, the memory chip, and for the same reason. Advancing production technology has crowded ever more transistors and other components onto a fragment of silicon a few millimetres square. The number of transistors per chip soon became hundreds of thousands, and today is hundreds of millions, as the typical size of components fell from a tenth the thickness of a human hair to something a hundred times thinner again. Processing speeds have risen a thousand-fold to billions of cycles per second (gigaHertz), though clock speed is no longer the sole definition of a chip's power; modern chips take in data 64 bits at a time.

A major innovation, coming to light in the 1980s, was Reduced Instruction Set Computing (RISC). With ever-more sophisticated architecture, processing chips could accept much simpler instructions. This made programming easier and increased competing speeds

Opposite: A new generation hybrid silicon 'laser' chip that combines electronics with the use of light in a technology called photonics. Ultimately, photonic devices will do everything today's chips can but even faster and cheaper.

Tens of thousands of processors like the CPU in your desktop run together to generate blinding speed.

yet again. Most modern microprocessors use RISC or something like it. Designs are now so complex and sophisticated that humans need help from computers to create them. The immense expenditure, amounting to hundreds of millions of dollars, required to design and test a next-generation microprocessor, and to build the high precision equipment ('foundries' or 'fabs') to make it by the million, has caused the smaller firms to drop out. The mass production of microchips is dominated by a handful of large companies such as Intel, IBM and Motorola.

Supercomputers FLOP
Seymour Cray

All computers have become much faster and more powerful over recent decades, but some are well ahead of the rest. A computer's speed can be measured in FLOPS (floating point operations per second), roughly the number of calculations done each second. The earliest computers did a few thousand (kilo-FLOPS), the first personal computers a few million (mega-FLOPs). Modern desktops manage a few billion (giga-FLOPS), which sounds quick, yet leading edge supercomputers are a thousand times faster again and more. In 2006 a machine at a leading American defence laboratory competed an astounding 200 million million calculations each second (200 peta-FLOPS). Upgrades should take it to 500.

The pioneer was Seymour Cray, who worked for Control Data Corporation (CDC) in the sixties, before moving out to start his own company. The first machine under his own name, the CRAY-1, reached the market in 1976. Building faster computers raised many issues. Cray figured that a number of smaller computers running side-by-side (aka 'parallel processing') would go faster than a single larger machine. To minimise the time taken to shift data from processor to processor, the connections were as short as possible. His most famous designs were cylindrical. Lastly, the data had to be moved in and out of the computer fast enough not to slow the calculations. As Cray said, 'Anyone can build a fast processor. What we need is a fast system.'

The success story

Microprocessors are not only unbelievably swift and sure, they are ubiquitous. Almost all machines or devices—from car ignitions to air conditioners and electric irons, from the office photocopiers and PABX telephone systems to your iPod and PDA (personal digital assistant), from traffic lights to industrial robots to probes in deep space—are under microprocessor control. CPUs assess a situation constantly and initiate required actions in accordance with programmed instructions, much as a human would but faster and more reliably, without losing concentration or needing a day off. Where will it all end?

Chapter 13: The computer story

Above: A computer on your desk. This early 1980s IBM PC was slow, with a monochrome screen, no mouse, no hard drive, little memory and primitive software, but it raised the curtain on a new age.

For several decades Cray-designed 'supercomputers' —a term he did not use—whipped the opposition for speed. Nothing else came close, though CDC was nearly bankrupted building the early models. Cray hit financial turbulence himself later, perhaps not surprising in so risky a business, in which one machine could cost $5 million. He was successful, selling 100 CRAY-1s and setting the early pace. Others have now taken over, using 'massively parallel' computing. Tens of thousands of processors like the CPU in your desktop run together to generate blinding speed.

The demand for speed comes from military agencies, universities and major research laboratories with big tough calculations to do. Tasks include forecasting the weather—always a rugged one— figuring the shapes of chemical molecules used as medical drugs, cracking codes, or simulating how new nuclear weapons or aircraft will perform—much cheaper than a wind tunnel. By the way, only the top machines can beat the best human chess players.

'The computer moves in'
The rise of personal computing

Early electronic computers were massive, expensive and, by today's standards, ponderously slow, even if faster than hand calculators or slide rules. New technology was beginning its march, with the use of transistors, and a decade later integrated circuits, hacking into computer size and weight. By the early 1970s, the typical minicomputer was only as big as a refrigerator or a grand piano, though still for collective rather than individual use. Anyway, who would want one for themselves?

Most people old enough to remember would say that the personal computer, such as most of us now have on our desks or in our briefcases and generally reserved for our use alone, entered our lives in the mid-1980s. That's true overall, but the early movers were already a decade down the track. In the early 1970s, the Xerox Alto, produced by the Xerox Corporation, was the typical fridge size but getting user-friendly—though very few got to use one. It was the first computer with two features later almost universal: a mouse, invented by Douglas Engelbart in 1963 to move a pointer around a display screen; and a graphical user interface—images and objects on the screen at which the user could 'point and click' and so tell the computer what to do next.

The 1970s were the heyday of computer hobbyists, the original geeks, happy to assemble computers like the 1975 Altair from a pile of parts with a soldering iron, and

Below: Personal computers have now become cheap enough for almost anyone to own. Even an entry level laptop, such as the Apple iBook can outperform the mightiest mainframe of 50 years ago, and they are user-friendly. Mostly.

Chapter 13: The computer story

The Commodore 64—with an unprecedented 64 kilobytes of inbuilt data memory—released in 1982, was to become the most popular single model ever; between 1982 and 1994 it sold 17 million units.

to meet with other nerds to swap war-stories. Among the people writing 'software' for the Altair were Paul Allen and his high-school friend Bill Gates, who later co-founded the leviathan Microsoft. Also active were the two Steves, Jobs and Wozniak, co-founders of Apple Computers. Their Apple II, released in 1977, quickly became the market leader, typifying the personal computer phenomenon then stirring. It had coloured graphics and the new floppy disk drive, first devised in the 1960s but now cheap and reliable enough to replace cassette tapes as a way of loading programs and data.

It was also innovative in 'software'—a relatively new word to describe the coded instructions that guided the computer's operations—particularly in supporting the Visi-Calc spreadsheet, invented by Dan Bricklin and Bob Frankston in 1979, for users needing to juggle figures. This 'killer application' made computers attractive to business, a major new market. Users today would think both its processing speed (around 2 megaHertz) and onboard memory (typically 4 kilobytes or 100 words) absurd, but such computers found buyers then, with more than 1 million Apple IIs sold for $1000 each, a lot of money at the time.

Released in the same year as the Apple II were the Commodore PET and the TRS-80, museum pieces now but with many devotees in their prime. The PET's successor, the Commodore 64—with an unprecedented 64 kilobytes of inbuilt data memory—released in 1982, was to become the most popular single model ever; between 1982 and 1994 it sold 17 million units.

The giant firm IBM ('Big Blue') had been very successful in the 1950s and 1960s with mainframe computers; these had a large central processing unit supporting many terminals. Now IBM was on the sidelines. It had seen no future in minicomputers and built none, but in 1980, spurred by the success of Apple, decided to enter the new microcomputer market. The first IBM-PC reached the stores in August 1981. Its impact, and that of Apple, was so great that *Time* magazine named the personal computer 'Person of the Year' for 1982, the first non-human so acknowledged. The 3 January 1983 cover of *Time* announced with foresight 'the computer moves in'.

In 1984 Apple released the first Macintosh, which would be IBM's major competition into the early 1990s, but IBM clones, with the same structure as the IBM-PC and using the same software but a variety of manufacturers, were soon proliferating, the competition exterminating some of the pioneers such as Commodore.

Into the 1990s new features and functionalities were emerging: miniaturised hard drives to store data and programs for quick access and use; compact discs as 'read only memories' (CD-ROMs) that loaded programs faster; systems that delivered CD quality sound; and more realistic graphics for computer games. Software proliferated and became more sophisticated, with programs for processing words and images, databases, spreadsheets. Computers in offices and even in homes were now sending email and accessing the Internet.

By 2000 most of the features of today's desktop computer were in place and the 'laptop' was a realistic description, with smaller, lighter batteries and more compact circuits. Personal computers were everywhere. In barely twenty years, they had become all but indispensable.

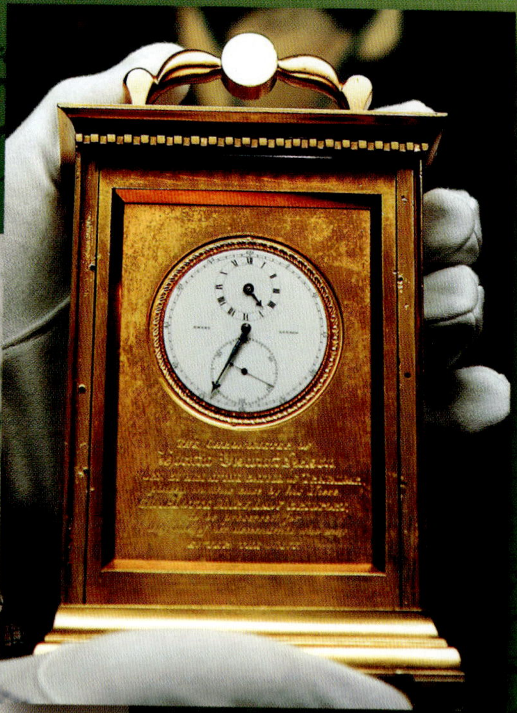

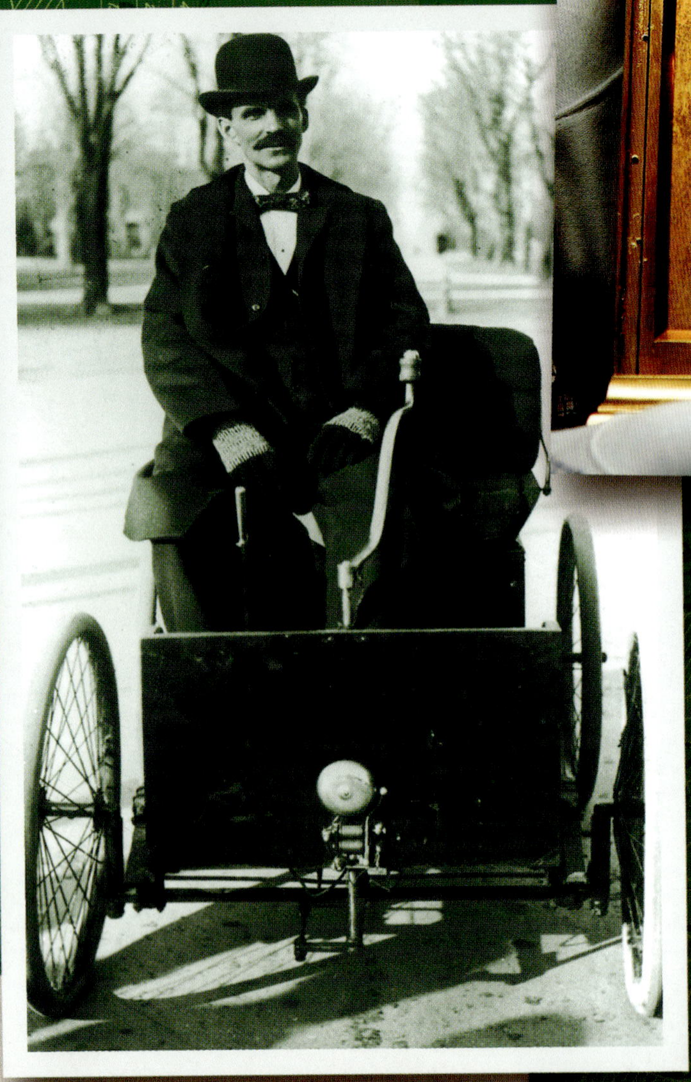

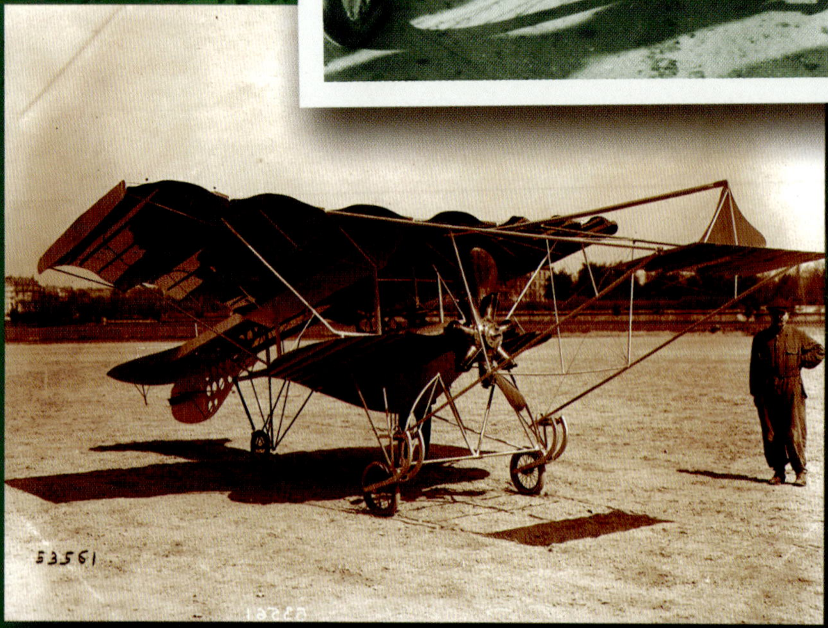
53561

03

PART THREE:

On the move

CHAPTER 14:
Running on iron rails

After most people ceased keeping horses and before the automobile took over the world, the preferred way to make a long journey was by train. By the standards of the day a train trip was cheap, safe, fast and reliable. Iron rails began connecting cities and town from the 1830s, with coal-fired steam engines to pull the carriages and freight wagons. Those iron rails, often running side-by-side with the copper wires of the equally new electric telegraph, drew the industrialised world together as never before, tethering the city and the country.

An old idea made new
Richard Trevithick

While the steam railway was a 19th-century invention, running trucks on rails is an old idea. Miners in Transylvania had used wagons running on wooden rails to carry out ore since the 16th century. These were much easier to push and pull than wagons running over rough ground, and the practice soon spread. Coalmines in England, Germany and other countries, increasingly important as the Industrial Revolution spread, were equipped with wooden tracks from the workface to the pithead. If there was room, horses were used to pull the wagons, and a hardy breed of 'pit ponies' developed. Otherwise the labour was human, unless gravity was able to help.

Increased use wore the wooden rails; their life was extended by covering them with removable strips of hardwood, later strips of iron, and finally by making the rails themselves of cast iron, which was strong and hard-wearing, though brittle.

What we now call railways became possible in the early 19th century. Improvements in power and efficiency allowed a steam engine mounted on wheels to pull not only its own weight but more besides. In 1804 the pioneering engineer Richard Trevithick first used a steam engine to pull trucks in a mine in South Wales. But the engines were still so heavy they often cracked the cast iron rails. So mine owners mostly shunned the 'iron horse' and stuck to the real ones. The notion of using steam engines and rail tracks to carry goods—and perhaps even people—over long distances seemed fanciful.

Above: Long before steam trains were pulling trucks on rails, hardy ponies were doing the same in mines. When locomotives later replaced equine power, their weight often broke the brittle cast iron rails.

Seeing the future
George Stephenson

As always, some visionaries could see the future of the new technology. Chief among these was Englishman George Stephenson. Coming from a mining family, he knew how valuable a steam 'locomotive' could be; skilled in mechanical matters since his youth and familiar with James Watt's steam engines, he thought he knew how to build one. The *Blucher*, which he trialled successfully in 1814, was the first practical locomotive, hauling coal at the Killingworth colliery in Northumberland. He soon improved it by use of the 'steam blast', directing steam exhausted from the cylinders up the boiler chimney. This made the fire

> **The notion of using steam engines and rail tracks to carry goods—and perhaps even people—over long distances seemed fanciful.**

hotter, increasing power and efficiency, and extracting more work from each tonne of coal.

By the mid-1820s railways were taking over Stephenson's life. He had had some successes in other fields, such as his 'safety lamp', able to be used in mines where explosive 'fire damp' (methane) seeped from the coal seams. In his lamp, the flame was shielded by metal gauze, which stopped it igniting the surrounding methane. Humphry Davy, the leading English scientist of his day, had the same idea around the same time, with resulting controversy over who had invented what when.

But trains were the thing. Stephenson's early success in steam locomotive-building resulted in his building a factory to make them. He built several railways for mines and coal haulage, including a public line between Stockton and Darlington. The engine he built for this line could reach the unheard-of speed of 25 kilometres per hour.

The big triumph came in 1829, when he was chief engineer on the proposed railway from the industrial city of Manchester to the port city of Liverpool. The sponsors of the railway had not been impressed by the performance of the early engines, and had a preference for horses to haul the carriages. On the Stockton–Darlington track, although goods travelled by steam, passengers still travelled in horse-drawn carriages.

Above: In the 19th century, steam engines too, not only to the rails, but also to the roads. This 'traction engine' could go most places, and a belt wrapped around its large driving wheels could power agricultural machinery.

Stranger than fiction

The grand opening of the Liverpool and Manchester Railway on 15 September 1830 was a triumph for Stephenson. However, the occasion was marked by tragedy; the local Member of Parliament, the Honourable William Huskisson, fell in front of the train and became the first person to be killed on the railways.

Under pressure from Stephenson, determined that steam be given a chance to compete, a £500 prize was offered for an engine that was at least as good as a horse. Stephenson's own entry, the *Rocket*—built with the assistance of his only son, also George—easily outpaced the opposition in the famous Rainhill trials, covering 100 kilometres at an average of 20 kilometres per hour. Among his other innovations, Stephenson used a new style of boiler, with the water sent through the heat of the fire in a number of tubes, so that it boiled faster and delivered higher steam pressure.

The Liverpool and Manchester Railway was opened in 1830 by the Duke of Wellington after five years of construction. The age of railways had come; enterprising businessmen were soon assembling capital, buying up land and putting labourers to work. Over the next few decades, railway lines appeared all over England. Travel times tumbled; once the track was laid, Edinburgh could be reached from London in 50 hours rather than twelve days by coach or horse. The nation's leading engineers were engaged. For example, Isambard Brunel, later the eminent builder of the massive steamship *Great Eastern*, was chief engineer on the Great Western Railway from London to Bristol.

Stephenson's innovations settled many features of railway technology still seen today. Needing a flange (a projecting rim) to keep the wheels on the rails, Stephenson had opted for a flanged wheel rather than flanged track; others followed suit. The gap between the tracks—the 'standard gauge' of 4 feet, 8½ inches (1.44 metres)—was taken from the traditional spacing of tracks in coalmines. By now the rails were made of wrought iron, less brittle than cast iron and allowing faster travel. Stephenson was also quick to promote the use of tunnels, cuttings and bridges to reduce gradients and curves on railways.

Trains go global
George Westinghouse among others

Railways appeared quickly elsewhere. A steam-powered passenger service was running on the South Carolina Railroad in the United States soon after the Liverpool to Manchester railway opened. Nearly 200 kilometres long, it was judged 'a wonder of the world'. North American trains soon developed their distinctive features: the 'cow catcher' to clear obstacles off the

Below: It took two or more to make a steam train move, one to drive, the other to stoke the insatiable fire under the boiler with broken coal. The 'fireman' or stoker certainly earned his passage.

Once the track was laid, Edinburgh could be reached from London in 50 hours rather than twelve days by coach or horse.

track, the bell and heavy whistle, the substantial cabin to protect the crew from severe cold, the sandbox to support traction on steep gradients, and small wheels in front of the main driving wheels to guide the train on a steeply curving track.

Other nations were soon active, though some of their early trains were horse-drawn. France had its first trains in 1829, Germany in 1835, Russia in 1837, Italy in 1839. Stephenson himself built railways in Belgium and Spain. An early symbol of the looming dominance of the 'railroad' was the completion of the track across America, coast-to-coast, in 1869. Railways would rule long-distance travel for most of the next century, though steam would eventually yield to other forms of power.

The triumph of the locomotive drove improvements in materials and manufacturing techniques, and promoted other innovations as well: the 'electric telegraph', first used for railway communications, and the development of 'standard time'—as opposed to 'local time'—so that timetables could be coordinated. There were technology losers too; the extensive networks of canals, built over the previous century to carry heavy goods, were too slow to compete.

One impediment to more and faster trains was concern over safety. In 1866 the train carrying 20-year-old New Yorker George Westinghouse, recently returned from the Civil War, narrowly avoided running into a train wreck. With the brakes applied by hand, it took hundreds of metres to stop.

The inventive Westinghouse already held a patent for a rotary steam engine. Wanting to stop trains more quickly and safely, he experimented with steam power, then with compressed air, connecting carriages with a flexible pipe so brakes could be applied to all of them from one location. Patented in 1869, the Westinghouse air-brake was 'failsafe' once fully developed. Air pressure was needed to release the brakes, not to apply them. Loss of pressure—for example, if a carriage uncoupled from the rest, it broke the line—automatically applied the brakes.

In 1893 automatic braking systems in trains were mandated by law. By 1905, 90,000 locomotives and 2 million freight cars and carriages in America carried the Westinghouse system. Trains could now travel

Left: A typical train in the mid-19th-century America. The elaborate smokestack caught sparks which might otherwise have set the prairie alight.

Opposite: A hundred years later, the famous *20th Century Special* was the apex of engineering excellence, passenger comfort and futuristic styling. There is nothing like it today.

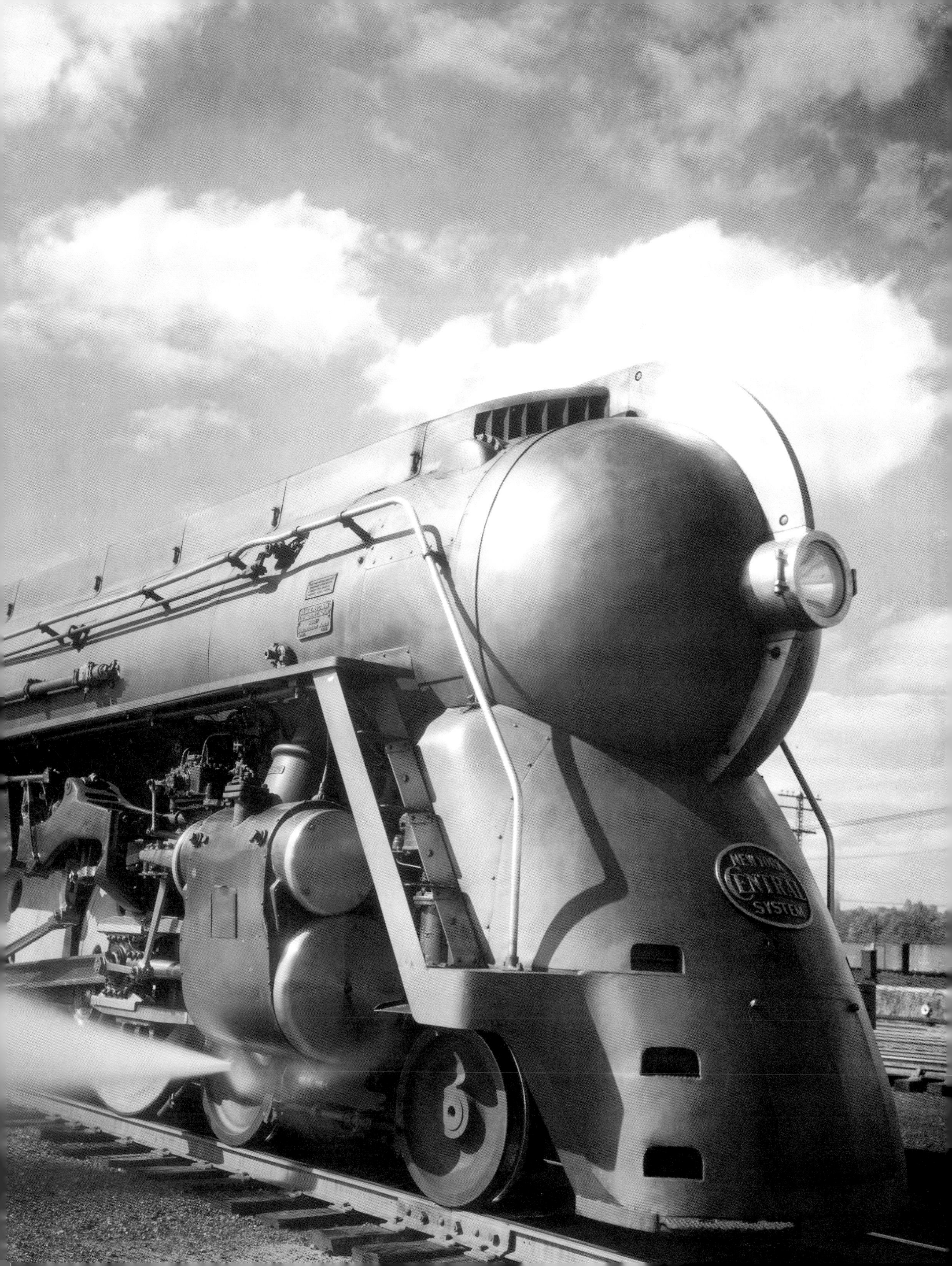
NEW YORK
CENTRAL
SYSTEM

safely at higher speeds, reducing travel times, increasing profitability and transforming the industry. Safety was further enhanced by Westinghouse's 1880s invention of automatic train signals that warned of blockages ahead.

How fast can a train go?
The Shinkansen and the TGV

The first successful steam locomotive, George Stephenson's *Rocket*, could manage 20 kilometres an hour. It outpaced any other mode of travel at the time over a long haul, and was quite fast enough, thank you, for most passengers. Over the next 150 years, steady, if incremental, improvements in engines, tracks and railway operations, combined with increased willingness by the travelling public to move at a faster pace, saw speeds reach five or eight times that of the *Rocket*.

France and Japan, however, thought seriously about new designs and systems that would go much faster still. Japan's *Shinkansen* ('new trunk line')—known as the Bullet Train from its shape—began running between Tokyo and Osaka in 1964, in time

Below: The future on rail? An experimental 'mag-lev' mono-rail train on trial. It is held above its one rail by electromagnetic fields to virtually eliminate friction. They also pull it along at speeds of up to 400 kmh.

Right: Advanced wheel-on-the-rail trains like the TGV and the Bullet Train can average 400 kmh on long hauls, completing a city centre to city centre journey in less time than you would take by plane, using less energy.

for the Tokyo Olympics. Seventeen years later, but still well ahead of anyone else, the first French *Train à Grand Vitesse* ('very fast train') began plying between Paris and Lyon.

Not everything about these trains was new. The newly-built rolling stock was streamlined to cut wind resistance, but engines were still electric-powered, picking up energy from overhead wires, and still carried by steel wheels on steel rails over conventional—if often newly-laid—track-bed. The French and Japanese were simply pushing the existing technology as far as it would go.

That was quite a way. The speeds were impressive, averaging over 250 kilometres per hour over a whole passenger journey of several hours and reaching over 500 kilometres per hour in trials. Travel was comfortable, departures punctual, environmental consequences minimal, and the safety records impressive, especially once the tracks were fenced off. Train lines operating on similar principles began to appear in other countries, and the lines could compete with air travel over short distances, being cheaper, taking less time door to door and consuming less energy.

Was that enough? Some thought not and were already experimenting with the next-generation technology, a major leap towards magnetic levitation ('mag-lev'). No longer would trains run on rails but float above them, held up, guided and even driven forwards by the interaction of electromagnetic fields in the track and the train. With no motors, the trains could be much lighter and economical to run. No mag-lev train is yet in service, but the Japanese, active in the area since 1970, have reached nearly 600 kilometres per hour in trials. Even higher speeds, approaching 1000 kilometres per hour, would be possible with the train running through an evacuated tube. Will we ever see that?

The speeds were impressive, averaging over 250 kilometres per hour over a whole passenger journey ... and reaching over 500 kilometres per hour in trials.

CHAPTER 15:
The world on wheels

In picking the top invention of the last 500 years, many would go for the car. Certainly it has provided extraordinary and all-but-universal personal mobility and transformed the we way we build and run our cities. Variations on the same idea have impacted massively on agriculture, through the tractor, the mass movement of people and goods, through trucks and buses, and land warfare, with tanks. Other elaborations speed us over snow and water.

The first efforts at a self-powered road vehicle, with an on-board engine replacing the traditional horse(s), go back nearly 250 years, but the 'horseless carriage' as a mass consumption item is a 20th-century phenomenon. Before automobiles arrived we had already begun to improve the roads, and had been through one personal transport revolution by the agency of the bicycle.

Chapter 15: The world on wheels

What happened when?

1770	Nicolas Cugnot produces a horseless carriage
1807	Isaac de Rivaz replaces the steam engine with an internal combustion engine
1816	John McAdam makes cambered roads of stone blocks with gravel in between
1816	Karl Drais introduces the hobbyhorse (Draisine)
1860	Etienne Lenoir introduces engines powered by coal gas
1861	Pierre Michaux proposes the velocipede, an early pedal-powered bicycle
1868	A traffic control device is introduced outside Parliament House, London
1876	Nicolaus Otto assembles a four-stroke internal combustion engine
1888	Bicycles gain inflatable rubber tyres, as developed by John Boyd Dunlop
1889	Daimler and Maybach build a complete automobile with a two-cylinder engine, able to reach 10 kph
1893	Karl Benz mass produces the Ideal car
1900	Ferdinand Porsche exhibits the first hybrid car
1908	The first Model T Ford comes off the production line
1912	Lester Wire introduces electric traffic lights
1916	The Mark One tank goes to war
1935	Parking meters are first installed in Oklahoma City, United States
1940s	Snowmobiles introduced by Joseph Bombardier
1954	William Hamilton trials the jet-boat
1974	Victor Wouk fits a Buick Skylark with a hybrid engine

Starting with steam
Nicolas-Joseph Cugnot, Isaac De Rivaz and Etienne Lenoir

Nicolas-Joseph Cugnot's *fardier à vapeur* (steam wagon) was probably the first 'horseless carriage'. It appears this French army officer was the first to put one of the still-new steam engines on wheels so it could pull loads, such as heavy artillery pieces, as horses had done for millennia. By 1770 he had a version, a three-wheeler, steered by tiller like a boat, able to pull four tonnes at four kilometres an hour.

Steam engines of the time lacked power and efficiency; important advances by James Watt and others still had not had much impact. So Cugnot's wonder regularly ran out of steam, needing to stop every twenty minutes to let the pressure build up again. The vehicle was also the subject of the first recorded motor vehicle accident, running off the road into a garden wall. A second, improved model was never seriously tested, though it has been on show since 1800 in a Paris museum.

Cugnot had personal troubles too. He lost his pension when the Revolution broke out in 1789, living in exile and poverty in Brussels. Napoleon pensioned him again, just before Cugnot died

Below: The beginnings of the bike. Early 19th-century onlookers see a man on a 'hobbyhorse' go by. Without any pedals, it was faster than walking, but not much. It did provide the wealthy some amusement though.

Right: Music to cycle by. By the early 20th century (before the automobile arrived) cycling was such a craze that songs were written about it. Apart from continual refinement, little has changed about the bicycle since.

in 1804, but showed little interest in the machine—surprising, given the Emperor's well-known technological awareness.

No-one in France followed up Cugnot's work. Across the Channel pioneers like William Murdock and Richard Trevithick were working with steam locomotives, but concentrated on railways. Steam engines were still too heavy for the poorly-made roads.

Around 1807 Swiss inventor Isaac de Rivaz tried another tack, successful in the long run. He replaced the boiler and fire of the steam engine with an 'internal combustion engine'—perhaps the first, other than the failed experiments of Huygens and Papin with engines driven by gunpowder. He drove the pistons in his engine not with introduced steam but with hot, expanding gas made by igniting a mixture of hydrogen and oxygen—of course, this is actually steam too. Around 1813 his last model, 6 metres long and weighing a tonne, could barely move itself.

Fifty years later French inventor Etienne Lenoir tried again, using coal gas as fuel. The engines themselves worked well and were popular with small factories not needing a large steam engine. From around 1860, he sold 500 in the Paris area alone. Again they were too heavy and inefficient for a practical vehicle, but the first successful 'automobiles' were not far away.

Roads fit for travel
John McAdam and Thomas Telford

Road-making is an ancient skill. The Roman conquerors built an excellent network of roads throughout Europe, but 2000 years later these had fallen into disuse and disrepair, and the construction techniques had been forgotten. Roads in 18th-century Britain were worse than any—merely winding, rutted cart tracks in most cases—so, appropriately, this is where the first modern road builders became active.

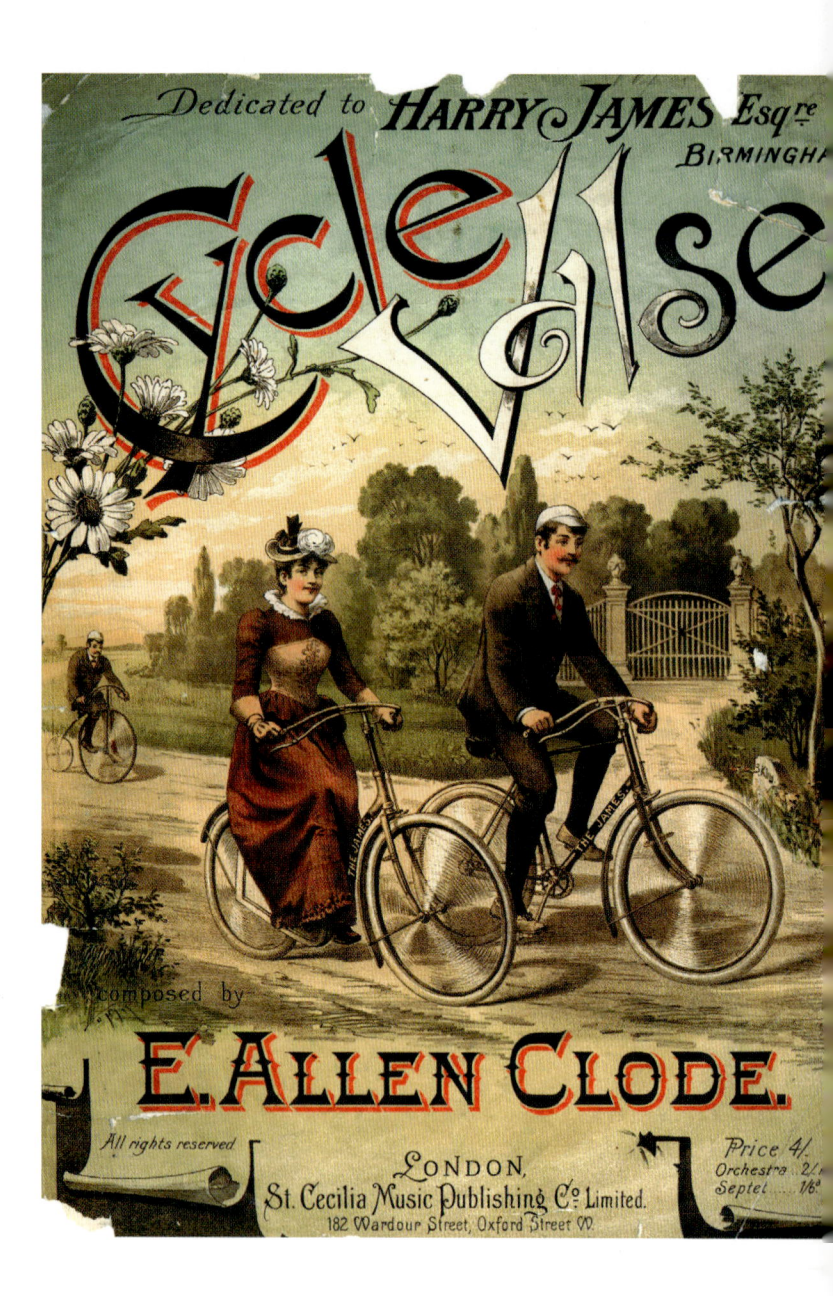

Scotsman John McAdam is most often mentioned; roads built according to his methods were said to be 'macadamised'. McAdam had made a small fortune in his 20s and 30s in his uncle's New York counting house. On returning to Britain in 1783, he purchased a substantial estate in Ayrshire; problems encountered in moving around his estate led him to consider how roads might be improved.

But others went before him and beside him, such as the blind Englishman John Metcalfe and the Scot Thomas Telford, McAdam's contemporary, a brilliant, versatile engineer. Clearly a solid foundation was needed; one solution was stone blocks. The gaps were filled with small durable stones or chips of granite rammed into place. For Metcalfe, this was enough; he built hundreds of kilometres of such roads in Lancashire and Cheshire in the late 18th century.

Telford and McAdam went further, completing the road with extra layers of stones of decreasing size, ending up with tiny fragments or gravel on top. All this was pressed into place with heavy rollers to make a smooth surface. Water was sprayed over it to increase bonding. Such roads needed a lot of manual labour, but were durable. The camber (curve) on the surface ensured that rain ran off into side drains and did not soak in. Roads were better built once the steam-powered road roller—better known as the 'steamroller'—was introduced by former farmer Thomas Aveling in 1865, 30 years after McAdam died.

By that time, Telford, McAdam and many others using similar methods had built thousands of kilometres of 'highways' throughout England and Scotland, speeding travel by coach. Similar efforts were underway throughout Europe, the Americas and elsewhere.

These 'unsealed' roads became dusty or soft depending on the weather. Spraying the surface with a liquid derived initially from coal tar made it waterproof, binding in the dust, and producing a 'tar-macadam' road or 'tarmac', a term commonly used today for aircraft landing strips even when these are made of concrete, as many modern roads are.

Going on two wheels
Karl Drais, Kirkpatrick Macmillan, Pierre Michaux and John Starley

Why did the invention of the bicycle take so long? For 5000 years, wheels had been used in various combinations, including two side-by-side. But no-one had put two wheels in the one line and sat between them until the French Comte de Sivrac around 1690. Yet it did not take on. For one thing, de Sivrac could not steer his machine.

Around 1816 the German inventor Karl Drais marketed his Draisine—a 'running machine', often dubbed a 'hobbyhorse'. Although steerable, it had no pedals; the rider propelled the machine by kicking the ground on either side. It was not a comfortable ride; the wooden wheels had iron rims. Yet, satisfied with his design, Drais made no efforts at improvement.

In 1839 the Scotsman Kirkpatrick Macmillan connected the back wheel to pedals with cranks, and a rider could easily move faster than walking pace. But the bicycle boom did not come until the French carriage-maker Pierre Michaux, asked to repair a Draisine, proposed that a pair of pedals should be

Left: Built around 1885, this three-wheeled vehicle, invented by Edward Butler and called the Petrol-Cycle, had an on-board petrol engine and somewhere to sit. It was slow, uncomfortable and open to the elements, but was probably the first motorbike.

A bicycling craze began; clubs were set up in the 1880s and a man called Thomas Stevens rode such a penny-farthing across the United States.

fixed to the front wheel, producing the 'velocipede'. That was in 1861.

The machine moved forwards by only the circumference of the front wheel at every turn of the pedals, so manufacturers made the front wheel larger and larger, producing the 'penny-farthing'. With the rider perched on top of a very large front wheel and a small trailing wheel for balance, these were not easy to ride, and tumbles were common. Yet a bicycling craze began; clubs were set up in the 1880s and a man called Thomas Stevens rode such a penny-farthing across the United States.

The propulsion breakthrough also came in the 1880s: the first 'bicycle chain' had metal links to drive the back wheel from pedals; different sized gear wheels made the back wheel turn faster than the pedals; and a 'freewheeling' mechanism let the rider stop pedalling while still moving.

The internal combustion engine: how does it work?

In Otto's invention, still the basis of nearly every car engine in the world, a piston in a cylinder first moved down to draw in the fuel (a rich mixture of air and petrol vapour, rather than the expensive coal gas Lenoir used). The piston then moved up again to compress and heat the fuel. A spark was fired to explode the fuel mixture, driving the piston down again. On the fourth stroke, the piston moved up again to drive out the exhaust fumes.

With the addition of inflatable rubber tyres in 1888, ball bearings to reduce friction in the works and many experiments with the form of the frame, ultimately made out of steel tubes, bicycles such as John Starley's Rover gained very much the form we see today, and unprecedented popularity. People now had the speed and freedom of movement they had enjoyed riding horses. Better roads ensured a smoother ride. Both people and roads were ready for the bicycle to be overtaken by the next big thing in personal transport—the 'automobile'.

A carriage without horses
Nicolaus Otto, Gottlieb Daimler, Wilhelm Maybach and Karl Benz

Etienne Lenoir's gas-powered engine may have been too heavy and wasteful to power a vehicle, but the possibility intrigued young German travelling salesman Nicolaus Otto. Fascinated by new technology, Otto thought he could improve on Lenoir's design and, with the help of sugar factory owner Eugene Langen, he assembled the first 'four stroke' internal combustion engine in 1876. A year later it won a Gold Medal at the Paris Exposition.

For the first time, the steam engine had a real rival. Over the next decade 30,000 'Otto Cycle' engines were built for use in factories and mills. Otto's engine was lighter and cheaper to run than Lenoir's and gave much less trouble. Otto himself attached an engine to a bicycle and pioneered the use of the 'magneto'—a small generator driven by the engine to provide the spark. He patented his design in 1877. But this patent was later

TO THE NOBILITY & GENTRY
MAY 1896

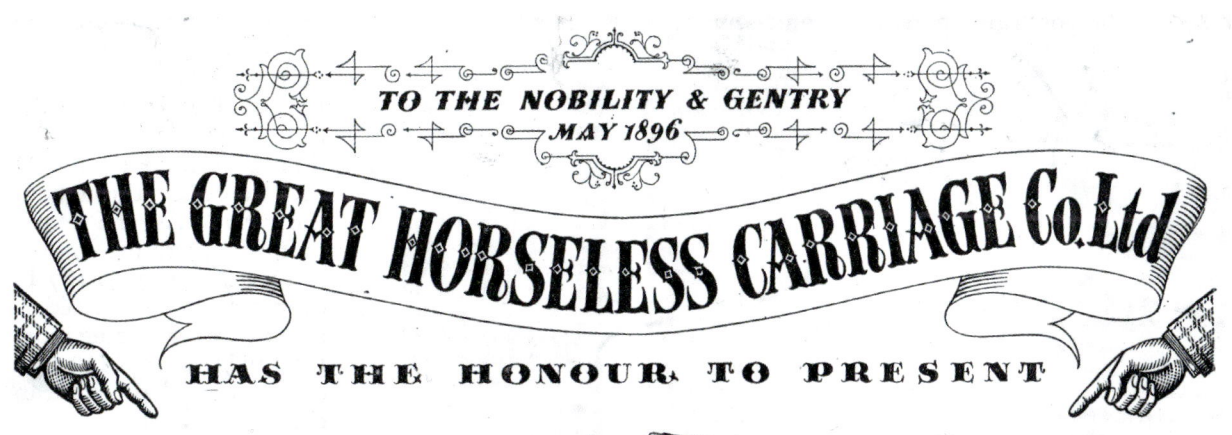

THE GREAT HORSELESS CARRIAGE Co. Ltd

HAS THE HONOUR TO PRESENT

This NOVEL vehicle is propelled by an **INTERNAL COMBUSTION ENGINE** OF 2 CYLINDERS AND 6 HORSE POWER relying on petroleum for its motive force

THE MECHANICAL carriage will attain the comfortable speed of **TWELVE MILES PER HOUR** on the level, while hills can be ascended and descended in safety

The Daimler Wagonette is admirably suited to the needs of the **SPORTSMAN** AND LOVER OF THE COUNTRYSIDE, giving as it does full facilities for the enjoyment of **FRESH AIR AND AN UNINTERRUPTED VIEW OF THE** Scenery

'A new mode of transport that has undoubtedly come to stay'
—VIDE DAILY PRESS

THE TWIN-CYLINDER 6 H.P. WAGONETTE {See Engraving}

---

overturned in favour of Frenchman Alphonse Beau de Rochas, who had had drawn up a specification for a four-stroke engine a decade earlier, though he had never built one. Otto's company, now called Deutz AG, still makes engines today, though he died aged only 59.

Among Otto's employees were Gottlieb Daimler and Wilhelm Maybach; they pushed the technology rapidly forwards. Their 1885 engine was light-weight and fast. A year later Daimler installed one in an existing stagecoach to make the first four-wheeled automobile. In 1889 the pair built a complete automobile from the ground up, with a two-cylinder engine and a top speed of 10 kilometres an hour.

The Daimler company was building cars to his designs by 1890. In 1894 the winner of the first-ever car race had a Daimler engine. Maybach parted company with Daimler soon after, going on to design the Mercedes—named after his wife—and later engines for the new Zeppelin airships.

Above: The power of advertising. In this 1896 pitch to sell the new Daimler, able to reach 12 miles an hour (on the level), the absence of any covering was a virtue—'fresh air and an uninterrupted view of the scenery'.

Daimler's rival was fellow German Karl Benz, an iron foundry owner from Mannheim. Seeking additional income, he marketed stationary engines, and in 1885 Benz married a scaled-down engine to a two-seater tricycle, producing the first gasoline-powered automobile. This had electric ignition, mechanically-operated engine valves, a water-cooled engine and differential gears—all found in cars more than a century later. Benz sold some, but later moved to the more stable four-wheeled format. In 1893 his Ideal model became the world first mass-produced car; 150 were made in 1895. A 3 horsepower motor allowed for a top speed of 14 kilometres per hour, but it struggled on hills.

> ### The success story
>
> Unlike one-idea Dunlop, the inventive Thompson's portfolio included a fountain pen; a steam-powered road tractor—with solid rubber tyres—able to pull a 40-tonne load and later to make the draught horse obsolete on farms; and an improved 'mangle' for squeezing the water from clothes, able to turn either way and invented when he was only 17. His technique for setting off explosive charges safely by an electric current saved lives in mining and construction, his improvements in sugar refining—undertaken while on assignment in Java—greatly enhanced productivity. Thompson, who had worked with railway genius Robert Stevenson, was inventing up to his death in 1873, aged only 50. His last patent—posthumous—was for an inflatable rubber cushion.

Supported by the business acumen and unflagging optimism of his wife Bertha, Benz was the biggest maker of cars in the world by 1900 but stuck to old designs and paid no attention to increasing speed. His vehicles were soon being overtaken—in both senses—by Daimler's more innovative designs. The Daimler and Benz companies merged in 1926.

Riding on rubber
John Dunlop and Robert Thompson

Today we travel smoothly by car, bus, motorbike or bicycle thanks to the pneumatic (air-filled) rubber tyres on the wheels. The name associated with this source of comfort is usually John Dunlop; certainly Dunlop tyres are still sold around the world. Like so many such linkages, it is only partly true.

Scots-born John Boyd Dunlop was a successful veterinarian, working in Belfast. One day in 1888, his young son wanted something to put around the solid wheels of his tricycle to smooth the ride over the roughly-cobbled streets. Dunlop had some skill working with rubber, so he made ring-shaped rubber tubes that fitted over the tricycle wheels and filled them with air.

The benefits were obvious. Dunlop sold some of these new tyres to local cyclists, who won more races using them—wheels with such tyres had less friction. The head of the Irish Cyclists' Association partnered Dunlop in marketing the tyres. The company prospered, but Dunlop sold out too early and did not leave much in his will.

All true but incomplete. Unknown to Dunlop, the pneumatic tyre already existed, invented by fellow-Scotsman Robert Thompson 40 years earlier. The idea was basically the same; an annular (ring-shaped) tube of rubberised canvas, filled with air and protected by an outer covering of leather. Bicycles hardly existed then and Thompson's tyres supported horse-drawn wagons.

The rubber industry was new, vulcanised rubber barely discovered, and Thompson could not secure enough rubber of the right quality and thickness. So he turned to solid rubber tyres, initially for wheelchairs. Four decades later Dunlop had the right rubber, and his tyres reached the market, but he could not gain a patent because Thompson had precedence.

Opposite: The pioneer of automobile mass production, Henry Ford and his son Edsel, pose in a Model F Ford in 1905, in front of their Detroit house. The car now has inflatable tyres and lights, but note the state of the street.

Automobiles by the million
Henry Ford

Into the new century, gasoline-powered cars began to outnumber steam vehicles, not that there were very many of either, given the cost. Frenchmen René Panhard and Emile Levassor set the pace with new designs, moving the engine and radiator to the front of the car, with power transmitted to the rear wheels to improve balance and steering.

The making of cars soon spread. The Duryea brothers were selling 'motor wagons' in Massachusetts in 1896. The Duryea was an expensive limousine, in production until the 1920s. America's first mass-produced car was the Oldsmobile, designed by Ransom Olds and manufactured in Detroit from 1901, soon the hub of the United States motor industry.

The big name, of course, was Henry Ford. Ford built his first car in 1896, using four bicycle wheels. In 1903, aged 40, he left the Edison Company to found his car-

The success story

Ford transformed the motor car from a luxury item into a workaday machine almost anyone could own, and implanted the 'assembly-line' firmly in the lexicon of manufacturing. The Ford Foundation, created from his immense fortune, currently disperses half a billion philanthropic dollars a year. It is a heady legacy for the dour Michigan farm boy, a tinkerer with machines since his youth.

making firm, declaring he would make 'a car for the great multitude', a promise he kept. The first of the iconic Model T Fords came off the production line in 1908, costing $1000. Over two decades he built 15 million of these pioneering machines: the price fell by 75 per cent.

The result was an irrevocable change in American—and later global—society. Cheap personal transport transformed the shape and operation of cities: creating the suburbs and, later, gridlock; linking widespread population centres; enriching lives, although at a cost to the environment—the first cars had, however, promised relief from streets knee-deep in horse manure.

Ford and his son Edsel made many improvements to automobiles, including the one-piece V8 engine in 1932 to drive a new generation of large, powerful and smooth-riding cars, but his greatest innovations were in the methods of manufacture. He did not invent mass production or the assembly-line; but he did bring them to a state of exquisite efficiency. His Highland Park plant in Michigan divided production into dozens of steps, undertaken at a relentless pace. As partly completed automobiles moved down a 300-metre conveyor belt, components were added by workers, each doing only one task repetitively, in a time dictated by the movement of the belt. At peak production, around 1914, a 'Tin Lizzie' took only 90 minutes to traverse the line, a completed vehicle emerging every 30 seconds.

These immense gains in productivity allowed Ford to treat his workers relatively well, paying double the wages on offer elsewhere for fewer hours. But they were essentially slaves to the line, which could be speeded up to further enhance production and ran 24 hours a day with three shifts. Though Aldous Huxley later made 'Our Ford' a subject of veneration in *Brave New World*, the endless repetition of simple tasks was satirised by Charlie Chaplin in *Modern Times*.

Making the traffic flow
Lester Wire

Traffic lights, now essential to keep our city streets moving at all, are a technology with many claims to paternity, making it hard to trace the story with certainty. If we mean the present red-amber-green system, its first appearance was in New York in 1918, or perhaps Detroit in 1920. In all probability, it was in the United States and in response to the growing volumes of traffic as the factories of Henry Ford and

The result of mass production of the Ford Model T was an irrevocable change in American—and later global—society.

others poured out motor cars cheap enough for the average person to buy.

Attempts to control traffic by signals went back at least 50 years earlier to a signal post with movable arms, like railway semaphores, which was set up outside Parliament House in London in 1868. Red and green gas lamps provided warnings at night, though, according to reports, the signal blew up after a year of use, killing a policeman and discouraging further experiment.

The first electric traffic signals, though with only red and green lights, were the work of Salt Lake City policemen Lester Wire in 1912. Placed in charge of the traffic squad, he sought to ease the chaos on the city roads caused by the increasing popularity of automobiles. He also wanted to save fellow policemen from standing in the rain to direct traffic. Instead they could control the lights from a shelter nearby. Wire refined his system but never patented it, and died having received no royalties.

In 1923 African-American inventor and sewing-machine repairman Garrett Morgan, also the inventor of a gas mask and a chemical to straighten hair, patented a cheap and simple traffic control system, using hand-cranked movable arms. This became quite

Below: Traffic lights battle to shine through an accumulation of snow. The booming population of motor vehicles in the early 20th century needed to be regulated to prevent gridlock on the city streets. Traffic signals helped … for a while.

popular, and the myth grew that Morgan had invented traffic signals, which is clearly not so. The US Patent Office had issued 50 patents before his.

Early systems could be operated either automatically, the lights changing at set time intervals, or manually, controlled by a policeman or other official able to view the state of the traffic and make appropriate adjustments. Not until the 1960s were electronic circuits inserted in the roads to monitor traffic flows, so that the cycling of the lights could be adjusted to minimise delays. From the 1980s information technology allowed a variety of light-change cycles to be applied at any particular intersection as traffic conditions changed, and the succession of lights down a stretch of road to be coordinated.

Below: Steam powered and fitted with 'caterpillar' tracks, this American tractor made light of the steepest farmland and inspired the 'tanks' that first saw service in the Great War (WWI).

The tank goes to war
Ernest Swinton

The military tank was invented more than once. Leonardo da Vinci drew one around 1500, and the Australian inventor Lancelot de Mole designed one driven by a petrol engine in 1912. Neither was ever built, largely because no-one saw the need.

Trench warfare in World War I saw calamitous losses but no sustained progress by either side. Tens of thousands were killed in futile advances over chaotic broken ground, through barbed wire and in the face of withering enemy machine-gun fire. British Colonel Ernest Swinton glimpsed a solution in an article on American farm tractors which ran on caterpillar treads. These could handle the battlefield terrain, he thought and—covered with armour-plate and equipped with guns—would make an unstoppable advance through 'no-man's land', a phrase he had invented as an official war correspondent.

Stranger than fiction

With secrecy paramount, workers on the new military machines were told they were building 'water carriers for Mesopotamia'. This was later simplified to 'tanks', the name that persists today.

He built a primitive prototype, which impressed military authorities, including Winston Churchill, and the first vehicles were ordered. Production began in 1916. Mark One, also called Big Willie, was box-shaped, weighed 26 tonnes, and had two side-mounted cannon, six machine-guns and a crew of eight.

Tanks first went into battle in September 1916, in an effort to break the murderous stalemate of the Battle of the Somme. Forty-seven tanks began the advance at dawn, terrifying the Germans as they emerged, rumbling relentlessly through the smoke and fog. Mechanical failures and fuel shortages soon halved the force, but the remaining tanks penetrated 10 kilometres through German defences before withdrawing. That first encounter was far from decisive, and the war would still run another two years, but the potential of the weapon was clear. Other versions soon followed. The Whippet could reach 16 kilometres per hour; the Hornet had a turret for all-round fire.

Much enhanced, tanks have remained prominent in land warfare ever since, though not without challenge. In 1940, seeking a decisive anti-tank weapon, US Army Colonel Leslie Skinner put an explosive charge on top of a small high-speed rocket and fired the combination from a metal tube that could be held over the shoulder. The Mark One Rocket Launcher was soon renamed the Bazooka after a bizarre home-made trombone played by popular comedian Bob Burns.

Paying to park
Carl McGee

Traffic and cities have long gone together, especially once the horseless carriage replaced the horse and buggy. As motor cars began to throng city streets, there was soon nowhere to park.

Carl McGee, newspaperman and civic leader, was chair of the Traffic Committee of the Chamber of Commerce in Oklahoma City in the United States. With his colleagues he faced the challenge of the 'all-day parkers'. Those who worked in the downtown area left their cars parked against the sidewalk right through working hours. This forced shoppers to park a long distance away, which did not impress retailers. Time limits were imposed, traffic police chalked tyres to keep track of time and handed out tickets, but the problem did not go away.

McGee decided that he needed an inexpensive machine that would be stationed adjacent to each kerbside space, and which could record how long each

Below: Are these the most hated machines ever? Mid-20th-century motor cars come face-to-face with a line of parking meters, which demanded payment for what many motorists took as a right. They still do.

In 1900 automobile pioneer Ferdinand Porsche exhibited an unusual vehicle at the Paris World Fair... the first 'hybrid' car ... decades ahead of its time.

car stayed parked. Helped by staff and students at Oklahoma State University, he devised a machine nicknamed the Black Maria, and in 1933 went looking for a manufacturer. The early machines were driven by clockwork, so the tender went to a firm that made clockwork fuses to set off nitro-glycerin in oil wells.

The first 175 parking meters were installed in 1935, and proved so successful they were soon installed all over town. The rest of the world followed, inspired by the threefold success of McGee's innovation: the machines straightened out the parking problem, helped by a $20 fine on those who stayed too long; the fines and the ten-cents-an-hour fee for legitimate parking brought extra revenue to the City; and the values of downtown properties went steeply up as the parking nightmare disappeared. Indeed meters quickly became so popular with retailers that everybody wanted one outside their store. Not everyone welcomed the meters; some motorists objected to paying for something they thought should be free and tried to disable the meters, a not-uncommon fate for innovations.

Parking meters today work and even look much like the ones McGee installed, but there has been progress, initially putting two meters on the one pole to reduce clutter, and more recently the introduction of meters able to print out tickets for four to six spaces, the motorists then being required to display the ticket behind their windscreen. This has cut costs and boosted further revenue.

The car of the future?
Ferdinand Porsche, Victor Wouk and David Arthur

In 1900 automobile pioneer Ferdinand Porsche exhibited an unusual vehicle at the Paris World Fair. Its gasoline-powered engine did not drive the wheels directly. Instead it turned a generator sending current to electric motors on the wheels. This was the first 'hybrid' car—more precisely, a 'series hybrid'—decades ahead of its time. Back then, and for decades to come, petrol was cheap, fuel efficiency unimportant, air pollution not a worry. Only in the 1970s would such issues start to matter, and hybrids again attract attention.

In 1974 American engineer Victor Wouk, brother of novelist Herman Wouk, refitted a 1972 Buick Skylark with a much smaller engine connected to an electric motor. The car could be driven by the engine, the motor or both. Wouk had been stimulated by the passing of the Clean Air Act of 1970. Such a hybrid, he believed, would cut tail-pipe emissions, of gases like carbon

Regenerative braking: what's that?

With regenerative braking the motor can also serve as a generator. When the car brakes, its energy of motion is converted back into electricity and stored in a battery. This improves fuel efficiency still more. Regenerative braking is found in many of the hybrid vehicles now being marketed by most of the major manufacturers.

Above: Cars and petrol have traditionally gone together, but pollution, global warming and fuel shortages suggest that nexus might soon have to end. Future cars might run on sunlight, but they won't look like this uncomfortable experimental solar 'car'.

monoxide, by 90 per cent, while also greatly improving the 'miles per gallon'. This proved correct. Testing by the US Environment Protection Agency found that its emissions were only 9 per cent of those coming from a purely gasoline-driven car of the time. It also doubled the fuel efficiency of the car before conversion. Wouk, too, was ahead of his time. The program that funded the research was cancelled in 1976, and interest in hybrids was left to hobbyists.

In the 1990s in the face of rising fuel prices and troublesome air pollution, the American government sought collaboration from industry in developing 'next-generation vehicles', with hybrids much in the frame. By then we had another vital piece of technology, 'regenerative braking', brainchild of electrical engineer David Arthur.

Into the 21st century, hybrids, with their many advantages, seem certain to loom even larger in the market, as costs and performance match those of traditional cars. But there are other options, such as cars powered by fuel cells burning hydrogen, for example. The demise of the petrol-fuelled car is still several years away, but we can see it coming.

> **The demise of the petrol-fuelled car is still several years away, but we can see it coming.**

CHAPTER 16:
Over and under the water

We humans first found our feet on land but the oceans have always fascinated and drawn us. Explorers and traders have been crossing the seas for millennia, relying on human muscle and capricious winds to power their journeys. Only over the last 200 years have we had the aid of other energy sources, starting with the steam engine, making travel faster and safer and allowing ships to grow in size.

Unlike fish or other marine creatures we die if we stay too long underwater without aid. Yet the desire to explore the underwater world, to travel freely for long periods under the surface, is an old one. The oceans cover 70 per cent of our planet. The sea is deeper than the land is high, and its resources are immense. Inventions to meet this challenge began centuries ago with the diving bell, yet much of the oceans' depths, and the creatures that dwell there, remain mysterious.

Living and working beneath the sea
Edmond Halley, William Beebe and the Piccards

Englishman Edmond Halley has a comet named after him and contributed massively to physics, mathematics, astronomy, technology and navigation. Less well known is his contribution to the technology of underwater salvage. Halley did not invent the diving bell, a 'metal barrel lowered into the water upside down', as his predecessor the philosopher Francis Bacon described it. One had been used to recover cannon from the sunken Swedish warship *Vasa* in 1628. However, Halley made some major improvements to the design and operation of the bell, and willingly went underwater himself to test out the dangerous device.

His diving bell was built of wood, weighted with lead, with glass windows, a platform to stand on and benches for seating. Of course, it did not have a floor. A rope suspended it from a spar on its parent ship. To keep the air in the bell fresh, Halley released stale air from a valve at the top of the bell and sent down new air in lead-weighted barrels.

Men could work 20 metres under water for several hours, carrying out salvage operations on underwater shipwrecks, recovering cargo and cannon. Engravings at the time show men working outside the bell connected to the air supply inside via a 'cap of maintenance', a sealed helmet with an air hose attached. Halley ran a profitable business doing such work, helping to restore his family fortunes after the ravages of the Great Fire of London.

Above: Edmond Halley's diving bell in operation, late in the 17th century. On the left, a diver outside the bell and wearing a primitive helmet watches salvaged cannon going up. On the right, a barrel of fresh air descends.

Even at that depth, external pressure would have forced water in to fill two-thirds of the bell. That obviously put a limit on depth of working. The water could be kept out in greater depths by pumping in more air and

Men could work 20 metres under water for several hours, carrying out salvage operations on underwater shipwrecks, recovering cargo and cannon.

According to reports, his craft made several voyages under the Thames around 1620, once with the King onboard.

increasing the pressure, but this put the occupants at risk of compression sickness, as it was later known.

Diving very much deeper required a closed vessel, able to withstand water pressure, which increases steadily with depth. Of course, the occupants would be unable to leave and return freely, and so would be limited to looking or operating mechanical arms. Not until 1934 did American explorer William Beebe dive 500 metres into the Mediterranean in his 'bathysphere', suspended, like Halley's bell, on a cable.

To go down further still, the vessel needed to float untethered. In 1953 the Swiss explorers Auguste and Jacques Piccard descended more than three kilometres in their bathyscaphe *Trieste*. In 1960 the *Trieste* reached the deepest part of the ocean floor, the 11,000-metre deep Marianas Trench near Guam.

Steam across the water
Robert Fulton

Very early in the development of steam engines, inventive spirits like the 17th-century Frenchman Denis Papin dreamed of—and experimented with—boats driven by steam rather than by the wind or human effort, but many decades passed before those visions became reality.

Jonathan Hulls took out patents on a steamboat in 1736, but it was to be driven by a Newcomen engine, heavy and inefficient, and would never be a success. In 1763 William Henry in Pennsylvania put a Watt steam engine in a boat, but it sank. Two decades later in France, a steam-powered paddle-wheeler managed to last fifteen minutes against the current on the River Saone but lacked the endurance for longer trips.

Developments elsewhere included a boat driven by a steam-powered water-jet and able to do six kilometres per hour, but all these were too slow or too expensive to run. For example, American John Fitch successfully trialled his first steamboat in 1787; a later version carried passengers. Fitch tried a number of designs and solved many technical challenges; one of his boats could travel at 13 kilometres per hour, yet he could never convince sceptics that steam-boats would pay.

The 19th century came before real success could be claimed. In Scotland in 1802, Lord Dundas launched the *Charlotte Dundee*, a double-hulled paddle-wheeler with an improved engine by William Symington. This pulled two 70-tonne barges 30 kilometres along the Forth and Clyde Canal to Glasgow.

Soon after, success came to American Robert Fulton, whose countrymen called him 'the father of the steam-boat'. Inspired by news of the *Charlotte Dundee*, Fulton ran trials on the Seine, where he was hoping to attract French support for his submarine *Nautilus* (see below). He later imported a Boulton-Watt steam engine and built a boat to take it. In 1807 the *Claremont* began a regular passenger service between New York and Albany, 250 kilometres up the Hudson River, taking 30 hours for the trip. Within a few years, steamers were running on the St Lawrence in Canada and would soon appear on other rivers and lakes, including the Mississippi River, a most famous venue for the paddle-wheelers.

Back in Scotland Fulton's ideas inspired Henry Bell, who launched his *Comet* in 1812 on the Clyde between Glasgow and Greenock. Inside a decade, dozens of smoking steamers were plying the rivers, lochs and canals of Scotland, carrying cargo and occasionally passengers. The age of steam-boats had come.

Chapter 16: Over and under the water

Left: The largest paddle-steamer ever built, Isambard Brunel's *Great Eastern* at dock in New York Harbour. Ultimately too expensive to run as a passenger ship the *Great Eastern* was leased to lay the first submarine telegraph cable from Europe to America.

Submarines become reality
David Bushnell, Robert Fulton and John Holland

The diving bell let adventurous souls spend time underwater, but it was mostly in the one place. How much better would it be to be able to move about? The dream of sailing under the sea, rather than on it, is an old one. In 1573 English mathematician William Bourne proposed an ingenious leather-covered boat, which could submerge and surface with the aid of leather 'buoyancy tanks'. The first man reputed to have built such a craft was the 'Court Inventor' to England's James I, Dutchman Cornelius Drebbel. According to reports, his craft made several voyages under the Thames around 1620, once with the King on board. Details are sketchy. We do not know how the boat submerged, how it was steered or how the crew got air.

A submarine first went to war in 1776, following the American Revolution. Looking like a coconut built of oak, the one-man *Turtle*, the brainchild of David Bushnell, was powered by hand-turned screws, raised and lowered by buoyancy tanks, and allowed vision from a 'conning tower'. Its armament comprised a time-bomb to be attached by a screw to an enemy warship, but its one attack, on the HMS *Eagle*, was unsuccessful.

Decades later, the American steamboat pioneer Robert Fulton designed the *Nautilus*—later the name of Captain Nemo's submarine in *20,000 Leagues Under the Sea* and later still of the first nuclear-powered

Left: The bathyscape *Trieste*, in which the Piccard brothers rode to the bottom of the ocean in 1960. The term 'bathyscape' comes from Ancient Greek words meaning 'a ship in the deep'.

Left: A modern nuclear-powered submarine, here seen travelling on the surface, but able to stay submerged for weeks at a time since its engines need no oxygen.

submarine. Like a modern submarine, this had wing-like planes on either side of the hull to direct the path of the craft higher or lower. Napoleon was interested for a while, but it seems the *Nautilus* never went to sea.

Real progress in submarines needed a way to power them—other than by hand-driven screws—and an efficient weapon for use against ships. Compressed air, steam and gasoline engines and the electric motor provided the first, the torpedo the second. First to pull these elements together was the Irish-American schoolmaster John Holland. His early experiments were paid for by the Irish-expatriate Fenian Society, with the hope of destroying the British Navy. In 1893 Holland beat off many rivals to secure the first contract from the US Navy for a submarine.

The ultimate outcome of his efforts was the 15-metre *Holland*, the first practical and reliable submarine, bought by the US Navy in 1900. Though lacking a periscope, it was a modern submarine in most respects, with 'dual propulsion'—petrol powered on the surface, electric powered when submerged—which was used by all submarines until nuclear power took over, 50 years later.

Steam across the oceans
Samuel Cunard and Isambard Kingdom Brunel

Once steam ships were carrying cargo and passengers along the inland waterways and sheltered coastlines of Europe, North America and elsewhere, the challenge became to send 'steamers' onto the open ocean, such as across the Atlantic, between the Old World and the New. Travelling under steam power alone would demand engines using less coal so the ship could stay at sea for several weeks. Replacing paddle wheels for propulsion with something less affected by the rolling of the ship was required to provide a reliable service.

The success story

Brunel's 1858 *Great Eastern* proved too large, at over 200 metres and 19,000 tonnes, to succeed commercially, though it did successfully lay the first transatlantic telegraph cable. However, it led the way in technology, as ship-building moved from iron to the more flexible steel for construction, and to 'double expansion' engines for greater efficiency. Steam exhausted from one cylinder drove a second larger one, so that the discarded steam contained less energy. This cut coal usage. Crossing times steadily fell, to below ten days and then below six, as shipping lines sought the coveted 'Blue Riband' for fastest passage.

Without waiting for such breakthroughs, crossings under a combination of steam and sail got underway in 1819 with the American ship *Savannah*. A regular service took another two decades and introduced the famous name of Cunard. Securing the British government contract for the mail service across the Atlantic, Samuel Cunard established a shipping line in 1840, soon carrying passengers as well and offering guaranteed sailing dates. Cunard's first ships used a sail–steam combination, but the era of the passenger liner, using steam alone, was drawing close.

When it came to building the ships, the versatile British engineer Isambard Kingdom Brunel set the pace. Brunel, brilliant and daring, had already built the Great Western Railway. He created ever bigger ships— faster, more luxurious and comfortable for passengers. The *Great Western*, launched in 1838, was 70 metres long and crossed from Bristol to New York in just fifteen days.

His 1853 *Great Britain*, nearly 100 metres long and luxuriously appointed, was the first ocean-going steamship made of iron, and the first to use the submerged 'screw propeller' for propulsion in place of paddle wheels. The idea of the screw had been around since the experiments of the American John Stevens in 1803, but only in 1838 had a large steamer used one, the riverboat *Archimedes* built by screw pioneer Francis Pettit Smith. Later ships had twin screws for reliability.

In many modern ships, steam turbines have replaced engines with pistons, with fuel oil instead of coal to fire the boilers. Diesel engines keep others moving. The largest ships afloat now would dwarf Brunel's *Great Eastern*; the Atlantic can be crossed in

Below: The *Lusitania*, sunk during World War I, was typical of the great liners of the early 20th century, carrying wealthy passengers in luxury across the Atlantic in less than a week.

only four days. But in whatever form, the offspring of the original visions of Denis Papin and Robert Fulton continue to ply the seas in vast numbers.

Breathing underwater
Jacques Cousteau and Emile Gagnan

To stay very long underwater you have to breathe—easy enough in a submarine or diving bell, but harder if you want to move freely and independently. The traditional diving suit, with lead-weighted boots and heavy metal helmet, allowed the wearer to walk around on the seabed, but he was still tethered to a ship by a hose supplying him with life-sustaining air.

By the early 1930s, daring divers were strapping to their backs tanks of compressed air attached to a face mask. They headed underwater, but there was a risk. Water pressure increases with depth and threatens to crush the lungs unless the air in them is at the same pressure. So divers had to the increase the pressure of the air they were breathing as they went deeper, turning a valve by hand. Unless this control was precise—not easy to ensure if something else was distracting the diver's attention—their lungs could collapse and the diver would drown.

Above: Aqualung pioneer Jacques Cousteau prepares for a dive. The equipment he devised with his colleague Emile Gagnan transformed underwater activities, both serious and recreational.

Among the keen exponents of this hazardous sport of 'free diving', was French naval officer Jacques-Yves Cousteau. He had already had some close escapes and knew that safety could come only through a system that automatically adjusted the pressure of the air being inhaled to the external water pressure. Cousteau recognised the problem; in 1943 Emile Gagnan provided the solution.

It was Gagnan's vital invention that made possible the 'aqualung' or 'scuba' equipment (standing for Self-Contained Underwater Breathing Apparatus), and he shared the financial rewards, but in the popular mind Cousteau got the credit. Indeed, through his use of the new equipment on many pioneering dives, the subject of numerous films and books, he became a household name, opening up awareness of the undersea world to millions of readers and viewers. In the 60 years since, 'scuba diving' has become a popular sport and a great aid to marine biology and underwater archaeology.

The aqualung: how does it work?

An expert in regulating valves for gas supplies, Gagnan devised a unit with a diaphragm able to balance pressures. When the diver breathed in, water pressure pushed on one side of the diaphragm, opening up the compressed air supply, which then pushed back on the other side of the diaphragm. Air flowed until the two pressures were equal and then cut off. Breathing out pushed air through an exhaust valve so that the cycle could begin again.

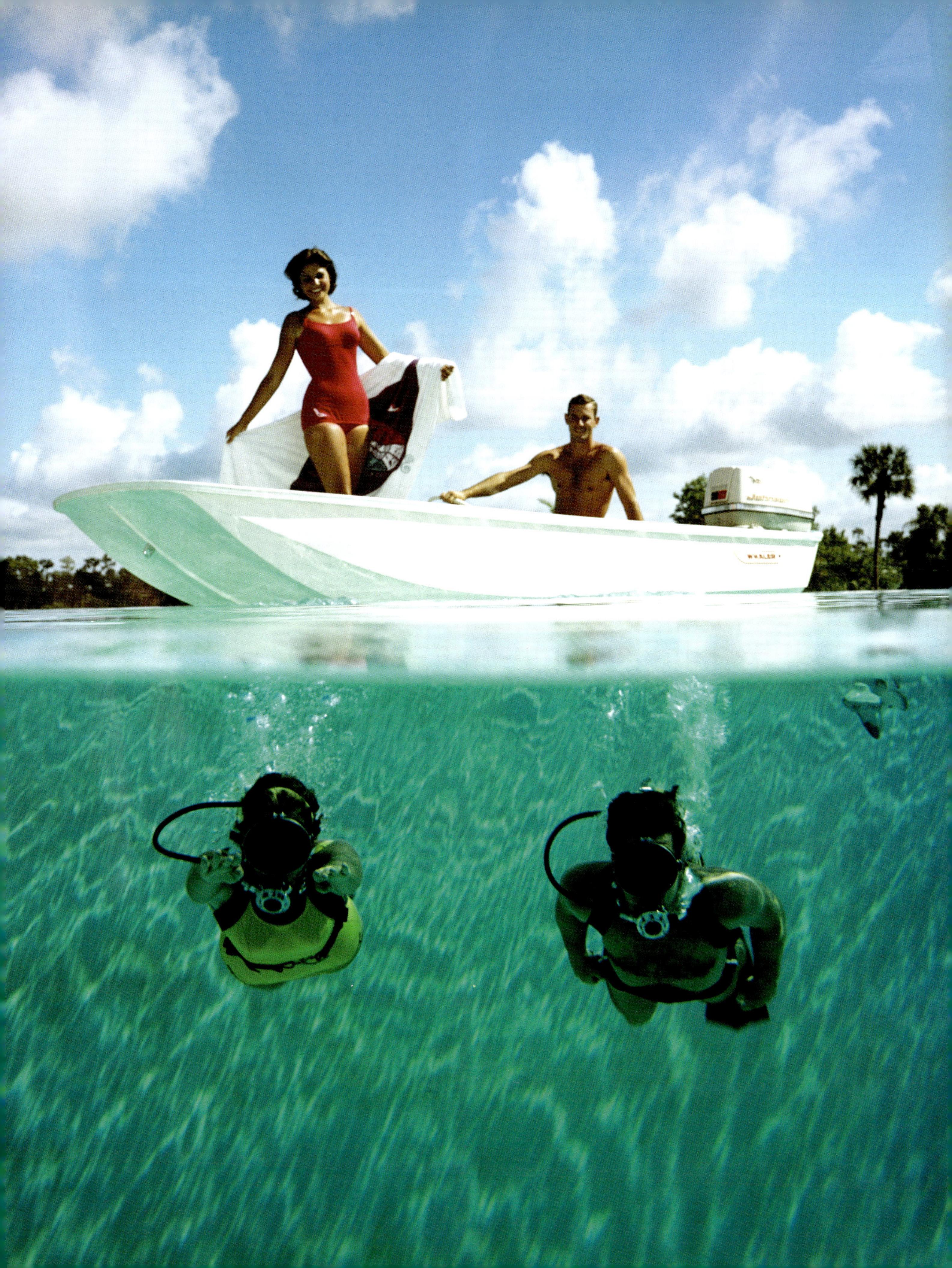

CHAPTER 17:
Taking to the air

Dreams of human flight are ancient, at least as old as the Greek myth of Icarus. Over hundreds of years, Leonardo da Vinci and others devised ways of lifting off the ground, at least in theory, by emulating the actions of birds. Humans actually flew for the first time in rather different circumstances, hanging beneath balloons. We have come a long way since then.

Balloons were not a new idea. Chinese paper lanterns, closed at the top, open at the bottom, and with a lighted candle hanging beneath, had risen aloft for millennia, buoyed up by the accumulation of hot air within them. Hot air is thinner than cold air. The balloon and its contents therefore weighs less than an equal volume of the surrounding air, so it floats. Modern aeroplanes are, of course, much heavier than air. It is still a wonder to many people that they stay up.

Up, up and away
The Montgolfier brothers and Jacques Charles

Who first thought that humans might fly with a hot air balloon, such as the Chinese had made for millennia, is unknowable, but a prime candidate is Frenchman Joseph Montgolfier. One story has him visualising soldier-carrying balloons storming the British-occupied Rock of Gibraltar. Montgolfier and his brother Jacques began experimenting in 1783 with balloons created from linen and paper and carrying a fire beneath.

Their first flight involved a hot air balloon alone. The next carried some passengers—a sheep, a duck and a rooster, according to the popular story. The duck broke its neck on landing after the fifteen-minute flight, but the other animals survived, proving that the air up there was fit to breathe. In November that year, in front of the King, Jean-Francois Pilatre de Roziére and Francois Laurent, brave men both, lifted off in a Montgolfier balloon. They reached 1000 metres and travelled eight kilometres in twenty minutes before landing safely. Human flight was possible.

The French Academy of Science asked rising star Jacques Charles to investigate. Charles reported that filling balloons with the light gas hydrogen—then newly discovered—would be safer than relying on a fickle fire. In December he went aloft himself in a balloon covered in varnished silk, the hydrogen provided by sulphuric acid reacting with iron. The balloon rose from where the Eiffel Tower now stands. Among the spectators was American inventor and statesman Benjamin Franklin, then his nation's ambassador to France.

Above: Getting a lift. The first flight of hot air balloon pioneers Joseph and Jacques Montgolfier in 1783 features in this classic advertisement for a food extract, intended to 'boost' health.

Two years later Frenchman Jean Blanchard and American John Jeffries crossed the English Channel in a hot air balloon, the Channel being the first obvious challenge. They barely made it. In the same year Pilatre de Roziére died attempting the same feat. He had—unwisely—thought a hydrogen balloon and a hot air

> **Their first flight involved a hot air balloon alone. The next carried some passengers—a sheep, a duck and a rooster.**

Above: Struts of the newly-invented duralumin form the spidery supports for a rigid airship or zeppelin. The brief Golden Age of the airship in the 1920s and 1930s was terminated by fiery accidents and competition from passenger-carrying aeroplanes.

The rise and fall of the airship
Ferdinand von Zeppelin

Pioneering airship builder Count Ferdinand von Zeppelin from Germany was not the first to tackle the main problem with passenger-carrying balloons. Lacking the power to steer, hot air balloons were at the mercy of the winds. Another term for airship, *dirigible*, from the French 'to steer', highlights the point. Frenchman Henry Giffard managed to fly 30 kilometres in a steam-powered airship in 1852; over the next 50 years others used petrol engines and electric motors.

Yet Zeppelin's is the name we remember. A lieutenant-general at 53, Zeppelin passionately promoted airships as war machines, for reconnaissance and the bombing of cities and ships at sea. He faced major engineering challenges. In larger airships the streamlined oval-shaped bag that provided lift was prone to instability and could collapse disastrously. Zeppelin's answer was to give his airships a rigid framework of girders to support an outer covering and enclose a number of separate gas cells. In later, larger models, the framework was made of the aluminium alloy duralumin.

Yet the vision failed the reality test when war came. While appearing huge and fearsome as they silently approached, zeppelins performed poorly as bombing platforms and did little damage. They were themselves vulnerable to attack from fighter aircraft and ground fire. By war's end, their roles were filled by airplanes.

Zeppelin died in 1917, missing the great age of civilian airships. In 1919 the British *R-34* was first across the Atlantic. Though many airships, including Germany's *Graf Zeppelin*, operated successfully for a decade or more, safer with non-flammable helium as the lifting gas, the fatal crashes of the British *R-101* in 1930 and the *Hindenburg* in 1937, among others, eroded public confidence. Through the 1930s airships could not handle the competition for passenger traffic from faster and more reliable aircraft, and were ultimately

balloon in tandem would provide better lift than either alone. The fire touched the adjacent hydrogen balloon, setting the gas alight and disaster followed.

Hot air balloons remain popular today for recreation and for feats such as crossing an ocean or orbiting the planet. For safety reasons few balloons now use hydrogen, other than the balloons sent regularly aloft by weather observers. Larger, gas-filled balloons now use helium—not as light, not renewable, but also not flammable. Such balloons have carried scientific and meteorological instruments—and humans in sealed capsules—to great heights, almost to the edge of space.

The success story

Luftschiff Zeppelin 1 made its mostly successful maiden voyage in 1900. Over the next decade, zeppelins proliferated in German skies, carrying mail and passengers, enjoying a remarkable safety record given that the lifting gas was flammable hydrogen. They were accepted into military service in 1908. By 1914 Germany had the largest airship fleet in the world, including 100 ready for military duties. Zeppelin had become a much-decorated hero, his creations a source of national pride.

withdrawn, though they played a major role in anti-submarine warfare during World War II.

The triumph at Kitty Hawk
George Cayley, Otto Lilienthal, Octave Chanute, Wilbur and Orville Wright

On 17 December 1903, American Orville Wright became first to fly in a heavier-than-air, powered machine, while his brother Wilbur watched from the ground. At Kitty Hawk, North Carolina, the *Flier*, equipped with a 12 horsepower petrol engine, covered about 40 metres in twelve seconds. The Wright boys did not start from scratch. Any number of experimenters had sought the conquest of the air, with work moving away from flapping wings, which never worked, to fixed ones able to glide as many birds do.

Around 1800 Englishman George Cayley began to study the forces that might help a man to fly, summing up his ideas in the influential *On Aerial Navigation*. From 50 years of work with kites and gliders, he knew the shape of the wings was critical in providing 'lift'. Cayley added a tail-plane for stability and tried a biplane design to add strength. He also knew lengthy flights would need some form of power. While his experimental craft never carried a human, many lifted the weight of one, showing it could be done.

Many came after Cayley with the same desire. The German engineer Otto Lilienthal boldly went aloft in his own machines; his 1889 book on aerodynamics set the Wright brothers on their course. In 1896, after 2500 flights, Lilienthal lost control of his glider in a strong wind gust, crashing to his death. American astronomer Samuel Langley, Secretary of the Smithsonian Institution in Washington, tried adding a steam engine to a glider. His first model flew a kilometre before running out of fuel; a later, much larger, version proved too heavy to fly. Langley gave up the hunt.

Many experimenters were now at work, achieving long hops across the ground, if not sustained powered flight. The American Octave Chanute summed up the situation in his 1894 *Progress in Flying Machines*, drawing on all the pioneers including the Australian Lawrence Hargrave, for whom he had a particular regard. The book, and Chanute himself, was another source of inspiration for the Wrights.

The big day at Kitty Hawk climaxed half a decade of methodical effort by Wilbur and Orville, printers and

Below: One of the many bizarre early designs for an 'heavier than air' flying machine intended to emulate a bird. Ultimate success came to those who understood aerofoils and the physical forces that can keep such a machine aloft.

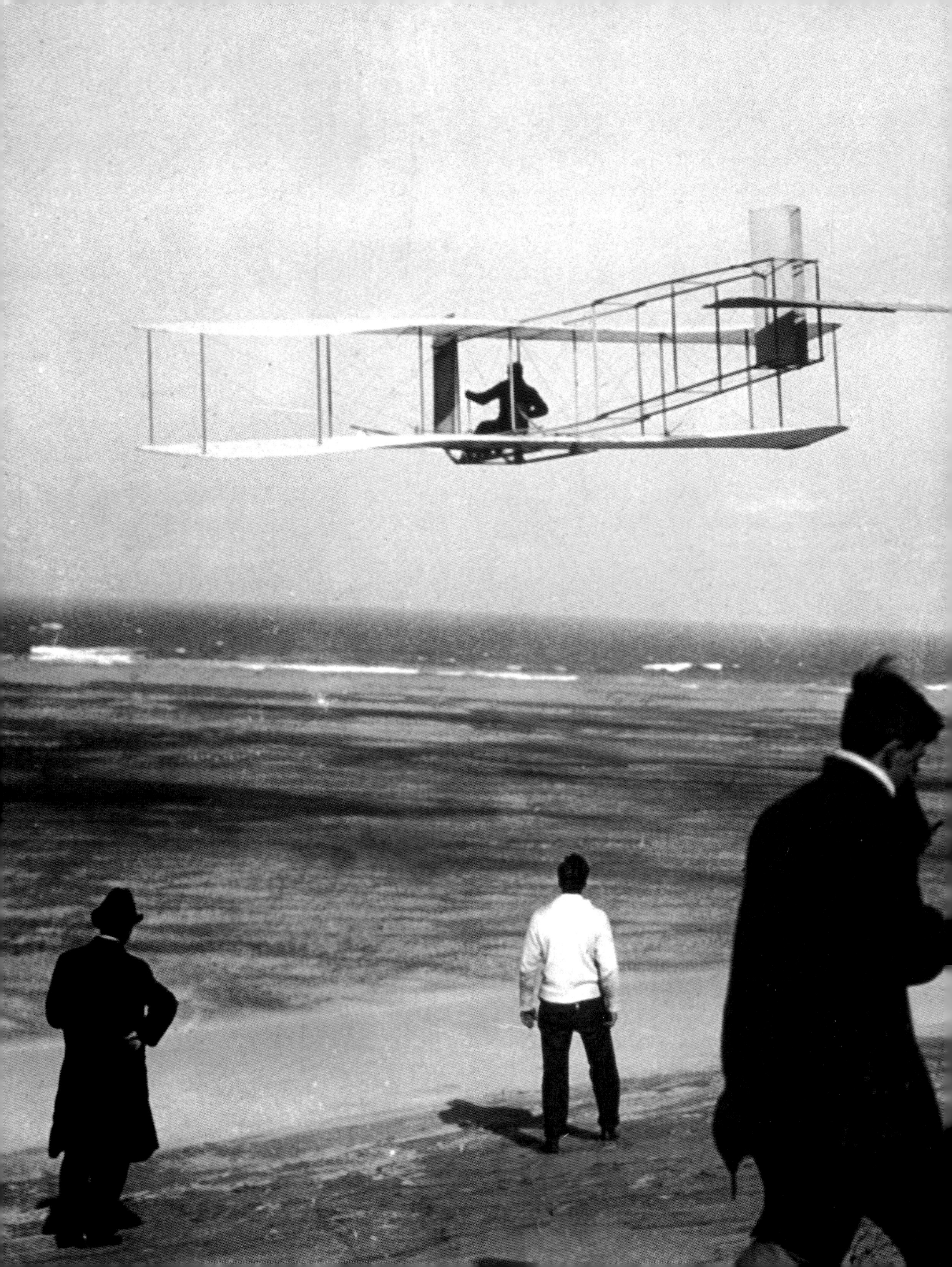

Opposite: The Air Age dawns. One of the Wright brothers soars above the beach at Kitty Hawk on 17 December 1903 in their fragile *Flyer*. Indeed, the machine was wrecked at the end of that great day by a sudden gust of wind.

bicycle builders from Dayton, Ohio. They had worked their way up from kites, through gliders and on to various models of powered aeroplane. In a key development, their craft could be controlled by the pilot about all three axes—turn, pitch and roll. Taking turns, they made four flights that first day, the longest lasting nearly a minute and covering 250 metres, before the plane was overturned and wrecked by a sudden wind gust. For two years they continued to work in secret, steadily increasing the distance and altitude of their flights, holding back public demonstrations until 1908.

By then daring inventors in several countries were taking to the skies. Aided by new designs and lighter petrol engines, their creations could stay aloft for half an hour and cover many kilometres, though the engines and the craft themselves remained unreliable and the flights literally death-defying. An early trophy was claimed in 1909 by French aviator, and successful maker of automobile headlamps, Louis Bleriot—a £1000 prize offered by London's *Daily Mail* for the first crossing of the English Channel. Reaching Dover in 37 minutes, Bleriot delighted his countrymen but worried the British. The bulwark of the Channel, which had halted the invasion plans of the Spanish Armada and of Napoleon, now seemed less secure. Politician David Lloyd George said with some concern, 'Flying machines are no longer toys.'

The Great War, erupting only five years later, showed the truth of that. Aircraft proved powerful in combat, for reconnaissance and for bombing, roles originally assigned—by the Germans at least—to airships. The combatant nations established embryonic air forces. Aircraft fought each other in the skies over France and elsewhere, in 'dog fights', and later using an innovation that let them shoot their machine-guns through the whirling blades of the propeller, increasing accuracy of

The aerofoil: what's that?

Cayley discovered that a wing designed with a curved upper surface could enable a craft to fly. Air travels faster over the curved surface of this aerofoil than over the flatter bottom surface. According to a well-known principle in physics, air pressure in the fast-moving air above the wing is significantly less than that below. This difference in air pressure creates an upward force that keeps the craft aloft.

fire. To survive such scraps, planes needed to be stronger, faster, more manoeuvrable; the science and technology of aircraft advanced rapidly. Without the goad of military need, aviation may have been much slower to take off.

1919 saw how military influence played out in peacetime. British aviators John Alcock and Arthur Brown flew a Vickers Vimy bomber, stripped of its wartime accessories and fitted with extra fuel tanks, in the first-ever crossing of the Atlantic by airplane. In quest of another *Daily Mail* prize, this time of £10,000, they took sixteen hours to fly from Newfoundland to Ireland, landing in a bog.

With the long-distance capacities of aircraft demonstrated, the way was open for regular flights to carry mail, cargo and soon paying passengers, all within twenty years of the Wright brothers' first achievement. Much more lay ahead.

Jet engines replace propellers
Frank Whittle and Hans Ohain

The jet engine is a great example of independent invention, an idea 'whose time had come'. Two military aviation engineers, Frank Whittle in Britain and Hans Ohain in Germany, unknown to each other, developed almost the same ideas through the 1930s. Since the time of the Wright brothers, aircraft had been driven by propellers. That technology peaked in military aircraft, fighters like the German Messerschmitt and the British Spitfire and Hurricane, but the limitations on how fast they could fly were already showing up. No propeller-driven plane, it was believed, could ever exceed the speed of sound (over 1000 kmh). A new approach was needed.

Frank Whittle patented his jet engine in 1932, though many doubted the idea had real merit. He struggled for funds and recognition, battling bureaucrats and potential industry partners, but the outbreak of war in 1939 spurred action, and his E-28 prototype engine flew successfully in 1941. The rewards came later, a knighthood, promotion to Air Commodore, a substantial ex-gratia payment from the government and the knowledge that, by war's end, Gloucester Meteor jets, using his engines, were destroying German V1 rocket bombs directed against London.

Like Whittle, Hans Ohain was only 22 years old when he first conceived an aircraft engine without a propeller. He patented his design in 1934, several years after Whittle, but with financial backing from leading aircraft designer Ernst Heinkel, he put his ideas to work much more quickly. His jet engine ran successfully on a test bench in 1937 and powered the Heinkel He178 in the air in August 1939, two years ahead of Whittle, and just a few weeks before war broke out. Germany maintained its lead when the Messerschmitt Me262 became operational before the Meteor.

The helicopter takes off
Igor Sikorsky

The idea for the helicopter, supported and driven by large horizontal blades rotating on top, was probably inspired by the spinning sails of a windmill. Instead of the movement of air pushing the sails around, powered sails would push on the air. Toys operating on this principle had a long history, originating perhaps in China, the rotors set spinning by pulling on a cord. Later models had on-board clockwork motors, even steam engines.

By the 19th century, screw propellers were pushing ships and featured in several early attempts at powered flight, but a practical load-carrying helicopter came only in the mid-20th century, mostly through the efforts of Russian-born engineer Igor Sikorsky. The son of a psychiatry professor and a doctor, Sikorsky's early interest in aviation was inspired by news of the Wright brothers and Zeppelin airships. He became a major innovator, building the first ever multi-engine aircraft and promoting revolutionary ideas like all-metal construction, enclosed cabins with upholstered chairs for passengers, even an on-board toilet. His designs were the forerunners of later airliners and his amphibious aircraft made possible airline routes into regions without landing fields.

The jet engine: how does it work?

Whittle and Ohain both designed what was essentially an air-breathing rocket. A traditional rocket had in its mix of chemical fuels one that supplied oxygen to make the rest burn (saltpetre in the case of gunpowder), the 'jet engine' takes its oxygen from the air it flies through, compressing and heating the air with a turbine, spraying in liquid fuel and igniting the mixture. Hot, high-speed gases injected from the rear cause the engine, and therefore the aircraft, to move forward, like a rocket. This makes more efficient use of the fuel than a propeller, generating more power and more speed, with fewer moving parts.

Above: A helicopter in its most familiar and welcome role, here helping rescue an injured crew member from a racing boat. Its unique capacity to hover can take the 'chopper' to places no other vehicle can reach.

Sikorsky had migrated to the United States after the Russian Revolution. In 1931 he returned to his early attempts at a helicopter, undertaken in Russia in 1909 aged only twenty. The weight of engines kept his creation on the ground. No experimental machine of the time went very far or very high, if it flew at all; control was a challenge. The first real success with horizontally-rotating blades came with the 'autogyro', devised by the Spaniard Juan de la Cierva in 1923. An ordinary air-screw pulled the plane forwards, supported by large unpowered rotors spinning freely.

As with early locomotives, progress depended crucially on the power to-weight ratio of the engines. When more efficient lightweight engines arrived, along with light strong construction materials, a number of inventors began to experiment again, including the German Heinrich Focke. Sikorsky had most success. To stop the whole machine turning in the opposite direction to the rotor, he added a second small rotor on the tail; other designers had used counter-rotating blades. His prototype VS300 first flew in May 1940, with the designer at the controls. A year later a production VS300 stayed aloft for 90 minutes. The larger R4 made the first-ever helicopter rescue in Burma in 1944. Sikorsky's single-rotor machine remains the dominant design today.

747s can carry more than 400 passengers at 85 per cent of the speed of sound, for distances in excess of 10,000 kilometres without refuelling.

In the wake of the Comet
The first jet airliners

Once World War II ended, designers and builders could turn to the possible use of the jet engines pioneered by Whittle and Ohain in civilian aircraft, for passengers and cargo, though new jet fighters and even bombers were soon in service and the 'sound barrier' was broken late in the 1940s. Jet airliners would become faster and quieter, and with the extra power would carry more passengers further. Costs could fall. The first such endeavour was the British-built de Havilland Comet, which went into commercial service in 1952, cutting the flight time from London to Singapore from two and a half days to one, and heralding the 'jet age'. Celebration was short-lived. Over the next two years, three Comets suffered catastrophic fuselage failure and crashed with great loss of life, though the tragedy had nothing to do with the engines.

The remaining Comets were withdrawn from service, but other jet airliners remained in development. In 1958 the American Boeing 707 first carried passengers. It was to prove acceptably safe and reliable. Propeller-driven airliners, flying since the 1920s, began to disappear from the skies, though there was later a place for the hybrid 'turboprop' engine in smaller planes, with a jet engine turning a propeller. Aircraft manufacturers concentrated on quieter and more efficient jet engines, fitted to larger and better-appointed aircraft. The cost of international air travel began to fall and demand rose.

Typical of the technology of air travel around the turn of the 21st century was the Boeing 747 or Jumbo, the most recognisable airliner, and until very recently the largest in service. First flown in 1970, and with 1400 operating worldwide as of 2006, 747s can carry more than 400 passengers at 85 per cent of the speed of sound, for distances in excess of 10,000 kilometres without refuelling. All these statistics would have seemed outrageous a few decades earlier. Still larger jet aircraft are being introduced by the two big manufacturers, Boeing and Airbus, as the anniversary approaches of Louis Bleriot just making it across the English Channel in 1909, though no change in technology comparable to the introduction of the jet engine is likely for some decades.

Below: A farewell in the early 1950s for a Comet, the first jet airliner to go into service. Three did not return. The strains on the fuselage of flying at unprecedented altitude caused them to crash with major loss of life.

Daredevils with parachutes
Pierre Blanchard, André-Jacques Garnerin and Charles Broadwick

Many people contributed to the invention of the parachute. Leonardo da Vinci is often credited with the idea. Certainly he sketched one around 1485. Others before da Vinci drew similar devices. Mediaeval documents tell of adventurous folk jumping and using something to arrest their fall, such as a loose cloak stiffened with wooden rods. Some survived; some did not.

The story picks up pace in the 1780s with the growing popularity of ballooning. Frenchman Sebastien Lenormand of France coined the name *parachute*, meaning 'to protect against a fall'. His creations, like earlier designs, used linen stretched over a frame.

Something more like the modern parachute appeared as inventors sought more compact and easily-used designs. Pierre Blanchard led the way; by the late 1790s he was using folded silk, which was both light and strong, but without a frame. In 1797 André Garnerin made the first jump with such a parachute. Perhaps this is the key date.

The parachute found a ready market with the arrival of powered flight. In Missouri in 1912, eight years after the Wright brothers, US Army Captain Albert Berry entered history as the first to parachute from a moving plane. Gender equity was restored in 1913, when crowds in Los Angeles saw Georgia Broadwick repeat the feat. The next year, Charles Broadwick patented the 'backpack' parachute, much as is used today, though others are credited with this discovery.

In World War I parachutes let crews escape from tethered observation balloons; hydrogen-filled, these were tempting targets for the early fighter planes. But the fighter pilots were not issued with parachutes, partly because the mass-produced canvas parachutes were too heavy for the primitive planes to carry, partly from a fear that having such an easy means of escape would promote cowardice among inexperienced flyers.

Stranger than fiction

The first parachute jump was made long before the invention of the aeroplane. One account has balloon pioneer and daredevil Pierre Blanchard first testing one using his dog. The dog survived, as did Blanchard himself in 1793 when his hot-air balloon caught fire and he had to 'bail out'.

Endeavour
United States

AeroVironment
ERAST
Pathfinder
NASA

Opposite: Way beyond the Wright brothers: the US Space Shuttle *Endeavour* lands after a successful mission. The stubby wings are useless in airless space but allow the craft to return safely to Earth for reuse.

Not until 1918 were German fighter pilots given parachutes. These were deployed by 'static lines', fixed to the plane, which caused problems when the plane was spinning. In 1919 American inventor Leslie Irvin introduced the 'rip cord', letting the pilot activate the parachute when clear of the plane. The first life so saved was reportedly that of William O'Connor in 1920.

The black box
David Warren

Whenever an aeroplane crashes, an early priority, second only to rescuing victims or retrieving bodies, is locating the 'black boxes', the flight data recorder and cockpit voice recorder, with their records of instrument readings and cockpit conversations. That information can be of immense value to investigators seeking a cause for the accident. Now every plane carries them. In 1967 Australia was the first nation to make them compulsory, following an unexplained crash in Queensland in 1960. There is some irony here. The devices had been invented by an Australian a decade earlier, but had initially generated very little interest.

David Warren worked as a chemist at the Australian Aeronautical Research Laboratories in Melbourne. Following the fatal crashes of several Comet airliners in the 1950s, which were initially unexplained, Warren explored the possibility of an onboard data and voice recorder, engineered to

Opposite: Solar soaring. Looking for new ways to stay aloft, this experimental aircraft reduces pollution by replacing conventional engines with electric motors powered by solar cells on its upper wings.

The instruments must survive very rapid deceleration, crushing, piercing, intense heat and deep-water immersion.

survive the crash. He reasoned that in many cases the crew would have known what was going wrong, or the instruments would have revealed it. By 1957 he and his colleagues had built a prototype 'flight memory unit', which retained up to four hours of data and voice recorded on steel wire, tape came later. Trials were successful, but he could not find a manufacturer and there was no support from local aviation safety officials.

Fortunately, his invention came to the attention of an official from the British aviation industry who was much more enthusiastic. Warren was invited to work in the United Kingdom with others to improve the device and house it in a crash-proof and fire-proof container. The first production model was dubbed the 'Red Egg' from its colour and shape. Sales were soon brisk, and over coming decades many lives would be saved through being able to determine the causes of accidents and take preventative action.

The term 'black box' was always misleading. To aid recovery, the recorders are usually painted red or orange. Requirements are stringent; the instruments must survive very rapid deceleration, crushing, piercing, intense heat and deep-water immersion. Most are now 'solid state', recording on microchips rather than on tape or wire. Up to 80 different parameters are logged, including speed, aircraft orientation, altitude, engine temperatures and the setting of the controls. Future versions are likely to capture a record from video cameras in the cockpit.

CHAPTER 18:
Time and place

Much technology has been invented to give us the information we need to organise our lives, for example, to know what the time is, or where we are. In some crucial cases we needed to know one before we could know the other. Other innovations improved on long-established aids like maps and the magnetic compass.

What time is it?
Peter Henlein

The first mechanical clocks, dating back to the 13th century, were large clunky pieces of machinery, adorning town hall steeples and church towers, driven by big falling weights. It was hard to imagine a timepiece small enough to be carried.

Around 1500 a locksmith from the medieval city of Nuremberg in modern southern Germany came up with a new power source for clocks. Others may have had the same idea, but Peter Henlein is the one credited with making it work. Henlein took a strip of thin iron sheeting, and coiled it up into a spring. Energy was stored in the spring as it was tightened ('wound up') and released bit by bit as the spring relaxed. Machines—such as model trains—powered by such coiled springs are still dubbed 'clockwork'. Timepieces could now be shrunk in size, set on a shelf or on a table, or even—and this was Henlein's greatest achievement around 1510—strung on a cord around the neck, portable if not yet a 'pocket watch'.

These early watches were the size and shape of a turnip, though increasingly they were made oval in shape and gained the name 'Nuremberg Eggs'. Often decorated with gems, they became symbols of wealth and fashion. However, as timekeepers they were poor, needing careful handling. The crude hand-made iron mechanism was stiff, needing a very strong mainspring to make time pass at all. As the spring 'ran down', the clock ran slower, and time began to drag.

Fixing that problem turned the early watch from a fashion item into a passably useful device. Around 1525 Jacob Zech, a Swiss-trained mechanic living in Prague, conceived the 'fusee'. A cord wrapped around a conical pulley connected the mainspring to the rest of the works, so that the spring maintained a constant pressure and the watch kept much better time.

By around 1600 watches were becoming commonplace. Malvolio in Shakespeare's *Twelfth Night* talks of his with pride. Brass and later steel replaced iron in the works, screws supplanted rivets, a glass face protected the 'hands', which now often recorded minutes as well as hours. Such watches were more than ornamental, but still far short of acceptable accuracy. Some way was needed to accurately measure the passing particles of time, count and record them. That took more than a century.

Above: A Nuremberg portable clock dating from 1525, with a bronze movement and a sundial face. Powered by the newly invented coiled spring, it is one of the earliest examples of its type.

> **These early watches were the size and shape of a turnip, though increasingly they were made oval in shape and gained the name 'Nuremberg Eggs'.**

King Louis XIV reportedly remarked that the survey had 'cost him much of his domain'—the new maps showed that Brittany extended west by 100 kilometres less than previously thought.

The art and science of surveying
Gemma Frisius, Leonard Digges and the Cassinis

The 16th century saw a boom in map-making, for very practical reasons. Rulers wanted to know how far their domains extended so they could plan the collection of taxes. Agreed national borders had to be firmly established. Lands were being newly discovered—or rediscovered—and the routes to them had to be charted. The crude maps that people had been using for centuries were not up to the task or simply out of date.

As a result we saw a surge of innovation in the technology of surveying, precisely locating major features in the landscape and using those to construct maps. Chief among the new techniques was the 'plane table', simply a large table on folding legs able to be made horizontal, with a sighting device that could be used to mark on a chart the direction from the observer to a distant object. When a number of such sightings had been made, the table was moved to another spot a known distance away and the sightings taken again.

When the two resulting charts were superimposed, the points at which the sighting lines crossed marked the relative positions of those objects. Initially the plot would give only relative distances, but measuring just one distance, for example, the 'baseline' between the two observing points, would immediately give the distances between all the landmarks plotted.

While this method, known as 'triangulation', probably had many inventors, it was first clearly described in 1533 by the Dutch mathematician, cartographer, physician and instrument maker—a versatility common at the time—Gemma Frisius. A key figure of his age, his students included the famous mapmaker Gerard Mercator, whose distortion-creating Mercator Projection—the flat 'map' of the world—controlled how people visualised the continents for centuries to come.

The plane table morphed into early forms of the 'theodolite', beloved of surveyors. The instrument—and its name, the meaning of which is uncertain—are usually ascribed to the Englishman Leonard Digges, who described one around 1571. The theodolite could measure vertical as well as horizontal angles and so

Left: British officers take theodolite readings before the flight of the *R-34* airship across the Atlantic. Measuring the angles between sight lines to known landmarks generated a grid of triangles over which detailed maps could be drawn to track the airship's movement.

find heights as well as distances. It often contained a compass to give magnetic directions, and a telescope, once they were invented.

Over 100 years elapsed before a large-scale survey using triangulation was undertaken, in an effort to correct the many errors in maps of the time. The Italian-born French astronomer Jean-Dominique Cassini began one of the whole of France, carried on by three generations of his sons and finished in 1740. Early on, King Louis XIV reportedly remarked that the survey had 'cost him much of his domain'—the new maps showed that Brittany extended west by 100 kilometres less than previously thought.

Supporting these grand surveys, and helping to pinpoint the ends of the baselines, were the evolving techniques to find latitude and longitude, though the main need for those was at sea rather than on land.

The secret of the sextant
John Hadley and John Campbell

For 18th-century science or commerce, no challenge was greater than letting a ship's captain establish his position once out of sight of land. Fortunes hung on the safe arrival of cargoes of raw materials, manufactured goods or slaves, not to mention the lives of the crew. Ships' captains and navigators needed to know where they were if they were to reach their destination safely.

Finding latitude was the lesser challenge. The earth is pretty much a sphere, so travelling north or south changes your view of the sky. It affects how high the sun, the moon or a 'fixed star' rises above the horizon to the north or south. For example, the sun can be right overhead at noon on the equator, though only on two days of the year. Navigators have to watch the calendar.

For centuries sea captains had used crude 'crossstaffs' to measure the altitude of the sun at midday, and so figure their distance north or south of the equator. In 1701 the English scientist Isaac Newton suggested that with the right equipment, the same idea could be used for any star whose position was precisely known.

But how to do it accurately? The old methods required a captain—often on a heaving deck—to keep one eye on the sun or star and the other on the horizon, and to measure the angle in between. It was a tough ask. In 1736 John Hadley of England simplified matters greatly with an instrument with internal mirrors. Move these to bring the horizon and the celestial object into the same line of sight, and the

Below: Beautifully crafted and elaborately engraved, this 16th century Persian navigational instrument combines a compass and a sundial. The tilt of the sundial can be changed with latitude.

A swinging pendulum a metre long marks off equal seconds. If these units could be tallied to mark the passage of minutes, hours and days, accuracy in timekeeping was assured.

angle between them could be read off a scale. His versions could measure angles up to 90 degrees (a 'quadrant') or 45 degrees (an 'octant'). A little later John Campbell settled on 60 degrees, and so we know the instrument as a 'sextant'.

Finding longitude proved a much tougher problem.

Keeping better time
Galileo Galilei and Christiaan Huygens

The sort of clocks known to the multi-talented Italian physicist Galileo Galilei could lose or gain many minutes in a day despite 50 years of development. He found the way forwards by chance. In 1581, when only 17, he was—so the story goes—standing in the cathedral in Pisa. He watched as the great lamp in front of the sanctuary swung from side to side.

As the swings got smaller, Galileo timed each one by counting the beats of his pulse. Each took the same time, whatever its size. Only the length of the chain appeared to affect the 'period' of the pendulum. A swinging pendulum a metre long marks off equal seconds. If these units could be tallied to mark the passage of minutes, hours and days, accuracy in timekeeping was assured. Such a clock should not run fast or slow at all if kept in the same location and the same conditions.

Galileo himself saw the possibility of a 'pendulum clock', but never built it. That honour belongs to the Dutch genius Christiaan Huygens, famous in astronomy and mathematics as well as clock-building. In 1656 he built a clock far more accurate and reliable than any built before. Each precise tick of the pendulum was passed through the mechanism known as an 'escapement' to the record the passage of time on the face of the clock. Scientists were probably the first to use such clocks; precise timing was becoming important in physics and astronomy. Precise timing then mattered less in everyday life, but Huygens patented accurate pocket watches as well.

Huygens had another vision, a solution to the vexed problem of finding the position of ships out of sight of land. If such a clock could be taken to sea, carrying within it for weeks at a time an accurate memory of the time at the home port, the captain or navigator could compare that time with local time as determined by the sun and so work out how far east

Latitude and longitude: what are they?

The 'global positioning system' in universal use in the 18th century had been devised 1500 years earlier by the venerable Greek geographer Ptolemy. Any point on earth lay at the intersection of two lines; a parallel of latitude running east to west aligned with the equator, and a meridian of longitude, running north to south from pole to pole. Identify the lines you are on, and you are where they cross.

Right: Admiral Horatio Nelson's pocket watch, rehoused in a carriage clock. Nelson was using it when he was killed at the Battle of Trafalgar in 1805.

or west the ship had sailed. Huygens strove to build such a clock, but it eluded him. Success in that, and the problem of 'the longitude', came in the next century and from across the Channel.

'Discovering the longitude'
John Harrison

In 1707 four Royal Navy ships and 2000 men were lost in a single shipwreck, mostly because they did not know where they were. A ship's captain could find his latitude quite easily but reliably finding longitude east or west of his home port had so far proved impossible. Taunted by such losses, the British Parliament passed the Longitude Act in 1714, offering to reward £20,000 to 'such person or persons as shall discover the Longitude' if the longitude of a ship at sea could be established within one degree, £40,000 if within half a degree. These were immense sums for the day, but the problem was urgent.

The solution had long been known, at least in principle. The captain simply needed to compare his local time set by the position of the sun in the sky with the time—at that same moment—back in the port he had come from. Since the earth turned through 360 degrees every 24 hours, each hour difference in time meant 15 degrees difference in longitude. But just what time was it back home?

There were two possible ways to find out; both had problems. You could take a clock with you from the place you had left, but would it be reliable enough? Champion clockmaker Christiaan Huygens had tried that route and failed.

Or you could use a clock that would be visible from both places, say something in the sky. For example, the

Chapter 18: Time and place

Left: A beautiful pocket or 'fob' watch that belonged to the British explorer Lawrence of Arabia.

movement of the moon in front of stars, though complex and not quite regular, could be likened to the hour hand against the numbers on a clock. The latter 'method of lunar distances', which used John Campbell's sextant, proved hard to compute and to use, but it let the average captain find his longitude to within a degree—about 100 kilometres.

Yorkshire carpenter-turned-clockmaker John Harrison did much better. He took the first route, and spent 40 years on the task. By 1765 he had done the impossible; built a clock so reliable that it lost or gained less than a minute in a six-week voyage. This gave the longitude well within the acceptable error. Envious manoeuvrings among the members of the 'Board of Longitude' scandalously denied him his reward for years, but with the intervention of the King, 'Longitude Harrison' finally got the money and his place in history.

Making a better compass
Elmer Sperry

The magnetic compass, invented by the Chinese, enhanced in Europe and used by every mariner from the great age of exploration onwards, has limitations. It does not point naturally towards the north and south geographic poles but towards the slightly different—and slowly moving—magnetic poles. You cannot trust it absolutely.

Left: A modern, waterproof, on-board compass for yachts. It 'floats' on gimbals to keep it level no matter how much the vessel rolls or pitches.

The gyroscope: what's that?

Any spinning object tends to stay spinning with its axis pointed in the same direction, a manifestation of 'rotational inertia'. So a spinning bullet from a rifle does not tumble, a moving bicycle stays upright on its two narrow wheels. A gyroscope is a metal ball or disc turning at high speed —today driven by an electric motor—and so able to stay pointing at the same spot in the sky. Three gyroscopes at right angles keep the *Hubble Space Telescope* directed very precisely to whatever distant object it is photographing.

Its modern replacement is the 'gyrocompass', invented by several people but most conclusively by the American Elmer Sperry in 1908. He started with the gyroscope, invented in the previous century by Frenchman Leon Foucault.

If you push and pull on the opposite ends of the gyroscope axis—so applying a 'torque' or 'couple'—in an effort to make it change direction, it responds in an odd way. It turns in a direction at right angles to both the direction of its own axis and the direction of the applied torque. If the couple is the result of earth's gravity, the gyroscope will slowly shift until its axis is pointing genuinely north to south, along the earth's axis, and there it will stay. The gyroscope becomes the gyrocompass.

Now any well-equipped aeroplane or ship has one, and usually more than one, since the movements of the 'master' gyrocompass can be transmitted to repeaters in other locations. Gyrocompasses are also totally immune to the effects of iron and steel nearby, another problem with the magnetic compass.

Sperry was typically prolific. He held 400 patents when he died in 1930, including for machines and for making fuse-wire. His gyrocompass was first installed in a ship, the US Navy battleship *Delaware*, in 1911. It found many uses in peace and war, including gyroscopic gunsights and bombsights, which kept pointing at the target however the plane or ship moved. The Sperry Corporation, founded in 1933, is now part of Unisys.

Where in the world am I?
Ivan Getting

The latest 'must-have' for your car is a talking roadmap, sitting on or installed in your dashboard. From time to time a gentle voice tells you what turn to take to reach your destination. This would be impossible without the Global Positioning System (GPS), another achievement of the Space Age. The godfather of GPS was Ivan Getting,

Right: The talking map: now coming into wide general use, a computer-readable map linked to signals from the Global Positioning System (GPS) can issue gentle instructions to drivers who might otherwise not know where they are going.

Defence authorities now access an enhanced system, which reputedly provides positions to the nearest centimetre.

a research engineer at Raytheon Corporation, very active in the developement of wartime radar and the first microwave ovens. His first customers were the American military. They wanted to be able to fire guided missiles from railway flat-cars to achieve security from attack through mobility. Their handlers needed to know precisely where their weapon was at the time of launch so it could be directed accurately to its target.

Borrowing from ideas first developed during World War II, Getting proposed a network of radio transmitters, each with a precise clock. A receiver at the missile launch-site, with a similar clock, would pick up several transmissions simultaneously. By noting how long each signal took to arrive, the missile's minders could figure their distance from each of the transmitters and so calculate their location.

In today's GPS, the radio transmitters are in space, circling the earth in precisely-known orbits 20,000 kilometres up. More than twenty such satellites, each the size of a small car, are aloft at any one time. A GPS receiver—now only as big as a mobile phone and costing a few hundred dollars, thanks to microelectronics—uses transmissions from three or four satellites to calculate its position in three dimensions (latitude, longitude and altitude) and is accurate to a few metres. It is an astounding capability.

The Americans launched the first GPS satellites in 1974. For a decade only their military could use them, but the government had promised civilian access once the system was fully operational. That came in the

Above: The *Hubble Space Telescope*—shown here being set free from the Space Shuttle during a maintenance mission—is kept very precisely in position by three gyroscopes arranged on mutually perpendicular axes.

1990s. Defence authorities now access an enhanced system, which reputedly provides positions to the nearest centimetre, used to guide 'smart' bombs and missiles. The systems cost about $6 billion in development and nearly $1 billion a year to operate and maintain, but economic benefits are piling up, especially from greater accuracy and safety of navigation of ships, planes and vehicles on land. One estimate sees savings reaching $50 billion a year by 2010. GPS already has competition, from systems launched by the European Union (Galileo) and the Russian Federation (Glonass).

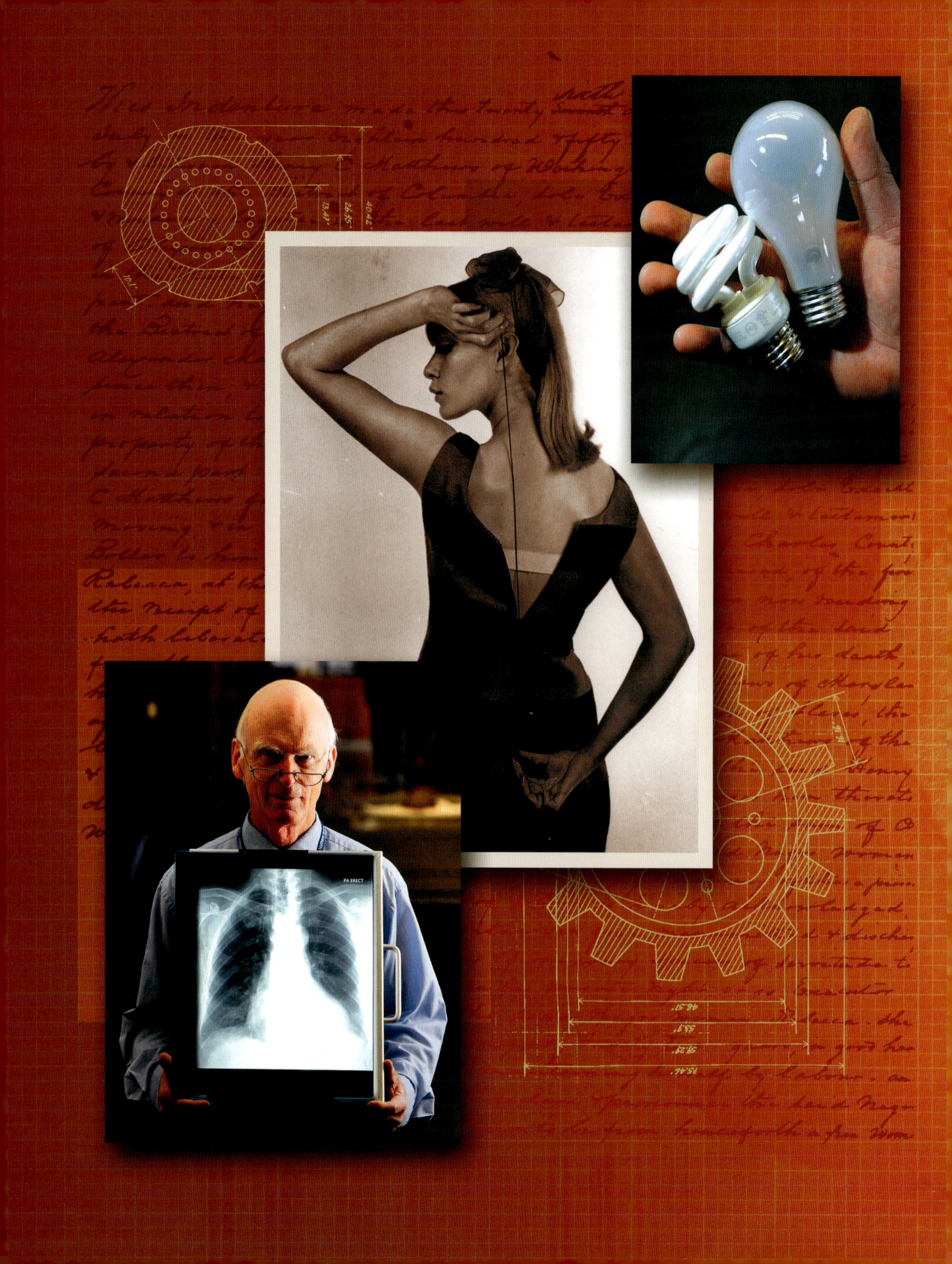
PA ERECT

04

PART FOUR:

Healthy and happy

CHAPTER 19:
Medical matters

We can find the fruits of invention all through the practice of medicine, in ways that keep us healthy and fight disease. That has always been so, even in early times when healers relied on folk remedies, but medical therapy today draws heavily on knowledge from biology, physics and chemistry and would be incomprehensible to physicians and surgeons of one or two centuries back. The transformation of medicine has been profound. Here are a few examples.

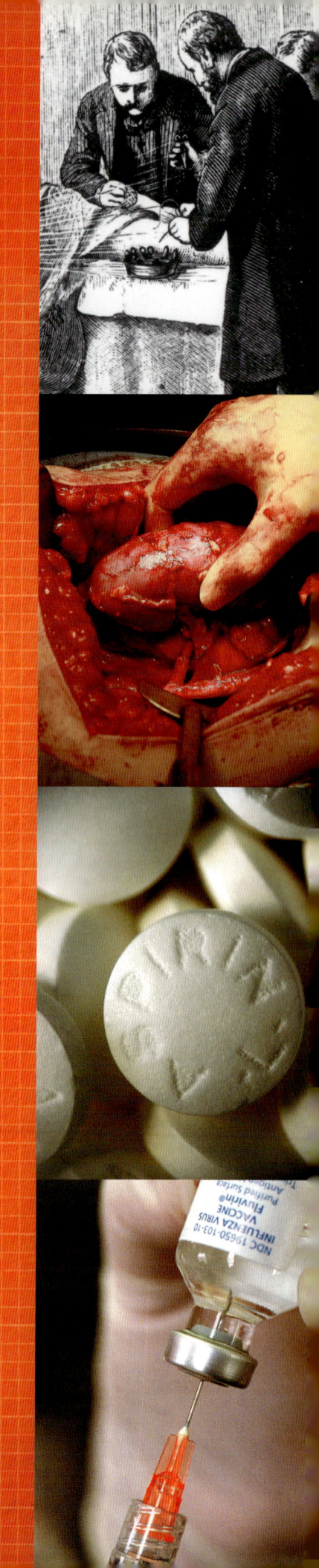

Chapter 19: Medical matters

What happened when?

1800	Humphry Davy discovers the relaxing effects of nitrous oxide
1844	Nitrous oxide used in operations to remove teeth
1846	Dr William Morton uses sulphuric ether to anaesthetise a patient during an operation
1850s	Hypodermic syringe used by Alexander Wood and Charles Pravaz
1867	Joseph Lister announces *The Principle of the Antiseptic Practice of Surgery*
1881	Louis Pasteur develops a vaccine against anthrax
1886	Pasteur successfully vaccinates a boy against rabies
1899	Bayer patents aspirin
1905	Robert Koch receives the Nobel Prize for his work on identifying micro-organisms
1910	Paul Ehrlich's Salvarsan arrives on the market to treat syphilis
1928	Alexander Fleming identifies *Penicillium notatum* as a potential treatment for bacteria
1937	William Gibbon pioneers the cardiopulmonary bypass machine
1941	John Hopps develops an artificial pacemaker
1943	William Kolff develops a continuously working kidney machine to clean blood
1950s	Disposable syringes first mass produced
	Paracetamol appears on the market
1953	First open-heart surgery
1955	Gregory Pincus reports the development of a successful contraceptive pill
1960	The Pill goes on sale
1967	Christian Barnard performs the first heart transplant
1984	Georges Kohler and Cesar Milstein win the Nobel Prize for developing 'magic bullets' to fight cancer

Anaesthetics against pain
William Morton and James Simpson

Anaesthesia, the controlled use of chemical compounds to reduce pain, was one of two 19th-century breakthroughs that has made modern surgery possible—the other being the control of infection.

Previously, pain could relieved by alcohol such as rum or by opium (laudanum), mandrake, cocaine, pressure on nerves, intense cold, perhaps even a blow to the head to render the patient unconscious—none very effective. Even into the 19th century, most operations took place without any anaesthetic, the surgeon relying on speed and skill to reduce the trauma of pain to the patient. There was little time for care or complexity.

The first chemical found to reduce pain, nitrous oxide, was initially used as entertainment at parties, hence its common name 'laughing gas'. English chemist Humphry Davy had discovered its exhilarating and relaxing effects in 1800, and predicted it would be used for minor operations such as removing teeth. But that did not happen until 1844, when first employed by Horace Wells, a Connecticut dentist.

Wells's work inspired a former student, Dr William Morton. He found that sulphuric ether, known since the 16th century, deadened pain when inhaled; it also put patients to sleep. After trials on animals, he first anaesthetised a human patient in 1846, during an operation on a tumour on the patient's jaw. Others had been

Stranger than fiction

Despite the contribution of others, it seems right to cite 1846 as the key date in the development of anaesthetics and William Morton as the pivotal central figure. Certainly his family seemed to think so. His headstone records: 'Inventor and revealer of inhalation anaesthesia; before whom, and in all-time, surgery was agony; by whom, pain in surgery was averted and annulled; since whom, science has control of pain.'

experimenting with ether; Crawford Long of Georgia used it on a patient in 1842, but kept his work quiet until Morton's became known.

Anaesthesia crossed the Atlantic in late 1846; Robert Liston, trained in the United States but working in London, used ether during an amputation. The news encouraged others, including Scottish midwifery specialist James Simpson. He preferred chloroform, more pleasant for the patient and able to be administered by a few drops on a gauze over the nose. He found it useful for both surgery and obstetrics; the practice gaining popularity after Queen Victoria accepted it during one of her confinements.

Since that time, anaesthesia has seen much technological improvement—new anaesthetics, new techniques to administer them, injectable painkillers, epidurals and the use of curare as a muscle relaxant.

The versatile hypodermic
Alexander Wood and Charles Pravaz

Hypodermic means 'under the skin'; a syringe with that name has a fine hollow needle able to inject drugs into tissues or blood vessels, or to withdraw blood if needed. Its invention appears shared by Scottish doctor Alexander Wood and French physician Charles Pravaz in the 1850s, working quite independently. Prior to this, syringes—such as used by Edward Jenner in vaccinating against smallpox in 1798—were crude and clumsy. Wood used the new device to inject painkillers such as morphine into the tissues of the face to ease the pain of neuralgia, though he was aware of other possible uses.

It did not take long for a downside of the use of hypodermics to emerge. Overuse of easy-to-administer morphine injections to control the pain of battle wounds during the American Civil War resulted in many soldiers returning home addicted. It is reported that Wood's own wife died from an overdose of self-injected

Below: The necessary needle. A hypodermic syringe is used to withdraw medication from a vial ready for injection into a patient. 'Hypodermics' are widely used in medicine and, less happily, to administer illegal drugs like heroin.

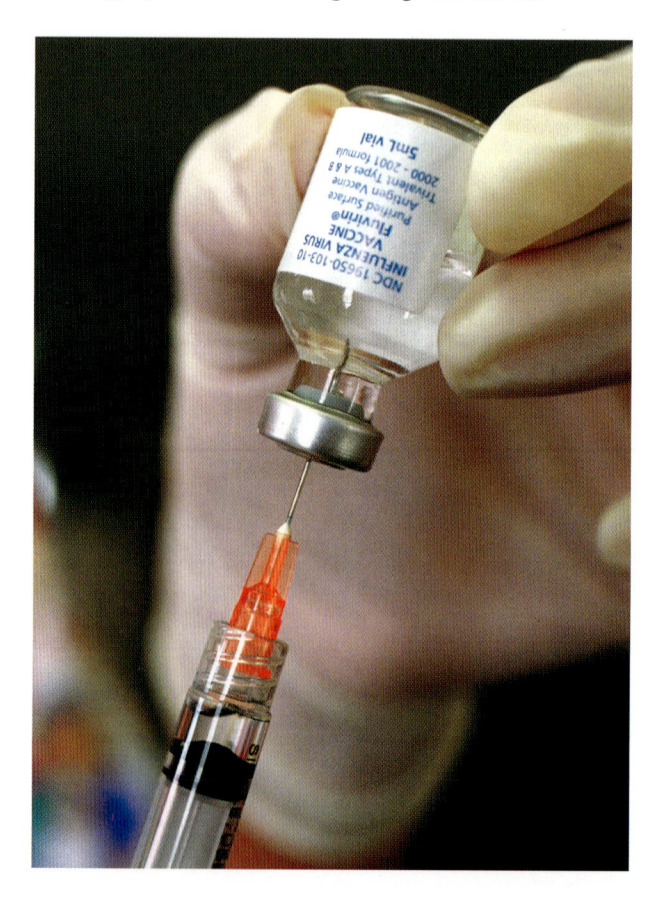

morphine, the first recorded such fatality.

Disposable (use-once-and-discard) syringes of glass were first mass-produced in the 1950s to support the large-scale vaccination of children against polio with the new Salk vaccine. Plastic disposables came soon after. Even more modern syringes must be discarded, since one use disables them. This has helped stem the spread of diseases such as AIDS, associated with injection of illegal drugs using contaminated needles.

Surgery without infection
Joseph Lister

Little more than a century ago, even successful surgery often ended in death because wounds became infected. The introduction of antiseptic surgery changed all that, and the great pioneer was English doctor Joseph Lister.

Lister's family were prosperous Quakers. After training and early experience, he became professor of surgery at the University of Glasgow. The all-too-frequent infection of surgical wounds was a constant challenge. At that time, septic wounds were blamed on 'miasmas' of bad air. Surgeons saw no reason to wash their hands before an operation; their bloodstained operating coats were never cleaned. Bed linen was rarely changed and usually dirty. The appalling smell of rotting wounds filled the wards.

Lister developed an alternative explanation for wound infection, based on the 'germ theory' pioneered by Louis Pasteur. The Frenchman had shown that fermentation and putrefaction was caused by micro-organisms, active even if deprived of oxygen. Lister confirmed this with the microbes he found responsible for the often deadly infection gangrene.

To kill the microbes, Pasteur had suggested heating, boiling or exposure to chemicals. Lister began to experiment with carbolic acid (phenol), distilled from coal tar and known to remove the smell from rotting sewage,

> **Surgeons saw no reason to wash their hands before an operation; their bloodstained operating coats were never cleaned.**

as an 'antiseptic' (able to 'act against' the micro-organisms of 'sepsis'). His methods were comprehensive. Swabs for cleaning wounds and bandages for covering them were soaked in a solution of carbolic acid. Surgical instruments were disinfected and the surgeons' hands similarly washed. The air around the operating table was sprayed with carbolic vapour.

Though the use of the irritating spray was later shown unnecessary, Lister obtained impressive results. Where carbolic was used, gangrene rates dropped markedly, many more patients lived. In 1867 Lister was able to announce 'The Principle of the Antiseptic Practice of Surgery' in the leading medical journal *The Lancet*. Despite much scepticism, indeed scorn and ridicule, among his colleagues, Lister had been on the right track. Surgeries now are designed to be both

The success story

For his work on antiseptics and other achievements, the shy, deeply religious Lister became a baron and one of the first members of the Order of Merit. He is also remembered through the antiseptic mouthwash Listerine, and the micro-organism *Listeria*, a common cause of food poisoning.

Left: Without the invention of sterile surgical procedures and safe anaesthetics, operations such as this one to transplant a living kidney and save a life were unimaginable.

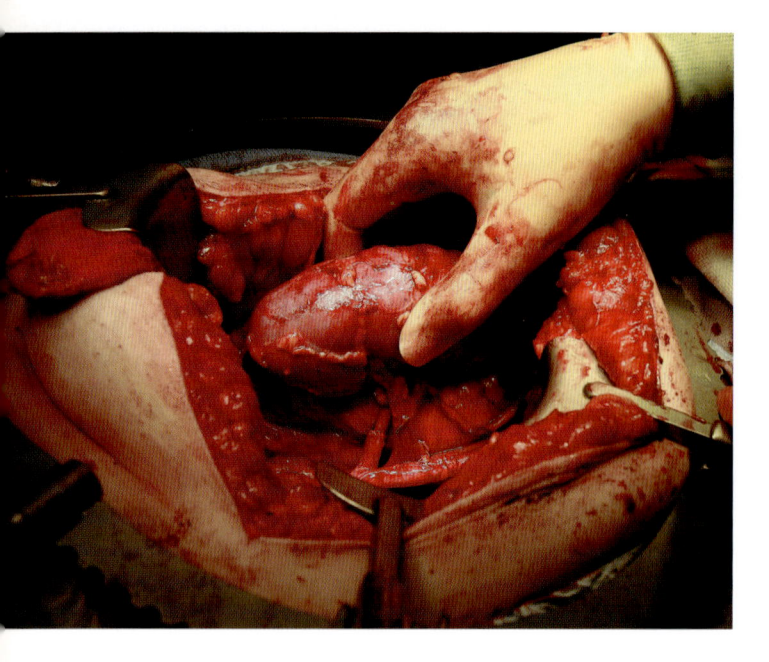

aseptic (free from bacteria) and antiseptic—though with more effective and convenient agents than Lister's carbolic.

Finding weapons against disease
Louis Pasteur and Robert Koch

By the time German doctor Robert Koch arrived on the scene around 1870, his illustrious French contemporary Louis Pasteur had won the early battles on behalf of the 'germ theory' of disease. It was increasingly accepted that diseases in humans, animals and plants, are due not to 'miasmas' of infected air, but to infectious micro-organisms or 'bacteria'.

Koch won his reputation through the nasty disease anthrax, suffered by animals and capable of spreading to humans. Although a busy medical officer in a rural area, he still found time for research using a microscope his wife had given him. Other scientists, inspired by Pasteur, had already found a rod-shaped micro-organism (a 'bacillus') living in the blood of infected animals. Koch placed samples of blood from animals that had died from anthrax in cuts in the skins of mice.

The mice quickly died of anthrax. Mice inoculated with blood from healthy animals—blood free from the anthrax bacillus—did not die.

A hard man to convince, Koch grew a 'culture' of the bacillus through several generations as free as possible from other contamination. Even though the micro-organisms had never directly contacted a diseased animal, he discovered that they could still spread the disease. The case appeared proven, but it was not until Pasteur found an anthrax vaccine in 1882 (see below) that Koch's findings were really accepted.

Koch added to his achievements by identifying beyond doubt the microbes that caused other diseases, including tuberculosis, cholera and septicaemia. Tuberculosis, known as 'consumption', killed one in every seven people in Europe at the time. Koch's work was therefore judged so important that he won the 1905 Nobel Prize.

Koch died in 1910, but his influence outlived him. He had developed ingenious and powerful methods to separate and culture various sorts of bacteria. Armed with his methods, Koch's students identified the germs that cause diphtheria, typhoid, bubonic plague, tetanus, syphilis and pneumonia.

When Koch found the anthrax bacillus, Louis Pasteur was one of the first to react. He quickly corroborated Koch but also noted that not all animals catch anthrax. Chickens are immune. Was this because their body temperature is around 43 degrees Celsius rather than the 37 degrees of other animals? Perhaps heat weakened or even killed the anthrax germs. Pasteur chilled the body of a living chicken to 37 degrees; it lost its immunity.

In a trial, publicly staged to answer his critics, Pasteur heated anthrax bacilli and then injected them into 25 sheep. Nothing obvious happened. He then

The boy seemed certain to die, so Pasteur treated him—despite having no licence to practise medicine.

injected the same sheep and 25 others with fresh anthrax. The first lot lived, the second lot died. Soon millions of sheep and cattle were being protected against anthrax. Someone estimated that the savings were equivalent to the reparations France had to pay after losing war against Prussia.

Pasteur repeated the methods successfully with chicken cholera, another costly animal disease. He called the treatment 'vaccination', so honouring Edward Jenner who 90 years before, in 1796, had pioneered something similar, using cowpox to protect people against the more virulent smallpox. Pasteur called the cultures of weakened bacteria 'vaccines'.

Pasteur's greatest triumph was over rabies—known then as 'hydrophobia'—which could kill humans bitten by an infected animal. He made a vaccine from the spinal cord of rabbits with the disease, though he had not yet clearly identified the bacteria. He could soon stop the disease from spreading from dog to dog, but would not try the treatment on humans for fear of something going wrong. He took the step only in 1886, when nine-year-old Joseph Miester was brought to his laboratory after being mauled by a rabid dog.

The boy seemed certain to die, so Pasteur treated him—despite having no licence to practise medicine. Miester made a complete recovery, as did a young shepherd similarly bitten a few months later. Publicity was intense; Pasteur became perhaps the most famous scientist in the world. Public gratitude supported the founding of the Pasteur Institute, where Joseph Miester served as gatekeeper for many years. Pasteur died from the complications of a stroke a decade later, aged 75.

More than 100 years later, many human and animal diseases, whether due to bacteria or the more recently discovered viruses, can be controlled by vaccination, and there have been some remarkable public health triumphs, including the total elimination of smallpox globally and almost the same success with polio. Deaths from regularly recurring new strains of influenza are gradually being reduced through the production of new vaccines. In a few diseases, however, most notably AIDS, a vaccine still eludes medical science.

Vaccination: how does it work?

When Pasteur injected the heated anthrax bacteria into sheep the dose had not been strong enough to cause the disease, but was enough to cause the animals' bodies to recognise an 'intruder' and build up their immune defences. When exposed a second time, the sheep were able to fight off the disease swiftly and efficiently by rapidly reproducing the antibodies they already had in stock, so to speak.

Controlling pain and fever
Aspirin and paracetamol

The most-widely used medical drug worldwide is probably aspirin, known scientifically as acetylsalicylic acid. It had been used for generations in an unrefined form in potions made from certain plants, including the bark of the white willow tree *(Salix alba)*. Aspirin is versatile, reducing pain and lowering fever by affecting body chemicals called prostaglandins. In more recent times aspirin has been found to thin the blood, reducing the risk of strokes and heart disease. In high doses it gives relief from arthritis.

An early form of aspirin, salicylic acid, was extracted from plants early in the 19th century, but its use generated some severe side-effects—bleeding in the stomach, even death. Consequently little interest was shown until 1897, when Felix Hoffman and a colleague at the large German chemical firm Bayer found how to make acetylsalicylic acid, a form of the chemical that did not exist in nature and which had much fewer side-effects. The first patient was Hoffman's own father, who suffered severely from arthritis and could not tolerate straight salicylic acid.

Bayer patented the drug in 1899 and the modern pharmaceutical industry was born. One other company was already selling it—a French chemist had devised a similar form 40 years earlier but thought it impractical for large-scale use. Bayer was first to mass-produce the drug, and it is best known by the Bayer brand name. The impact of the non-addictive aspirin was ultimately global. During World War I various countries, including Germany, devised their own ways to make the drug synthetically, from the chemical phenol found in coal tar. In Australia, the result was called Aspro.

At the time aspirin came into use, the only other antipyretics (drugs to reduce fever) came from the bark of the cinchona tree, along with the anti-malarial quinine. Supplies of the bark ran short in the 1880s, and the hunt was soon on for alternatives. This quickly led to the identification of the antipyretic phenacetin and then of paracetamol, though the latter's importance was not recognised for half a century. Not until the 1950s was

Below: The world's most-used medicine? Aspirin has controlled pain and fever for generations in many countries, but side-effects, such as bleeding in the stomach, have seen it often replaced by other treatments like paracetamol.

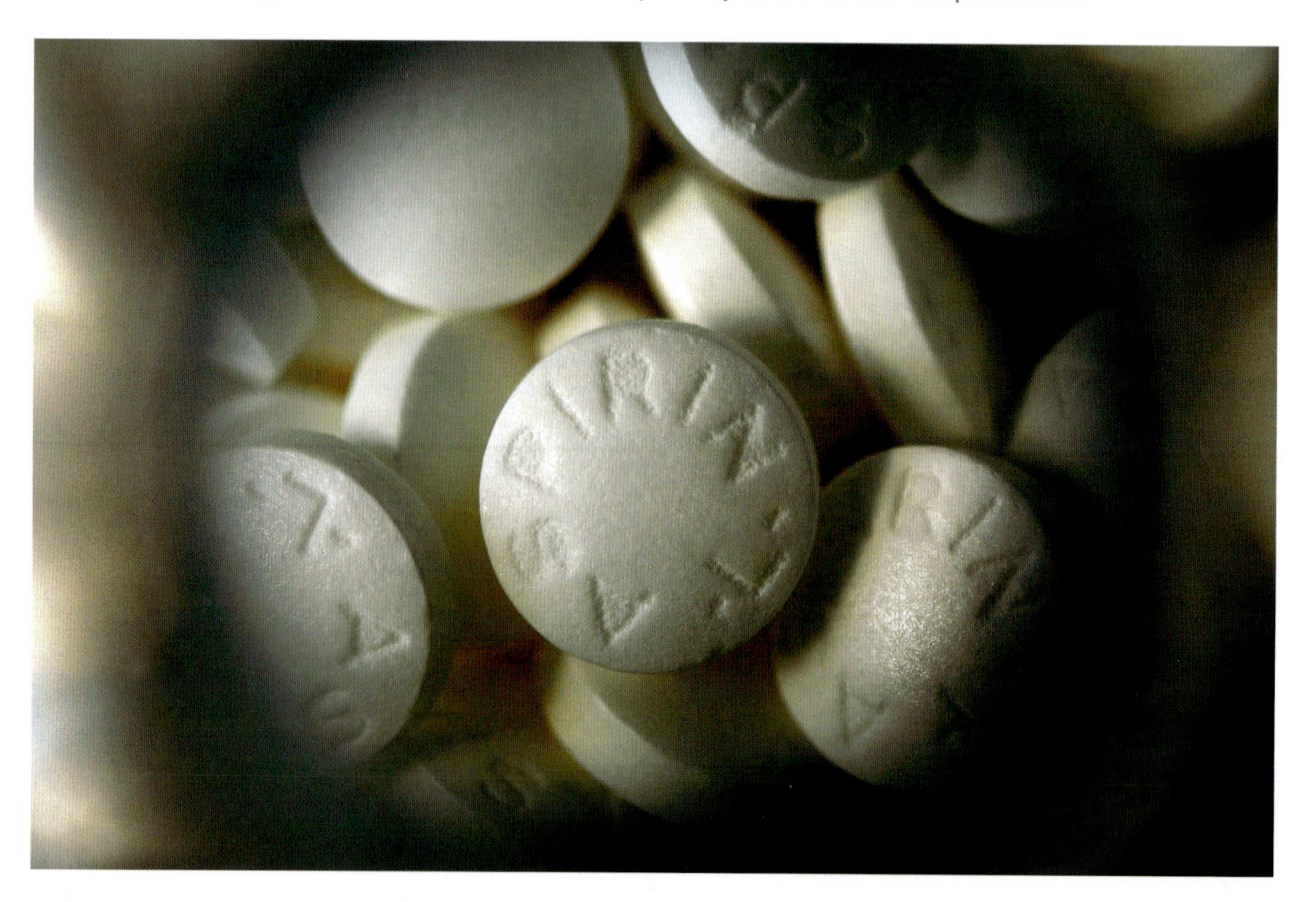

paracetamol on the market, widely known by the brand names Tylenol in the United States and Panadol in the United Kingdom, where it was at first available only on prescription.

Inventing chemical cures
Paul Ehrlich

When William Perkin found a beautiful purple dye in coal tar, previously merely a waste product from making coal gas and coke, he started a gold rush among chemists. They began to identify systematically all sorts of chemical compounds, many of them organic (containing carbon). As methods improved, thousands were extracted and named. Since these chemicals, or at least the possibility of them, had always existed, we cannot say most were invented. Innovation came through finding uses for them in industry or medicine, as dyes, drugs or synthetic materials, and through careful reworking of the compounds to maximise their effectiveness.

The idea that some of these new chemicals might help to control disease and promote health motivated Paul Ehrlich, a German doctor who spent his early career working with Robert Koch. By the early 20th century, Ehrlich was deeply involved in the search for chemicals that could be targeted against specific diseases or conditions without harmful side-effects—'magic bullets—as he called them. Hundreds of candidate chemicals were tested in search of those that worked; the hundreds chosen from thousands on the basis that they were constructed in a way that made them more likely to succeed. An early development was a drug to treat sleeping sickness. The parasites that caused this disease became known as trypanosomes and could be killed with the dye Trypan.

Ehrlich's public fame rests mostly on Salvarsan, the arsenic-containing chemical that killed the organism causing syphilis. The successful drug was number 606

> **Stranger than fiction**
>
> Ehrlich, who ate little, was sustained through these labours by a diet of 25 strong cigars every day; he always had a cigar box under his arm.

on his list of trials—though it had been initially judged ineffective—but number 914, Neosalvarsan, proved even better in use.

Ehrlich had to battle opposition to his new ideas and methods, but he persevered, benefiting public health and cementing his own reputation; he had already shared the 1908 Nobel Prize for Medicine. The chemotherapy he pioneered is a vast enterprise and industry nowadays. The invention and marketing of chemicals as pharmaceuticals for almost every known ailment, physical and psychological, is possible because we much better understand how the structure of the chemicals causes them to interact with the chemical workings of the body.

The first antibiotics
Alexander Fleming, Howard Florey and Ernest Chain

In the 1920s Scottish physician Alexander Fleming, working at St Mary's Hospital, London, was keen to find better ways to control infection in wounds, including surgical ones. He knew from experience in the trenches of the Great War that infection could kill more soldiers than bullets or shells. Thanks to Louis Pasteur, people already knew that bacteria caused infection and that some bacteria could be controlled chemically, but such control required cleanliness, hard to maintain on the battlefield or at accident sites.

Fleming's famous discovery occurred mostly by

chance. Returning to his laboratory from a holiday in 1928, he examined some unwashed glass dishes on which he had been growing the sometimes deadly bacteria *Staphylococcus aureus* ('golden staph'). He noted that a blue-green mould was growing on one dish, and that the growth medium on the dish was free of bacteria around each cluster of mould. Was something seeping from the mould that stopped the bacteria growing?

The mould had apparently come from soil in the garden, blown in through an open window. Fleming identified the mould as *Penicillium notatum*, so he called the bacteria-killing substance penicillin. There was scant evidence of its ultimate value. Fleming reported that it was unstable and difficult to purify. Fleming was a doctor, not a chemist and so not the man for the job. After a few years, the interest in penicillin petered out.

A decade later German-born chemist Ernest Chain, working in Oxford with Australian Howard Florey, came across a report of Fleming's discovery. The pair decided to find a way to mass-produce penicillin, a goal that required significant innovation. The results were startling. Penicillin cured bacterial infections in mice, then in human patients, and by the end of the war was saving the lives of seriously wounded soldiers. Soon half a ton of penicillin was being produced every month.

Available to the general public once the war ended, penicillin was the first of the 'antibiotics'; different moulds would later produce other drugs such as streptomycin. Together with the 'sulpha' drugs, also quite new, antibiotics proved effective against many infections. Ominously, strains of bacteria not killed by these 'wonder drugs' began to emerge within a few

Above: Alexander Fleming, discoverer of penicillin, holding one of the plates that revealed the presence the bacteria-fighting drug. The mass production of penicillin was stimulated a decade later by the outbreak of war.

years. Such drug resistance keeps ingenious chemists busy devising new antibiotics to this day. Often the bacteria seem to be winning.

Mustard gas; good from evil?
The start of cancer chemotherapy

Late in 1917 at Ypres in northern France, the technology of war entered a new phase with the first deployment of a poison gas specifically refined to serve as a weapon. Mustard gas—also called Yperite—was very nasty stuff. It was powerful—low concentrations were effective—and

Penicillin cured bacterial infections in mice, then in human patients, and by the end of the war was saving the lives of seriously wounded soldiers.

very persistent, remaining active in the soil or on a soldier's equipment for days or weeks. It had been known since the 1860s, though it does not occur in nature. Its use required the invention of equipment to manufacture and disperse it in large amounts. The other side was not slow to make use of the new weapon; it was a major factor in breaking through the Hindenburg Line in 1918.

Since the gas was absorbed directly through the skin, wearing a gas mask made little difference. Despite a faint if distinctive smell, soldiers were commonly unaware they had been exposed until symptoms developed a few hours later. The first organs affected were the eyes, which swelled up and closed over, then the corneas became ulcerated and died, so victims went blind. Large blisters appeared on the skin. Mustard gas, actually a liquid dispersed as an aerosol, was more incapacitating than life-threatening, since it was believed sick soldiers would more impede an army advance than dead ones. Only one per cent of exposures resulted in death, mostly from damaged breathing.

Despite international action to outlaw biological weapons, there have been consistent reports of mustard gas being used in conflicts since that time, such as by Japan against China during World War II and by the Iraqi government against Kurds in 1988. Despite this grim history, some unexpected good came from mustard gas.

After it had been discovered accidentally during the war that mustard gas could also lower white blood cell counts, an analogue (something that has the same chemical make-up) called mustine was developed, and was the first drug ever tried for cancer chemotherapy.

> **Stranger than fiction**
>
> When a stockpile of mustard gas maintained by US forces in Italy was bombed in 1943, civilians exposed to the gas experienced a noticeable drop in their white blood cell counts. Scientists began to think that the gas, or something like it, could treat lymphoma, a cancer in which white blood cells proliferate uncontrollably.

Combined with oestrogen it was also used treat prostate cancer. This extended the use of chemicals against disease first promoted by Paul Erlich.

Machines to replace organs
John Gibbon and William Kolff

Some people have defective kidneys, unable to cleanse their blood of wastes adequately, so threatening their health and even life. Others must manage without their hearts for a time, so they can be operated on. The triumphs of biomedical engineering include machines to mimic those organs: the dialysis machine which supports ailing kidneys and the heart-lung machine which is able to do the work of both for a time.

Dutch doctor William Kolff created the 'kidney machine'. In 1939 he filled a 50-centimetre tube of cellophane—actually an artificial sausage skin—with blood containing a lot of urea (the waste product from the breakdown of proteins). He then shook the sealed tube vigorously in saline (salt) solution. Much of the urea passed through the walls of the tube into the saline, leaving the blood greatly purified. In 1943, working under difficulties during the Nazi occupation, Kolff mechanised the process and made it continuous. A long tube around a drum rotating in saline drew blood from his patients' arteries and returned it to their veins. During the war, he treated 15 kidney patients in that way, though only one survived long-term.

Migrating to the United States at the end of the war, Kolff made many improvements; dialysis machines became smaller, cheaper, simpler to use. But the real impact did not come until the 1960s and the development of techniques to let patients be connected and disconnected quickly and without the need for surgery.

The 'cardiopulmonary bypass machine' was pioneered by American cardiac surgeon William Gibbon in 1937, working at the same time as Kolff and motivated by the death of a heart patient some years earlier. Gibbon argued that bypassing the heart would allow time for careful heart surgery. His experimental devices pumped blood around the body and also oxygenated it, so duplicating the operation of the lungs. Doing both greatly simplified the procedure. However, he got little support for the idea.

After early experiments on cats and dogs, Gibbon gained financial support from IBM Chairman Thomas Watson, improving the machine to the point where human use was safe, for example by reducing damage it did to the blood. In the first-ever 'open-heart' operation in 1953, he closed a hole in the heart of 18-year-old Cecilia Bovalek. During the operation the machine replaced her own heart and lungs for 26 minutes. In 1967 South African surgeon Christian Barnard used a similar machine during the first heart transplant.

Keeping the heart on pace
Wilson Greatbatch

We rely on our hearts; if they stop for long enough, we die. Yet the heart is not complicated, being essentially a muscle which serves as a pump. Activated by nerve impulses, it sends a pulse of blood into our arteries every second or so. Some heart problems can be traced to a faulty heartbeat. Nervous impulses, generated in

Chapter 19: Medical matters   241

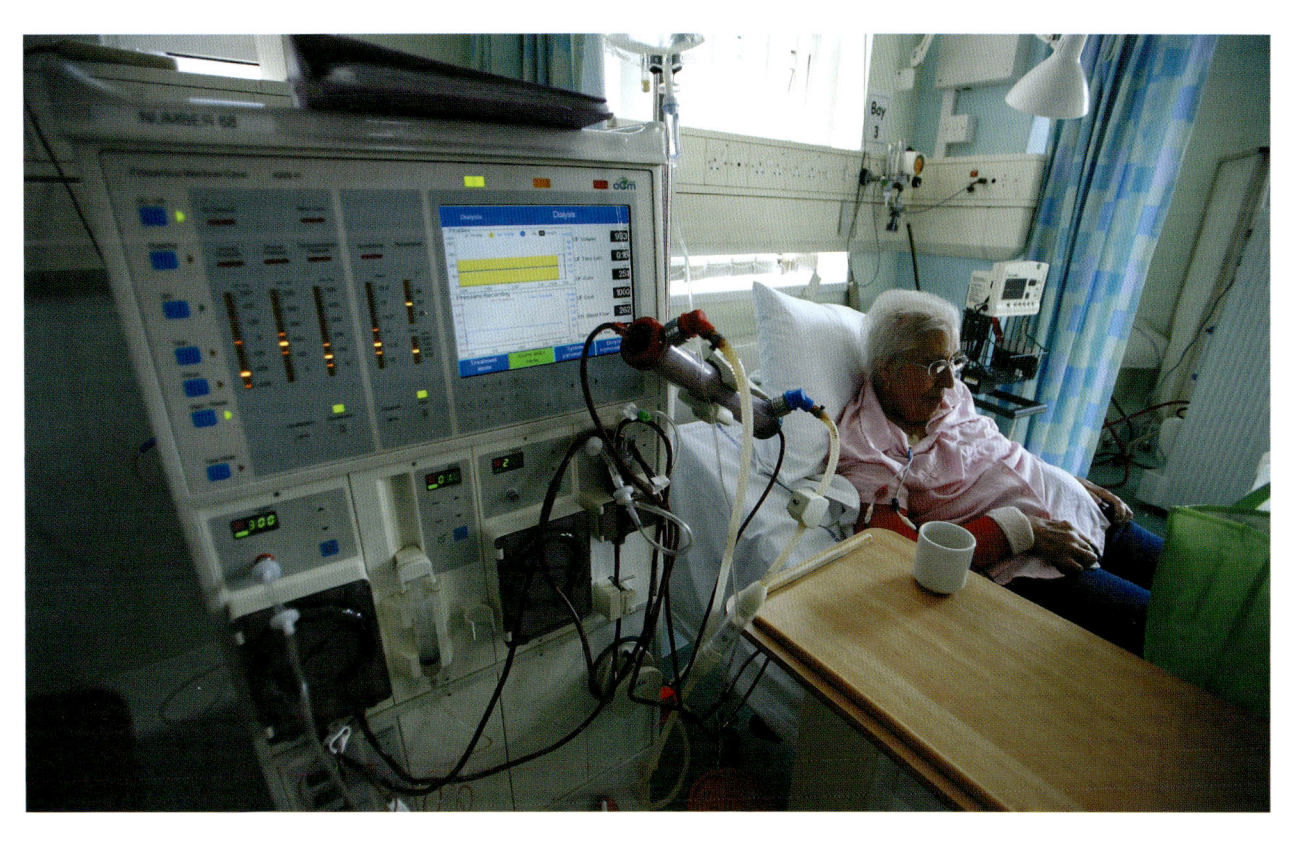

Above: Patients, like this woman with life-threatening kidney disease, now enjoy much better quality of life, thanks to technology. A dialysis machine filters her blood, but the process takes several hours every week.

heart tissue called the pacemaker, fail to drive the heart in a regular rhythm. The engineering solution was the artificial pacemaker, able to send regular pulses of electric current into the natural one if it is failing.

The idea of using electricity to shock the heart muscle back into regular action is an old one, with 19th-century physicians trying to induce the necessary currents from outside the body. The first to try it internally was probably Canadian engineer John Hopps around 1941. Cables were run through the chest wall into the heart tissue, but the pulse had to be generated from outside by equipment the size of a modern television, and the shocks delivered were painful.

The implantable artificial pacemaker, which is now carried by millions, was invented by American bio-engineer Wilson Greatbatch around 1958. There was some serendipity in his invention. While building a device to record heart sounds, he stumbled upon a circuit that produced regular pulses like heart beats. It was small enough to fit inside the patient's chest. Having generated interest among heart surgeons, Greatbatch began handcrafting pacemakers in his barn using his own funds, later quitting his job to concentrate on the device.

The success story

In 1983 the American National Society of Professional Engineers rated the pacemaker as one of the ten most important engineering contributions to human society in the previous 50 years.

The success story

Few, if any, chemical inventions have had the social impact of the Pill. Some commentators place it alongside fire and the wheel in importance in human history. By disconnecting sex and reproduction, it overturned long-standing mores. By giving women control over reproduction, and reducing the burden of unwanted children, it meshed with the women's rights and feminist movements in the 1960s and beyond. Relationships between the sexes were changed forever. There have been downsides. Sexually-transmitted diseases increased, as more people, especially younger ones, now had multiple partners and used condoms less often, relying on the Pill to prevent pregnancy.

Once the pacemaker was in place, the user was soon unaware of its operation. Around 1970 Greatbatch matched his new device with a small, lightweight lithium battery to power it, similar to the batteries now widely used in mobile phones and laptop computers. With circuits driven by transistors, the pacemaker drew so little current that batteries lasted for years.

Modern pacemakers now have the added advantage of microprocessor control, which allows them to respond more flexibly to changes in the heartbeat, including detecting and dealing with fibrillation (uncoordinated quivering) or tachycardia (when the heart beats too quickly).

The power of the Pill
Russell Marker and Gregory Pincus

Birth-control is an ancient human need, for example to limit population growth in hard times. Solutions have included abstinence and interrupted intercourse, intra-uterine devices, 'tube-tying', condoms, even abortion. At one time women in Europe soaked woollen tampons in vinegar or lemon juice to increase the acidity of the vagina and so kill sperm.

A more scientific approach required understanding the chemistry of female reproduction and began in the late 1930s with studies in rabbits. We soon knew that doses of hormones like progesterone at the right time could block the release of eggs by the ovaries, preventing conception. The first challenge was getting enough of the hormones for widespread use. In 1947 American chemist Russell Marker began extracting large amounts of progesterone from a wild Mexican yam, but there was another immediate problem. The natural hormone could not be swallowed, since digestion destroyed it. To use it for human birth-control required several painful injections daily.

Forms of progesterone that could be taken by mouth needed further chemical manipulation. Chemists at the Chicago firm of GD Searle and others came up with variants like norethindrone and norethynodrel. These would find use in early forms of the contraceptive pill, though at the time their discoverers had other uses in mind. American doctor Gregory Pincus made the connection to contraception in the early 1950s, encouraged by birth-control activist Margaret Sanger—who actually coined the term 'birth control'. Pincus first reported success in 1955. Following large-scale trials in Puerto Rico the next year, the first effective birth-control medication went on sale in America in 1960. Early high-dose pills raised health concerns, particularly for their effect on blood vessels, and lower doses of the hormones were soon found to be both safer and equally effective.

Making the magic bullets
Georges Kohler and Cesar Milstein

The development of monoclonal antibodies spans the sometimes-unclear boundary between discovery and invention, between science and technology. Based on a better understanding of how our bodies work, it also required innovative techniques if that knowledge was to improve human health. Monoclonal antibodies (MABs) give form to Paul Ehrlich's vision of 'magic bullets'. If chemicals could be found or created that selectively targeted a disease-carrying organism, something to kill that organism could be delivered at the same time, without damaging surrounding tissues.

Our bodies already make one form of such bullets. Particular white blood cells called B-cells manufacture antibodies when we are invaded by disease-causing bacteria or viruses. Certain antibodies—thousands of varieties are known—recognise cells that have turned cancerous, others can target our own healthy tissues, misjudging them as foreign and causing auto-immune diseases such as rheumatoid arthritis. Using antibodies deliberately for diagnosis and therapy requires their manufacture on demand, in large amounts, in the laboratory. Medical researchers Georges Kohler and Cesar Milstein, both working at Cambridge University, won a Nobel Prize in 1984 for inventing a way to do this.

The initial work used mice, and MABs are often still made that way, though since the 1980s, hybridomas can contain human genes as well. This reduces side-effects in human patients. Successful applications have been slow to emerge. When antibodies taken from mice are injected, the patient's immune system mobilises against them. This limits effectiveness. Nevertheless, dozens of MABs have been approved since the mid-1980s. They are used to act against transplanted tissue rejection and auto-immune diseases, by targeting the immune system cells that cause them; blood clots following heart surgery, by targeting platelets; and a range of cancers, by locking onto some identifier on the cancer cells and stirring the rest of the immune system into action. There is still a long way to go, however, to make the most out of MABs.

Monoclonal antibodies: how do they work?

They began in 1975 with B-cells that had become cancerous (a condition known as myeloma). Myeloma cells multiply uncontrollably, becoming essentially immortal. The researchers fused the myeloma cells, which had lost the power to secrete antibodies, with healthy B-cells that produced the antibody in which they were interested—each variety of B-cell produces only one pure antibody. The resulting hybridoma (a hybrid cell) divided continuously into a vast number of clones (exact copies), all making the same (monoclonal) antibody.

CHAPTER 20:
Looking within

We have sought to secure and maintain good health for countless generations, but only for the last 200 years or so have we had ways to look inside our bodies to see what might be amiss. Previously doctors had to rely on external signs, so often diagnosed wrongly. Now, from the stethoscope to the CAT scan, we have technologies that probe within, making our bodies all but transparent. We can identify and treat diseases and other conditions with greater ease and effectiveness, enhancing our health and lengthening our lives.

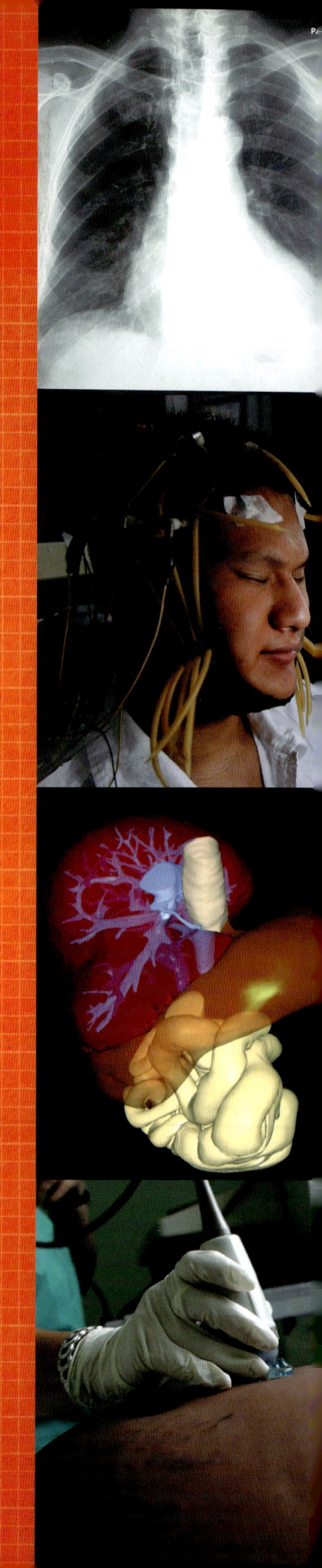

Noises in the chest
Rene Laennec

The name 'stethoscope' comes from Greek and means 'looking at the chest', but the organs of 'looking' are the ears. It was named by its inventor, French doctor René Laennec. Various myths surround the discovery: a former student reported that Laennec could not hear a patient's heart because street urchins were making a racket outside, and/or that the urchins were scraping on one end of a wooden fence and listening to the sound at the other end.

Laennec gave his version of events in an article in 1821. His patient that day was a young woman with a diseased heart, but he could not gain any information simply by tapping on her chest due to, as he put it, 'the great degree of fatness'. Because of her age and sex it would have been improper to put his ear directly to her chest.

Suddenly he recalled how the scratching of a pin on one end of a piece of wood could be heard clearly by placing an ear to the other end. Laennec did not use wood, but rather made a tube from twenty sheets of paper. He placed one end of the tube on the fat girl's chest, the other to his ear. 'J'entendre!' ('I hear!') he reputedly cried, as the sounds of her ailing heart reached him with unprecedented clarity.

Laennec treated patients with heart and lung diseases at the Necked Hospital in Paris, some with TB—then called 'consumption'. With his new device, he was soon able to develop new methods of diagnosis, 'a set of new signs of diseases of the chest, for the most part, certain, simple and prominent'. It was a major step forward in medical practice.

Laennec experimented with various materials, seeking the best transmission of sound. Surprisingly, solid materials like glass, metals and wood did not work as well as the column of air in his paper tube, though he quickly moved to wooden tubes 30 centimetres long, turned, reportedly, on his own lathe, with one end flared out into a bell. Some decades later the availability of flexible rubber tubes allowed the development of a 'two-eared' stethoscope, much as we have today.

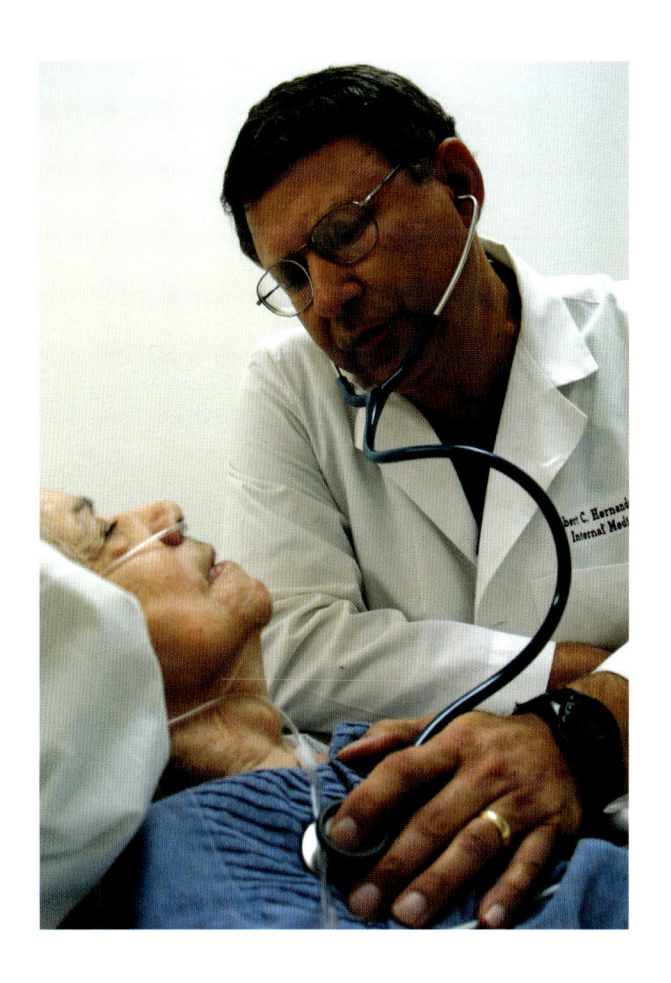

Laennec himself died of consumption aged only 45, a decade after his great invention and six weeks after returning to his native Brittany.

> **He placed one end of the tube on the fat girl's chest, the other to his ear. 'J'entendre!' ('I hear!') he reputedly cried.**

The X-factor
Wilhelm Roentgen

Late in the 19th century, German Wilhelm Roentgen, like many other physicists, was interested in the newly discovered 'cathode rays'. These were generated when high voltage electricity was passed between two electrodes inside a glass tube almost empty of air. It would soon be shown that cathode rays consisted of tiny fragments of electric charge—later called electrons. The phenomenon would give rise in due course to television and other uses of the 'cathode ray tube'.

Various researchers had noted, to their annoyance, that photographic plates stored near such 'discharge tubes' were inexplicably fogged, as if they had been exposed to light. Roentgen wondered if discharge tubes gave off invisible radiation that could leak through the packaging around the photographic plates. The walls of the tubes certainly glowed under the impact of the cathode rays.

One November evening in 1895, he found this was so. The radiation passed easily through a light-proof shroud around the discharge tube, and made a plate covered with a fluorescent chemical glow several metres away. For something so unexpected and mysterious, the term 'X-rays' seemed appropriate. They were also called Roentgen rays for a time.

X-rays were stopped by a thin sheet of metal, but could pass through human tissue and emerge with enough energy to leave an image on a photographic plate. The first 'X-ray photograph' ever taken was of Roentgen's wife's hand, clearly showing her bones and a ring she wearing, surrounded by the shadow of her flesh.

Roentgen did not 'invent' X-rays; they had always existed. But he did show how they might be used. X-ray images have proved of incomparable value to medicine, surgery and dentistry, as well as engineering

The success story

Roentgen became, justly, one of the most famous scientists of his age, winning the first Nobel Prize for Physics ever awarded. He remained strikingly reticent and modest, preferring to work alone, building his equipment with his own hands. A lover of nature and a daring mountaineer, he was finally taken by bowel cancer at age 78.

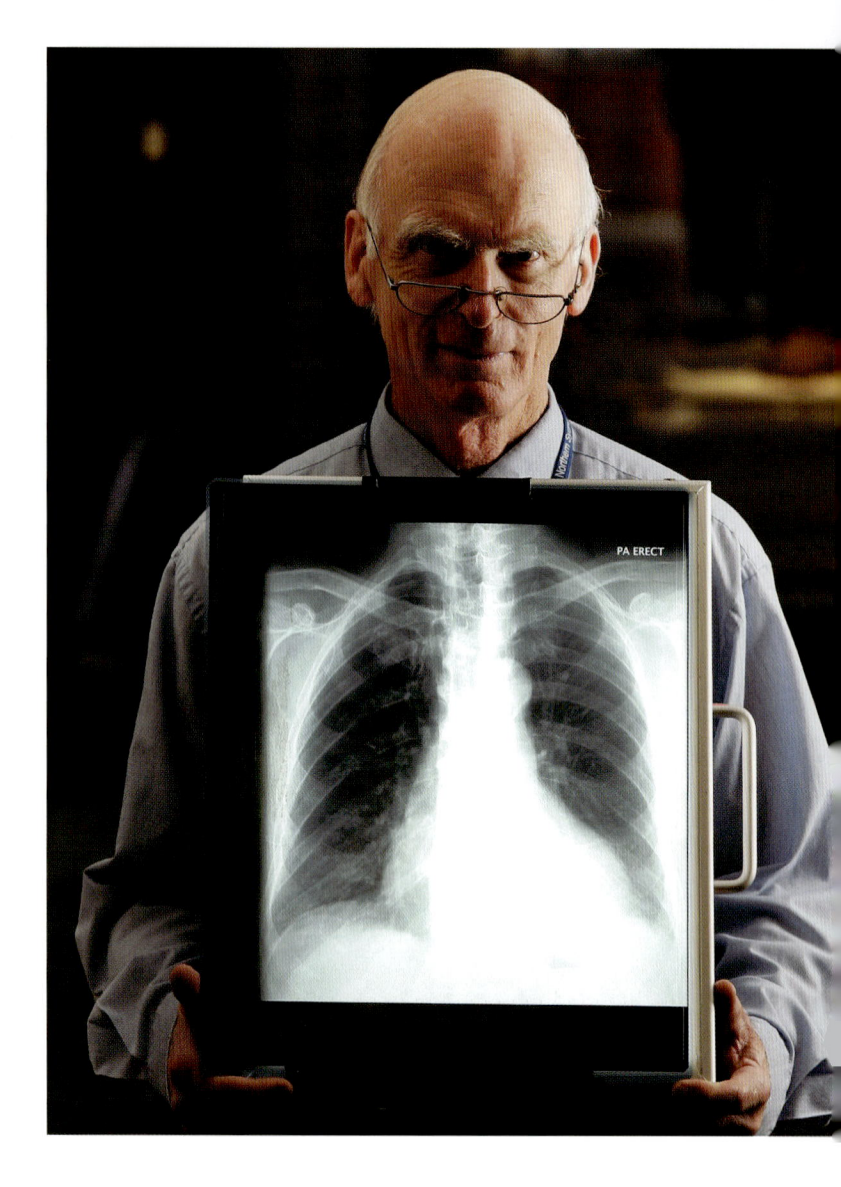

Opposite: Inner workings. An X-ray of the human chest clearly reveals the state of the lungs and the location of the heart. In most cases, the value of such knowledge far outweighs the risks associated it with obtaining it.

and industry. More powerful and precise X-ray tubes were soon invented, particularly by the American William Coolidge from General Electric. The hazards of the technology, such as its capacity to cause cancer, were discovered only later.

Electricity, electronics and medicine
William Einthoven and Hans Berger

From his discovery of 'animal electricity' in the 1790s, Italian anatomist Luigi Galvani may claim some part in the development of two key medical devices in the early 20th century: the electrocardiograph (ECG) and the electroencephalograph (EEG). Both use electrical currents associated with nerve impulses to analyse the workings of our bodies.

From the mid-19th century, we knew that the heart contained electric currents varying in synchrony with the pulse, and that these could be detected with a sensitive meter. In 1903 Javanese-born Dutch engineer William Einthoven invented the highly sensitive 'string galvanometer', which could detect and record on a paper chart electrical currents picked up at various points on the patient's chest. Later models used the new electronic amplifiers to further increase sensitivity.

Einthoven identified half a dozen distinct patterns of currents, some associated with a normally beating

Below: I can read your mind. In an electroencephalograph (EEC), a number of electrodes on the patient's head quickly reveal the pattern of tiny electric currents that reflect the functioning of the brain.

heart, others indicating abnormality and disease. His first machines weighed half a tonne and needed five operators, but a modern electrocardiograph (the name combines 'electricity', 'heart' and 'writing') fits easily on a tabletop. It may be combined with an oscilloscope to display the heart's electrical impulses on a screen, such as we find in every emergency ward. For his invention, Einthoven won the Nobel Prize for Medicine in 1924.

The human brain, centrepiece of the nervous system, also pulsates with detectable electrical currents. Here the pioneer was German physician Hans Berger. He had first tried to examine the workings of the brain by studying its blood flow, but in the 1920s turned to recording the minute electrical currents detected with electrodes attached to the skull. His first electroencephalogram ('electricity', 'head' and 'writing') was of his own son in 1924. Studies by Berger and others soon revealed a variety of brainwaves (patterns of currents), such as the ten-times-a-second alpha waves associated with relaxation.

Berger's invention of the EEG would revolutionise neuro-medicine, as the ECG had done for the study of heart disease. Strict, reserved and increasingly isolated, Berger worked on his researches in his spare time and in secret, and gained recognition only in 1937. Despair at the rise of Nazism drove him to hang himself in 1941.

Below: Ultrasound, the technique of probing inside the human body with high-frequency sound waves, is routinely used to check the health of the unborn, but other organs may be similarly scanned, the bladder for example.

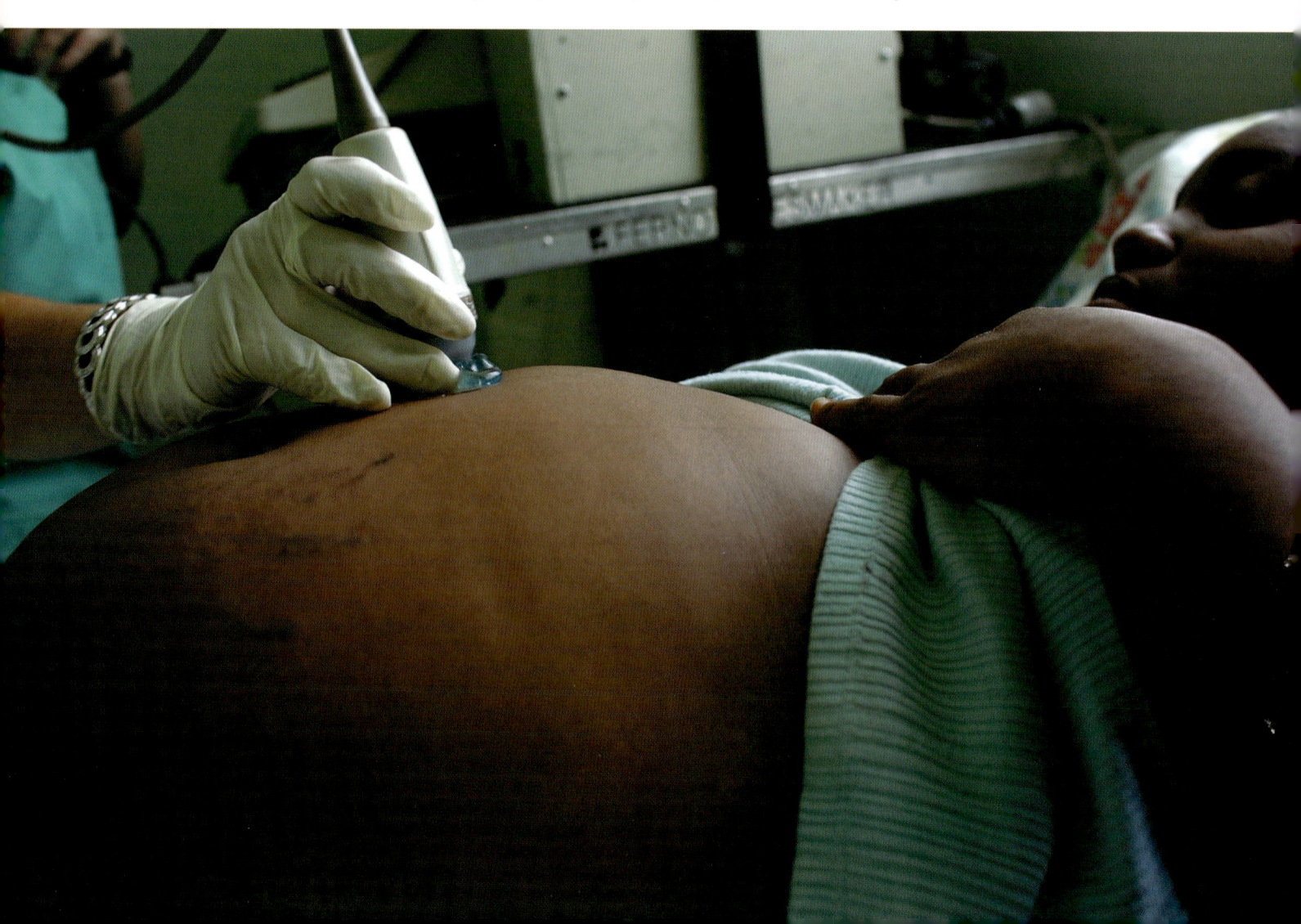

Chapter 20: Looking within

A multitude of uses

Ultrasound waves have applications in nature and in industry. Bats use them to navigate in the dark. Artificially generated by using electric currents to make crystals of quartz or other minerals vibrate rapidly, ultrasound can find hidden defects in metal structures. It is also the basis of sonar, which can detect underwater objects such as submarines or reefs, by the soundwaves they reflect.

Scanning the womb with sound
Ian Donald

'Ultrasound' refers to soundwaves of 20,000 cycles per second (hertz) or more, too high a frequency for us to hear. In medicine ultrasound was first applied as therapy, using the capacity of the waves to heat tissues or to break up structures like gallstones. Then a few researchers started to measure how soundwaves passed through regions like the skull. By collecting the reflected soundwaves, which were bent or absorbed by soft tissues including tumours, they slowly developed a capacity to make crude images.

Ultrasound has had a profound influence in obstetrics and gynaecology, particularly in imaging and monitoring the child in utero (in the womb). Here the key figure was Scotsman Ian Donald, newly appointed to the Chair of Midwifery in Glasgow in the 1950s. His inspiration came from several directions: war service, which taught him something about radar and sonar; and a long-standing interest in gadgets and machines. A lecture in London given by one of the other ultrasound pioneers quickly alerted him to what ultrasound scanning could do in obstetrics and gynaecology. His first investigations imaged anatomical specimens such as fibroids, cysts and tumours using an ultrasound metal-flaw detector at a shipyard—the visit had been made possible by the grateful husband of a patient.

Early trials on patients were disappointing, the work greeted by colleagues with a mixture of scepticism and ridicule. He turned the corner when his primitive scanner found a large operable ovarian cyst in a woman previously diagnosed with inoperable cancer. Attitudes quickly changed. In 1958 Donald was able to announce his findings in *The Lancet* under the title 'Investigations of Abdominal Masses using Pulsed Ultrasound'. From then there was no turning back. Ultrasound in obstetrics, gynaecology and elsewhere in medicine soon became mainstream.

Donald was strongly opposed to abortion on other than strictly medical grounds. Showing a mother images of her unborn child was one way he sought to discourage termination of the pregnancy.

Slices of information
CAT scans and MRIs

In modern science and technology, significant advances are often the work of more than one mind and heart. Two inventions in the 1970s and early 1980s show this. One of them built upon an existing

His first investigations imaged anatomical specimens such as fibroids, cysts and tumours using an ultrasound metal-flaw detector at a shipyard.

Left: A high resolution three-dimensional (3D) image of a human liver, constructed from computer tomography (CT) scans. Images like this allow surgeons to plan complex surgical procedures and visualise potential complications beforehand.

The resulting 3D image is usually presented as a series of slices across the body, from top to toe if need be, each slice being about a centimetre thick. The images are also much sharper and more detailed than traditional X-rays, able, for example, to pinpoint small tumours so they can be precisely attacked with radiotherapy. The CAT scan, as it became known, was first in the news in 1972, when Hounsfields's machine located a tumour in the brain of a living, conscious patient.

CAT stands for computerised (axial) tomography. A tomogram is an image of the internal structure of some object. While Hounsfield had the first real success with a CAT scan, others had already been active. These included South African-born American physicist Allan Cormack, who was to share the Nobel Prize with Hounsfield in 1979, and American-born William Oldendorf who did not, despite writing many papers to promote the idea and securing the first patents.

Magnetic resonance imaging (MRI) works on very different principles. It was not initially used in medicine, but to study various chemical compounds. Its 1930s discovery won the Nobel Prize for Physics. Instead of

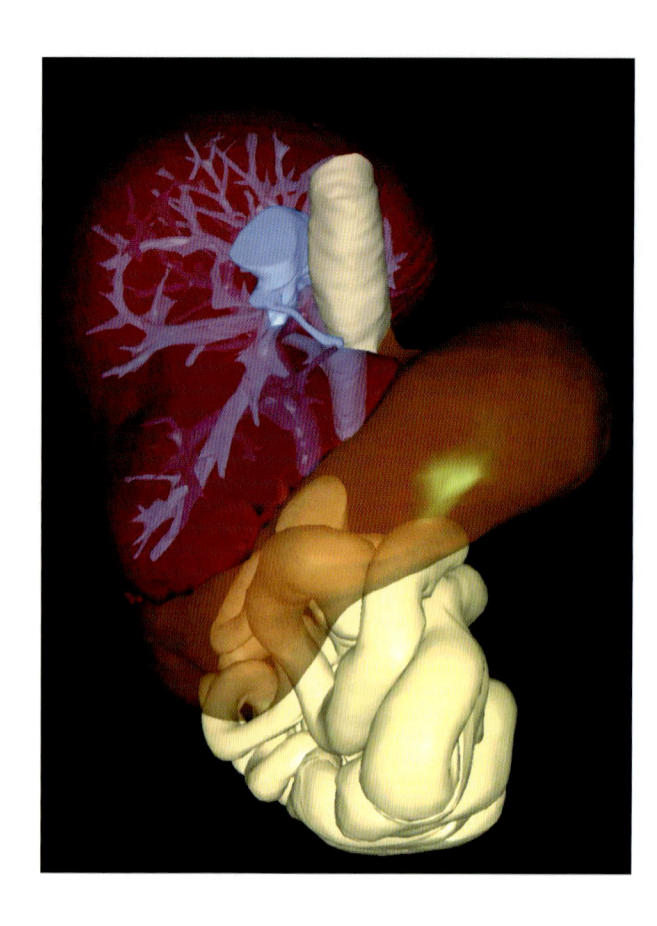

technology, the other found medical uses for a recent scientific discovery. Together, CAT scans and MRI scans have transformed much of medical diagnosis and therapy, and proved especially valuable in working out what is going on a patient's brain and nervous system. Both depended greatly on the development of high-speed electronic computers; both won Nobel Prizes for their inventors.

The discovery and rapid application of X-rays late in the 19th century brought immense benefits, such as the ability to image broken bones. Those were compounded 80 years later when British engineer Geoffrey Hounsfield first used the rapidly evolving computer technology to integrate a number of X-ray views of the body, initially taken one after the other from different angles. Hounsfield replaced the usual X-ray film with crystal detectors and electronic amplifiers.

Stranger than fiction

You can argue that British pop group The Beatles played a major role in the development of CAT scans. Their music publishing firm was EMI. Hounsfield also worked for EMI. The massive profits from The Beatles' records allowed EMI to fund the research needed to perfect the CAT scanner.

MRI: how does it work?

An MRI scanner generates a powerful magnetic field. This field flips the nuclei of water molecules (atoms of hydrogen) into new orientations. They give off distinctive radio signals when they 'relax' back to their old state. The human body is two-thirds water. The organs and soft tissues have different water contents, which make them show up distinctively on the scan. Bones, which contain little water, show up black. The location and density of water content can reveal diseases in organs.

sending beams of potentially damaging radiation from the outside, MRI effectively induces the body to scan itself from within.

To make all this happen, the patient must be subjected to a powerful magnetic field, more than 10,000 times as strong as that of the earth. Generally, however, this does not cause any harm, except, for example, to people with artificial pacemakers. MRI has spread rapidly since scanners went onto the market in the early 1980s. Despite their high cost, more than 20,000 of these expensive machines are now in use and more than 60 million scans have been completed.

It took much longer for MRI to be recognised with a Nobel Prize; more than twenty years, as opposed to the ten it took for the CAT scan. That was perhaps at least partly due to ongoing controversy as to who the actual inventor was. Some claim that American doctor Raymond Damadian should have shared some of the prize. His early work showed that cancerous tissues and normal tissues respond differently to the MRI technique. He also secured the first patents, and built a whole-body MRI scanner in 1977. Early histories of MRI attributed many of the early milestones to him, but later assessments claimed his technique had not been proven to reveal cancers.

When the prize was finally awarded in 2003, it went elsewhere, to American Paul Lauterbur and Briton Peter Mansfield; they had found how to process the signals with a computer to create 2D and 3D images, something which Damadian had not done. Yet many in the field, including the man himself, thought he had played a significant role in the creation of MRI scans, and ought to have been acknowledged.

Right: The remarkable technology called MRI makes the body all but transparent. Here, the details of a patient's brain are revealed as if the skull had been cut through. Any part of the body can be similarly sliced, bloodlessly.

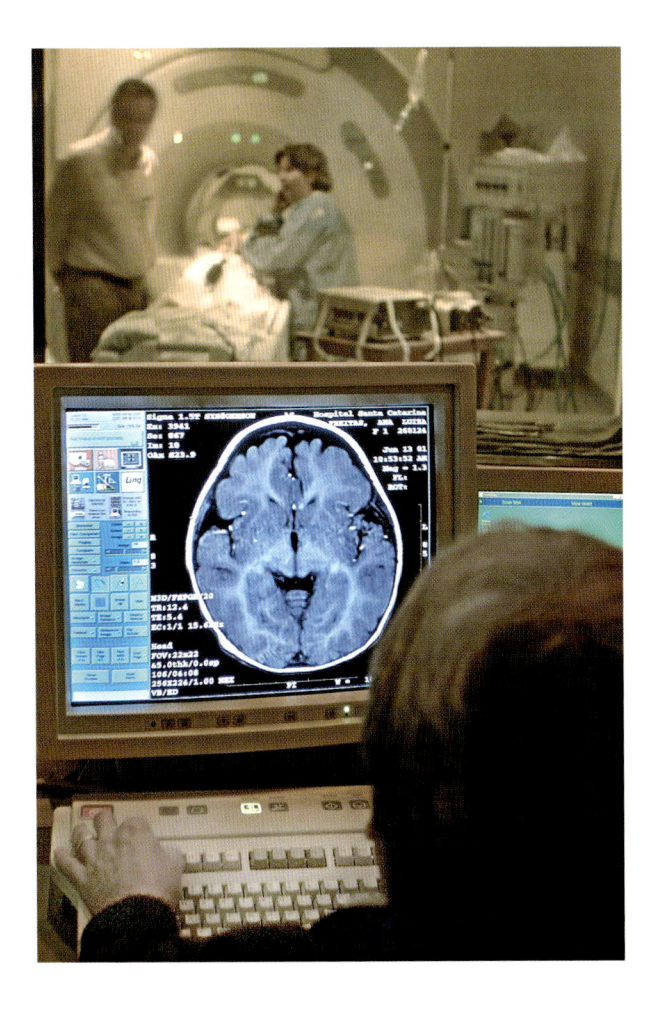

252

CHAPTER 21:
Lifestyle

Not all inventions are massive devices like jumbo jets and power stations and atom bombs, affecting hundreds or thousands of people at once. Many are small, even discreet, and impact on us individually, one by one, as we make use of them. Yet most of us could not imagine life without them. Of the hundreds of such technologies you could chose from, here are just a few, and the stories behind them.

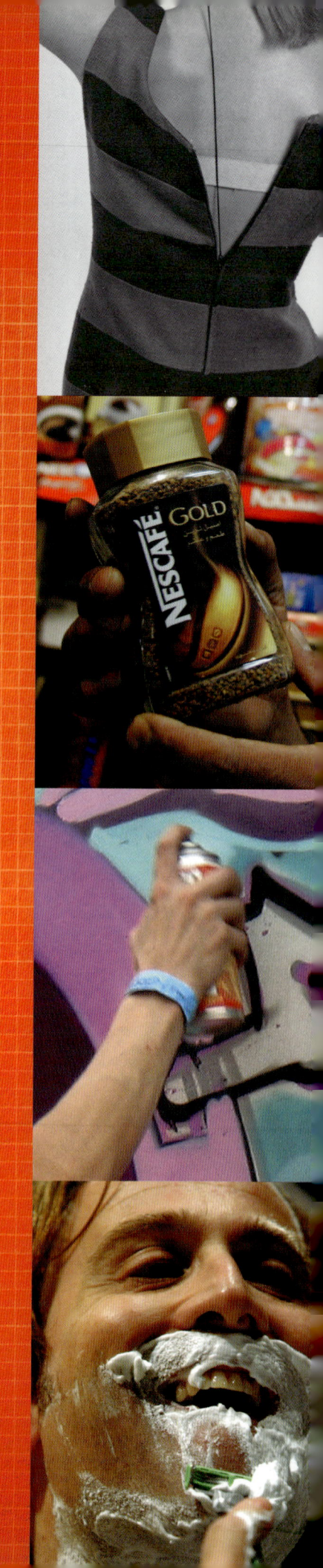

The bubbles in the wine
Dom Perignon

When the French monk and cellar master Dom Perignon first tasted his new brew, a wine with deliberately-induced bubbles, he is reputed to have so enjoyed the experience that he cried out to his fellow monks, 'I have been drinking the stars.'

Not quite so. Initially the good monk had been concerned with keeping the bubbles out of the wine, since they were unwanted and surely a sign of poor winemaking practice, at least in France. In England fizzy wines—such as 'perry' made from pears in the West Country—were popular. Here they often added molasses or sugar, particularly to wines imported from the Champagne region of France, to restart fermentation and fill the liquid with fresh bubbles of carbon dioxide. In the initial fermentation, in which yeast acts on sugar to produce carbon dioxide and alcohol, the gas escapes.

The English had two technological advantages in dealing with 'sparkling' wines: they sealed their wine bottles with cork, which trapped the gas inside—the French were still using wooden stoppers wrapped in hemp; and they used stronger glass in their bottles, so they were less likely to burst.

It was an accident of climate that first caused bubbles to show up naturally in wine from Champagne, where Dom Perignon's abbey vineyard of Hautvilliers lay. For some decades, the weather had turned very cold, shortening the growing season and ending fermentation early. In the spring the warmer weather caused fermentation to start again in the bottled wine, so producing the bubbles.

Initially thought undrinkable by the nobility, 'champagne' later became very popular in royal courts, especially in London. Dom Perignon, who had been sent into the region to try and save it from the plague of bubbles, now needed to make more bubbles, not fewer. By trial and error, he developed an early version

Above: The first name in champagne. A bottle of Dom Perignon and a phalanx of waiting glasses. The old monk did not know what he had started when he first tried to take the bubbles OUT of the wine.

of the *méthode champenoise*, putting extra sugar, and later extra yeast, to the bottled wine. He started using corks and stronger bottles, blending wine from various parts of his vineyard to achieve a more consistent quality. He was also made white wine from red grapes, which no-one had done before. Soon everyone was copying his methods.

And what is the appeal of bubbly wine? It seems that the bursting of the bubbles enhances the release of flavours from the wine, making it a treat for the nose as well as the tongue. Thank you, Dom Perignon.

Two pieces of paper
George McGill, Johann Vaaler and Walter Hunt

How do you hold two pieces of paper together so that they won't separate unless you want them to? This simple question, a vital one in our paper-driven age, has produced many answers: passing string or ribbon through holes or slots cut in the page—time-consuming; or pushing a straight pin again and again back and forth through the leaves—not very secure. Elaborations included brass fasteners with two legs passed through a hole and spread out on the other side.

Two more recent responses are the staple and the paperclip—simple, even obvious, but needing to be invented. American inventor George McGill—already the inventor of the brass fastener and its associated hole-punch—devised the first around 1870, though others probably had a similar idea. A U-shaped piece of fine wire was pushed through the pages and the free ends flattened back to hold them together. Now very familiar, then very new.

Early staples were fed into the machine one by one or formed from a long piece of wire. Only around 1895 did we first see the now familiar strip of staples connected together, struck off individually and punched in as needed. 'Stapling'—the term was not used till around 1900—became popular, particularly as a way of holding the pages of a magazine together along the spine, and demolished most of its competitors.

Coming close behind was the paperclip, which did not make holes in the paper and was very easy to remove. Norwegian inventor Johann Vaaler usually gets the credit; certainly he secured the first patent. It is not Vaaler's 1899 design we recognise now, rather the double oval 'gem' model, devised but never patented by an English firm, also in 1899. The extra loop prevents the steel point scraping across the page.

You could use a safety pin to fasten pages, though it usually holds other things. The 4000-year-old straight pin has well-known problems: it slips out and it can pick your fingers—or the baby. The answer probably came in stages. Someone bent a pin double so that its head and point could be latched together. Then in 1849 prolific American inventor Walter Hunt wound the pin into a spiral spring, exerting pressure on the latch to hold it closed. Hunt reportedly sold his newly patented idea, perfected in three hours of wire bending, to a friend for $400 to settle a small debt.

Seated on the throne
John Harrington and Alexander Cummings

Nothing perhaps better represents an advanced civilisation than the flush toilet. Disposing of human body wastes, and even controlling the smell, has been a challenge since forever, all the more since cities grew more crowded, and evidence was found that contact with excrement could spread disease.

Many people credit 19th-century British plumber Thomas Crapper with the invention of the loo as we know it today, not least because of the sound of his name, which is really just a coincidence. But key innovations lie much further back. Some archaeologists claim that remains of flush toilets can be found in the ruins of ancient Crete and the cities in the Indus Valley, thousands of years old.

By Tudor times in England, all that refinement had vanished; 'privies', where people might do their business into a bucket, were not nice places to be. The most up-to-date equipment, popular with the rich and powerful, was the 'close-stool', simply a comfortable padded seat above a metal or porcelain container, which of course still had to be emptied.

Some archaeologists claim that remains of flush toilets can be found in the ruins of ancient Crete and the cities in the Indus Valley, thousands of years old.

The man who moved us on was Sir John Harrington, courtier to Queen Elizabeth I and often in trouble for telling risqué jokes. He is reputed to have invented the 'Ajax' flush toilet for the Queen in 1596. This 'water closet' did not need emptying. Rushing water from a cistern set above the unit washed excrement and urine into a trap beneath the bowl, then via a drain to some convenient repository, like the castle moat or a nearby stream—that issue would not be addressed for several centuries yet. The invention did not really catch on.

Various designs for valves to separate the bowl from the trap and drain followed, but the key innovation, patented by Englishman Alexander Cummings in 1775, was the 'S bend', brilliantly simple. This curved tube kept an air and water seal between the bowl and the sewer so preventing any backflow or nasty smell. That's still what we use today.

And what of Thomas Crapper? It seems he did not really invent anything, but rather manufactured S-bend type bowls to someone else's design. He did, however, improve the operation of the cistern and made the units so well he received a royal warrant to provide toilets for the monarchy. Fame, and the use of his name as a familiar designation for a WC, followed inevitably.

The search for a safer match
John Walker and the Lundstrom brothers

Traditional ways to start a fire, always a priority, have included rubbing one piece of wood against another to make heat through friction, and striking a piece of iron on flint to liberate sparks. Now we use matches.

A Frenchman called Chancel developed a 'match' in 1805, with the head a mixture of sulphur, potassium chlorate, sugar and rubber. To be set the match alight, you dipped it into a small bottle of sulphuric acid—not a safe thing to carry. In the 'Prometheus' match, on sale in England from 1830, a tiny glass vial of acid on each match was crushed with pliers to set the match alight; again, not very convenient and, like Chancel's match, it did not catch on.

A more recognisable 'friction match', ignited by striking against a surface, was the brainchild of English chemist John Walker in 1827. Patented by Samuel Jones—Walker did not patent the idea though urged to do so—the matches were called 'lucifers' ('light-bearers'), a term used for many decades even when the technology changed. The smell of burning was unpleasant, like fireworks, and ignition could be violent or even explosive, but the matches were popular and apparently increased the number of tobacco smokers, previously denied a convenient 'light'.

Frenchman Charles Sauria added white phosphorus to the mixture on the head in 1831 to remove the smell. The matches, called 'congreaves' after a popular military rocket and marketed in a sealed box, were

The safety match: how does it work?

In the matches sold by the Lundstrom brothers 'safety' had two meanings: much safer red phosphorus replaced the white version; and the various chemicals were distributed between the head of the match (antimony sulphide and potassium chlorate) and the striking surface (a mix of powdered glass and red phosphorus). Heat from friction generated by striking turns a small amount of red phosphorus into white. This immediately ignites and sets the head burning. This is basically the match we use today.

cheap and sold well, but workers making them were exposed to phosphorus fumes and developed serious bone deformities, including 'phossy jaw'. Users were at risk too; white phosphorus is poisonous. Ultimately, white phosphorus in matches was banned.

The modern 'safety match' came to light in 1844, later commercialised by the Swedish Lundstrom brothers. Combining the matches and the striking surface into a 'matchbook' dates from the 1890s in the United States.

Women escape the corset
Marie Tucek and Mary (Polly) Jacobs

In search of an appealing body shape, women for centuries forced themselves into uncomfortable, even dangerous, undergarments. The corset (from Old French for 'small body') was reputedly invented by the French Queen Catherine de Medici in the 1550s, to ensure her ladies in waiting had narrow waists. Rods of whalebone or even metal were sown into the garments, and the restraint enforced by back-lacing.

By the late 19th century, women were beginning to rebel against the ordeal of the corset, and inventive minds were devising alternatives. Any number of women, most of them French, helped to invent the brassiere; some took out patents. Marie Tucek designed one in 1893 that looked much like a modern bra, with separate pockets for the breasts and over-the-shoulder straps. The word *brassière* is Old French for 'upper arm', the breasts being supported from above rather than squeezed up from below, as in the corset. But Marie failed to market her invention successfully.

Across the Atlantic in 1910, New England socialite Mary (Polly) Jacobs sewed together two silk handkerchiefs and some ribbons to make a suitable undergarment for wearing beneath a sheer and revealing gown. She took out a patent in 1914 and started selling her invention under the business name

Stranger than fiction

In 1917 the US War Industries Board entreated women not to wear steel-reinforced corsets for the sake of the war effort. Women responded, reputedly liberating steel sufficient for two battleships.

of Caresse Crosby. Though soft and light to wear, the Jacobs' bra was more flattening than flattering—conveniently that proved to be the preferred style in the Jazz Age of the 1920s. No businesswoman, Jacobs soon sold out to a corset company called Warners, which was to make 1000 times the patent purchase price over coming decades.

World War I speeded the demise of the corset and the rise of the bra. With their menfolk at sea or at the Front, women were taking their places in factories and offices and needed comfortable undergarments.

Through the 1930s and 1940s, new designs for brassieres often emphasised uplift, as the corset had. Some major engineering minds were at work, including the aircraft builder Howard Hughes; his ingenuity emphasised the talents of actresses like Jane Russell. In 1928 Russian immigrant Ida Rosenthal of Maidenform defined cup sizes, so bras fitted better; specialised bras for nursing mothers came in the 1930s; the 1970s sports bra successfully combined restraint with comfort.

The bandaid and the tampon
Earle Dickson and Earle Haas

In 1921 Earle Dickson was a cotton buyer for the leading American firm Johnson and Johnson, which made bandages and medical adhesive tape among other things. Dickson's young wife Josephine liked to

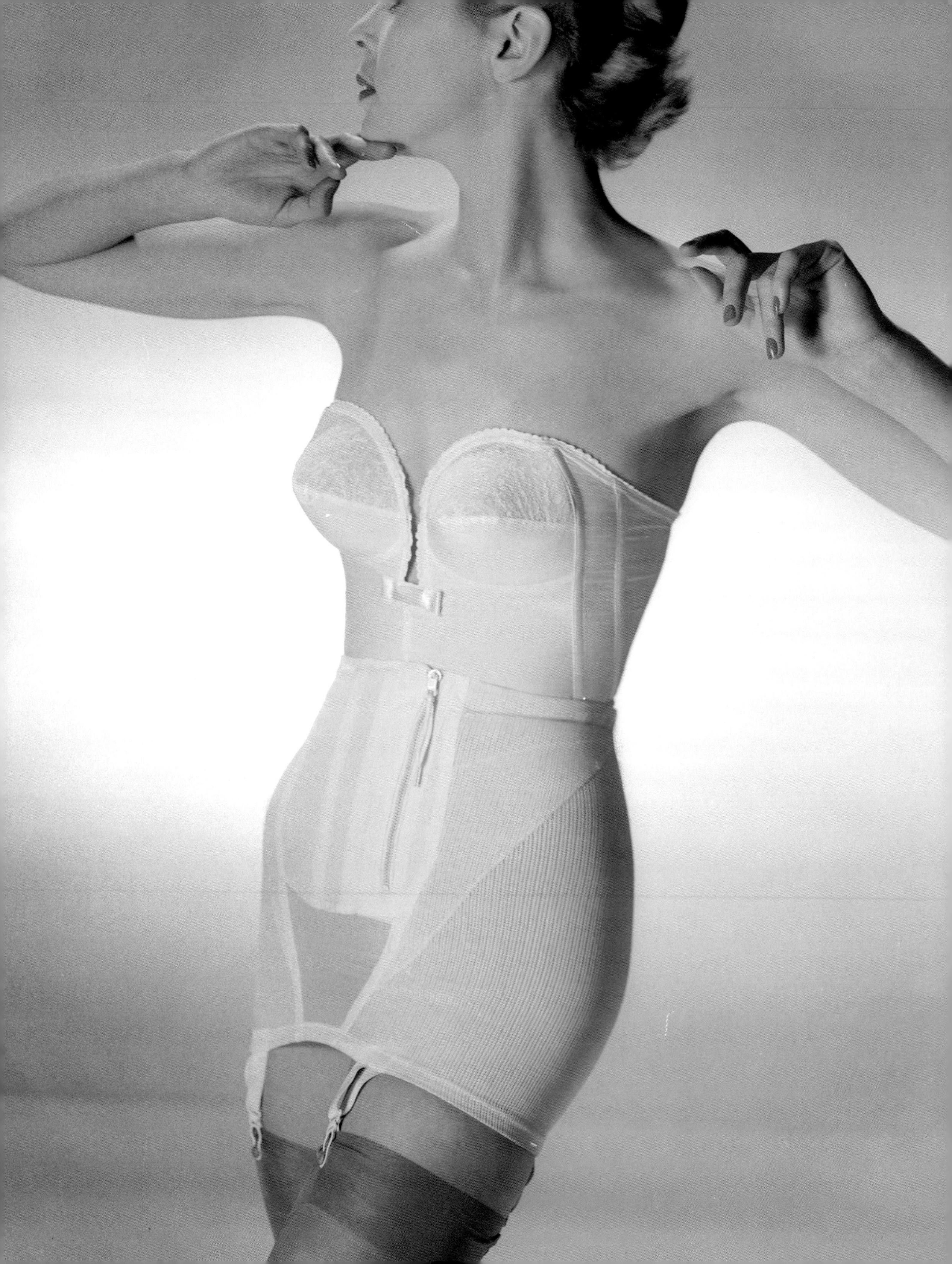

> **These stringent requirements can be met by an espresso machine, judged by some to be the centrepiece of civilisation.**

cook but was rather accident-prone, regularly cutting and burning her hands. Dressing the wounds was a two-stage process, with first a piece of gauze and then some tape to hold the gauze on place. Both had to be cut to size, and dressing wounds on your own hands was a trial. The dressings often came off. Like every innovator, Dickson wondered it there was a better way.

The task proved easy. He cut a long piece of tape and placed small pieces of gauze at intervals down the middle. He covered it with fine cloth (crinoline) to keep it sterile. All that his wife had to do was cut off a piece of the tape, pull off the crinoline and stick the dressing on. Dickson showed his invention to his boss, one of the Johnsons, who liked it and proposed marketing what would become J&J's most famous product. Dickson was promoted to vice president. Early sales of the hand-made dressings were slow until samples were given to the Boy Scouts to generate some publicity. Three years later, individually wrapped bandaids were being made by machine. From 1939 they were sold sterilised. Vinyl replaced cloth tape in 1958.

Unlike the bandaid, the tampon has a long history. Ancient Egyptians and Greeks used softened papyrus or lint wrapped around a little piece of wood. Later they were made of paper, wool or cotton, but only one at a time. The 20th-century approach to most things was mass production. Denver doctor Earle Haas, keen to find a replacement for the traditional 'rags' his wife and other menstruating women used, came up with a design and patented it in 1931. Basically a plug of compressed cotton sewn into a cotton bag, the tampon had an applicator for insertion, two cardboard tubes sliding one inside the other. No other tampon had that.

Seeking a manufacturer, Haas was turned down several times—including by Johnson and Johnson—so he sold his patent to a local entrepreneur, Gertrude Tenderich, an ambitious women recently migrated from Germany. She brought the Tampax tampon to the market in 1936, running up early batches on her home sewing machine.

Many ways to a good cup of coffee
Luigi Bezzera, Satori Kato, James Mason and Melitta Bentz

Coffee has been popular for centuries, originating in Ethiopia and spreading into the Middle East and beyond. Currently half a trillion cups are drunk around the planet each year. Discerning coffee-drinkers generally like it made on demand, tailored to their own tastes and produced quickly so they can get on with drinking. These stringent requirements can be met by an espresso machine, judged by some to be the centrepiece of civilisation.

The big name here was Italian Luigi Bezzera. In 1901 he patented a machine using steam pressure to force hot water through ground coffee held in metal filters. The coffee was stronger and more quickly made than with earlier low-pressure machines, but results were variable. The operator had to judge the pressure and temperature of the water and the time it took to

Opposite: In the quest for the ideal shape, women have tried all sorts of strategies, from steel-stayed corsets to more comfortable, but no less effective, 'foundation garments' like these. Polly Jacobs can take a lot of the credit.

Instant coffee: what's that?

To make coffee granules, water has to be extracted from the concentrated coffee. This was initially achieved by heating, later by spraying, now freeze-drying under vacuum. Frozen concentrate is exposed to low pressure. The ice evaporates without turning to water, leaving the soluble coffee power behind.

run through the grounds, since these factors all affected strength and flavour. Water hot enough to generate the steam tended to damage the taste. Modern machines have largely automated the process, with better ways to develop the necessary pressure: spring-loaded pistons in 1948, electric pumps in 1960.

With no espresso machine at hand, a quick alternative is 'instant' coffee. Japanese-American chemist Satori Kato devised a process in 1901, but sales were minimal until 1938, when the Nestlé company marketed Nescafé, firstly in Switzerland. Instant coffee was a hit with American soldiers during World War II, and was found in the ration packs of every GI. The American government bought up most of the production for several years; civilians had to wait until peace broke out before they too could freely enjoy instant coffee.

American James Mason and German housewife Melitta Bentz were other coffee innovators. He invented the coffee percolator in 1865; she patented coffee filter paper in 1908. To avoid over-brewing and to remove the grounds, she fashioned a conical filter from her children's blotting paper, placed ground coffee in it, and poured in boiling water. The coffee brewed as the water ran through. She and her husband Hugo were soon manufacturing filter papers under the still-familiar Melitta brand.

Done and undone with speed
Whitcomb Judson, Gideon Sundback and Georges de Mestral

For centuries ancient techniques have held our clothes together and on our bodies—buttons and button holes, hooks and eyes, laces. Chicago engineer Whitcomb Judson tried something different. In 1893 he patented a 'slide fastener'. Two rows of metal teeth faced each other and were made to interlock and then unfasten again by moving a slide one way or the other.

The cunning mechanism is little changed today. Moving the slide slightly widens the space between adjacent teeth, allowing opposite teeth to mesh as they are brought into contact. The initial market was for high boots; fastening those took time. But the device sold slowly. The crude initial models were likely to jam or come undone.

Swedish designer Gideon Sundback made the operation smoother and more reliable. Orders came for fasteners for army uniforms during World War I, but the clothing trade as a whole was uncertain. It appears the great clothes designer Schiaparelli made the new fasteners acceptable by incorporating them into garments from 1930, giving apparent approval to a device that let someone exit their clothes quickly. It was now called the 'zip-fastener' or the 'zipper', named by a PR man or a writer summing up its smooth speed as 'zip'.

We don't know where Judson got his basic idea from, but the story of the zipper's more recent rival is well known. Swiss engineer Georges de Mestral regularly returned from walking his dog in the Alps with his dog's fur and his own clothes covered in burdock seeds. Under a microscope, he saw how the seeds' shape let them hook onto clothing fibres and dog hair. He determined to emulate it, though others laughed.

In 'hook and loop' fastening (a generic name), which de Mestral invented in 1948, one side is covered with thousands of tiny plastic hooks, the other with even smaller loops. The hooks grab the loops, and with so many linkages over a wide area, a large force is needed to break them apart. Pulling the sides apart with a peeling action requires only a small number of links to be undone at a time.

The product is now called Velcro, from French words for 'velvet' and 'hook'. Like the zipper, it has drawbacks—accumulating lint and dirt, attaching to clothing—yet it has proved very versatile and is widely used in footwear and clothing, replacing buttons, zippers and laces, as well as enjoying some more advanced applications.

It's in the can
Erik Rotheim and Robert Aplanalp

The idea of forcing liquid out of a container using gas pressure is an old one. The first liquid so dispensed was soda water. In 1790, soon after its invention, Parisians could buy soda water in a device not unlike a modern soda-siphon. The soda water was driven out by its own carbon dioxide.

> **It was now called the 'zip-fastener' or the 'zipper', named by a PR man or a writer summing up its smooth speed as 'zip'.**

1927 saw the first patent on something like the modern spray-can, taken out by Norwegian engineer Erik Rotheim. His invention was commemorated in 1998 by a special stamp issue in his home country. Rotheim's metal cans could be filled with a range of liquids and used a variety of propellants such as methyl chloride. The name 'aerosol can' indicated that the liquid came out in 'aerosol' (a fine mist). Over the next 20 years, the cans were put to use as car fire-extinguishers, by General Motors, and to spray World War II soldiers in the South Pacific with insecticides to control the spread of malaria.

1949 was the big year for spray-cans. American Edward Seymour began marketing paint in spray-cans—reputedly his wife's idea. Graffiti artists have been thanking him every since. Perhaps more importantly, American engineer Robert Aplanalp—later a good friend of President Richard Nixon—patented a valve that did not clog. In 1950 his Precision Valve Company made 15 million of them. It is essentially the same valve found on billions of spray-cans today.

Among the early propellants were the man-made CFCs (chlorofluorocarbons). These seemed ideal—the right boiling point and inert—non-corrosive, non-toxic, non-flammable. They had been developed for use in refrigerators. However, in 1974 CFCs were first suspected of damaging the ozone layer, which protects us from the sun's ultraviolet rays. By 1985 the ozone layer over Antarctica was thinning and CFCs were blamed. By international agreement, the Montreal Protocol, CFCs were phased out over a decade, and mostly replaced with flammable hydrocarbons like

The spray-can: how does it work?

In a spray-can the propellant is not a gas, which would have to be at dangerously high pressure, but a liquid that boils at just below room temperature. The pressure inside the can is, however, slightly higher than outside air pressure. This raises the boiling point so only a little of the propellant evaporates inside the can. When the valve is opened the evaporated propellant is released—along with whatever is being sprayed. The internal pressure drops, evaporation starts again and the can repressurises.

butane and propane. Food products such as whipped cream are usually driven out by nitrous oxide.

Making it stick
Harry Coover and Spencer Silver

Making something stick to something else has been a challenge to craftspeople, and after them to engineers and manufacturers, for millennia. The many solutions have included natural gums and resins, boiled starch ('flour and water paste'), fish glue—patented in 1750—and adhesives made from casein (a protein found in milk). For tougher jobs adhesives dissolved in acetone (nail-polish remover) were popular, such as the long-familiar Tarzans Grip. Such drying adhesives set as the solvent evaporated.

In the 1940s Harry Coover of the photographic firm Eastman Kodak was busy with war work. He wanted to use threads of a chemical called cyanoacrylate as a sort of artificial spider silk to make cross-hairs for a gun-sight. It was a disaster; the cyanoacrylate stuck to everything. Years later he realised what he had. In 1958 Eastman Kodak went to the market with Eastman's 910 Adhesive—not an appealing name—later called Flash Glue, because it stuck so quickly—within a minute, even less with some additive—and then Super Glue, because it stuck so strongly. A patch just 2 centimetres square could hold up a tonne. Soon every home had a tube and the demand from industry was high.

Superglue—aka Crazy Glue and other names—works because the monomers (small chemical units) it contains join together when water is present—as it always is—to form long, strong chains of polymer. These intertwine to make an all-but-unbreakable adhesive mat. Since 1998 advanced forms of the glue have been approved for medical use, such as closing wounds and surgical incisions, rather than using stitches.

At the other end of the scale, Spencer Silver, employed by Minnesota Mining and Manufacturing (3M), was trying to make a strong glue but ended up with a weak ('low-tack') one, barely enough to hold two pieces of paper together. But it did not dry, however long you left it; it still stuck—sort of. He wondered what use there might be for it, if any. His colleague Arthur Fry found one, to put on bookmarks in his hymnal at choir practice. Ungummed slips of

A successful invention was one that would be bought over and over again by a satisfied customer.

paper fell out; those with the usual glues could not be removed. The new glue was ideal. Today we call slips of paper edged with a strip of Silver's glue Post-It Notes. They hit the market in 1980. Where would we be without them?

'Use it a few times and throw it away'
King Gillette

Acceptable facial hairiness in the 19th century changed with fashion, from the rigorously clean-shaven Beau Brummel to the elaborate whiskers of late Victorian times. Some shaving was always needed, increasingly done at home as implements improved.

The typical razor had a tempered steel blade, much like that of a sword; hence the term 'cut-throat razor'. The user could easily hurt himself and razors were expensive, requiring regular sharpening. In the 1880s the Kampfe Brothers in the United States reduced shaving injuries by fitting a metal wire grid on the side of the blade that pressed against the skin; only the hairs sticking through the guard were cut. However, the blades in this first-ever 'safety razor' remained expensive pieces of quality steel.

American King Gillette's family had suffered in the Chicago fire of 1871, the fire that set back the fortunes of reaper inventor Cyrus McCormick. Working as a travelling salesman to keep his family going, Gillette simultaneously hit on two momentous ideas: an improved safety razor, with only the very edge of the blade exposed, making skin cuts less common, and a new business model.

By selling the main device cheaply, good profits could be made from selling an element of it that required regular replacement. William Painter, inventor of the Crown bottle-cap, reputedly told Gillette that a successful invention was one that would be bought over and over again by a satisfied customer. Gillette succeeded by making the relatively cheap razor blade disposable, designed to go blunt and be replaced after only a few uses.

Experts assured Gillette that steel blades thin, strong and cheap enough for his purposes could not be made. Gillette thought otherwise and, with the aid of MIT graduate William Nickerson, triumphed after six years of work.

Despite his company's success, Gillette himself was financially ruined by the stock-market crash of 1929, an ironic fate since he was an opponent of rampant competition and capitalism, writing books promoting a socialist utopia.

The success story

Gillette's American Safety Razor Company, founded in 1901, went on to sell disposable razor blades by the billion. World War I provided a boost; the American government supplied his razors to every soldier. Sales soared further when changing fashions required women to shave their armpits. In time the company now bearing his name led the way in making the whole razor disposable, not just the blade, and in multi-blade razors.

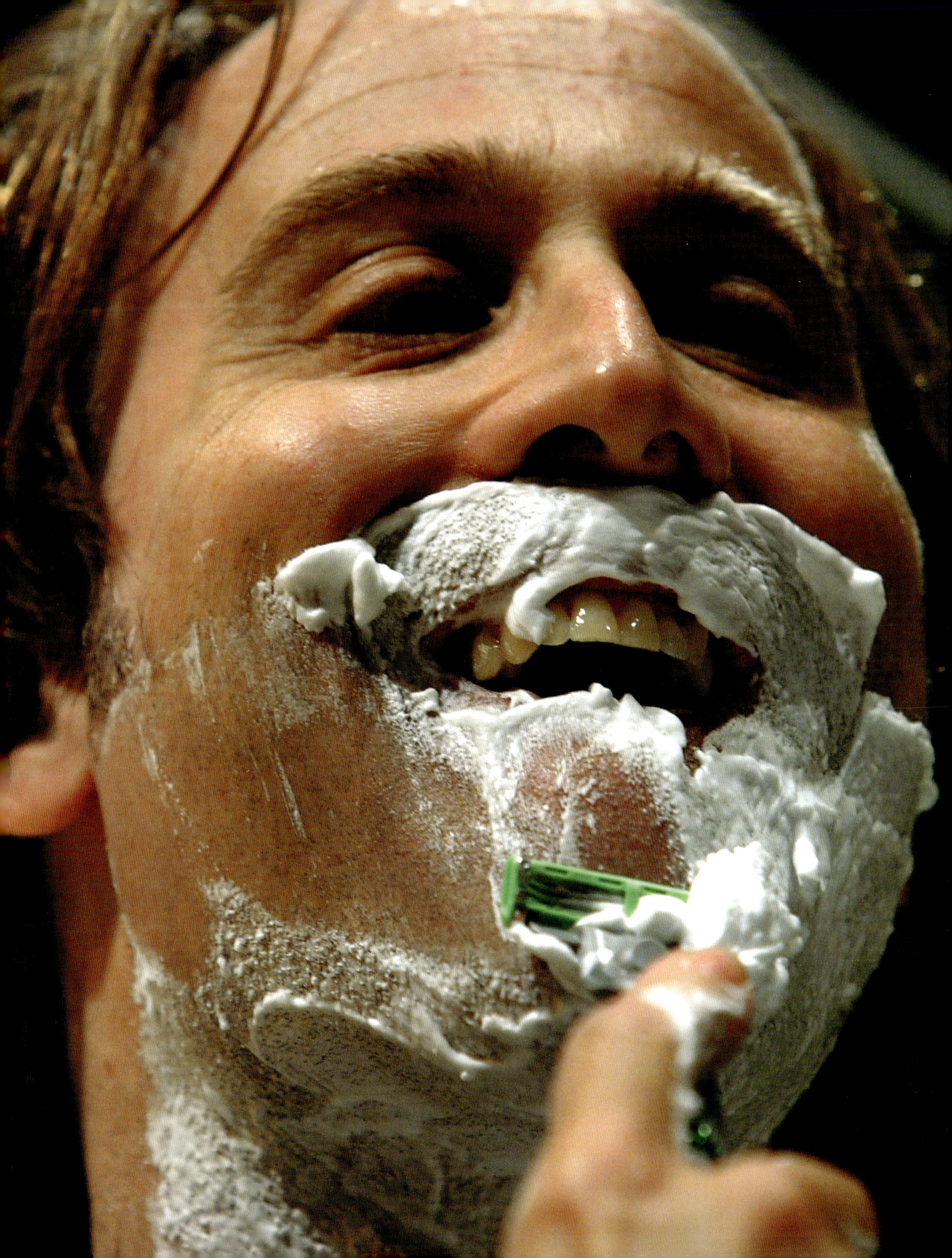

CHAPTER 22:
Having fun

Not all technology has a serious purpose. Many inventions have provided new ways to amuse or entertain ourselves, to fill up spare time enjoyably, indoors or out, alone or in the company of others. Some we call sports; others are games or pastimes. The need for such diversion has been with us a long time, but most of these entertainments are relatively new.

A stick and a ball
A multitude of new sports

Games in which a ball is struck with a stick are popular worldwide, and very diverse. The 'stick' can also be called a bat, a club, a racquet or a paddle; the objective may be to score a 'goal', to win a 'point', to hit a 'run'. Matches can be between individuals or teams. The ancestors of these games may be associated with ancient religious rites marking, say, the start of spring. Yet most of them were 'invented'—or at least took their modern forms—only in the 19th century.

Legend claims that Abner Doubleday invented baseball in Cooperstown, New York, in 1839, but in fact the 'the great American pastime' evolved from existing English games such as rounders. In 1846 the Knickerbocker Rules established the basic elements: nine-player teams, the four-base diamond. Over the next few decades, contests became limited to nine innings, national leagues were founded, professional teams took the field. The last established that sports could be as much for entertainment as for participation, for spectators as much as for players. It was a new idea.

The origins of cricket lie many centuries back, and games between teams of eleven were played for high stakes in the 17th century, but the key date is 1787, with the founding of the Marylebone Cricket Club (MCC) in London. The MCC drafted a set of rules—with the 22-yard (20.12-metre) pitch and specific ways for players to be dismissed—and remains dominant in rulemaking today. In the 19th century games involving county teams proliferated; Australia beat England in the first international 'Test Match' in 1877. The game spread through the colonies of the British Empire and remains strongest in those nations today.

Right: Like most modern games, hockey took its current form only in the 19th century. Today the ground beneath the players' feet is usually synthetic turf, while sticks and balls have benefited from modern technology.

Of the 'score-a-goal' games, hockey has ancient origins in Egypt, Persia and Greece, but developed its modern form in Europe, mostly in England. The Blackheath Club in London, the first ever, was established mid-19th century. Like cricket, hockey spread throughout the British colonies; nations on the subcontinent have a particular reputation. In the 1870s WF Robertson and RF Smith, students at McGill University in Canada, set down the rules of a new type of hockey played on frozen ponds with players on skates. Today, some see ice hockey as a game with few rules.

Canada also gave us lacrosse. An existing Indian game was renamed from the supposed likeness of the netted stick to a bishop's crossed staff (called *la Croix* in

> **Sports could be as much for entertainment as for participation, for spectators as much as for players. It was a new idea.**

French). The rules were first written down in 1863. The similar-looking game of hurling is barely known outside Ireland, but adherents declare its origins ancient. Rules were first agreed in the 1880s and the annual All-Ireland Championship has been held since that time.

Among individual stick-and-ball sports, two have massive global impact today—golf and tennis. Golf is perhaps unique in needing no opponent, though there usually is one. The contest can be just with the course or with your previous best score. With antecedents dating back to the *paganica* played by Roman emperors, 'golfe' emerged in Scotland in the 15th century. The Royal and Ancient Golf Club of St Andrews, established in the 1750s, maintains a dominant influence, but the global expansion of the game began only in the late 19th century, when the United States established its first golf courses and national championships.

Something akin to tennis was popular with 17th-century French aristocrats, though often played with the palm of the hand rather than a racket. The name comes from the Anglo-French word *tenetz* (meaning 'receive'), which the serving player was required to shout. A century later 'royal' tennis was popular in England, played indoors with some rules like squash. Modern 'lawn' tennis was invented in 1873—and

Below: Australian Evonne Goolagong Cawley at Wimbledon, 1976. The technology of tennis racquets has advanced since her day. Wherever a big match is played, modern communications bring the live action to the world.

patented a year later—by Major Walter Wingfield as a modification of royal tennis. It was first played at a garden party at his home in Wales, initially on an hourglass-shaped court with a very high net. Quickly established among the well-to-do in the United States, its popularity boomed and has never waned.

Among other racket sports, squash—originally 'squash rackets'—was devised at Harrow School, England, before 1850. Table tennis emerged late in the 19th century, but its origins are obscure. Badminton, also called 'shuttlecock', was brought from India by returning British Army officers, including the Duke of Beaufort. He played it at his country estate at Badminton, hence its name. The Badminton Association of England, established in 1893, still controls the rules.

Evolving technology has benefitted all these sports. Sticks and balls may contain advanced materials such as carbon fibres; scientific design now enhances performance. Information technology often speeds the scoring and helps to control the progress of matches—for example, line-calls in tennis; the increasing reach of mass communications can bring a game 'live' to billions of listeners and viewers across the globe.

The many forms of football
Rugby, soccer and American football

Games in which a round ball—originally an inflated pig's bladder—is kicked about by two teams have a long history, back to ancient China and Rome at least. The rules varied widely with time, place and circumstance. Modern codes arose within a few decades in the late 19th century, out of efforts to standardise rules for games in schools and elsewhere.

According to the popular story, in 1823 a student at Rugby school, William Webb Ellis, 'took the ball in his arms and ran with it,' defying the local rules. This probably apocryphal act was not reported for 50 years.

Above: Football (soccer or the 'world game') is the most widely-played of all team sports, but only one of several games that began through kicking around the inflated bladder of a pig. Their rules are now so diverse their common ancestry is obscured.

In any case the rules were constantly changing, so Ellis may not have acted illegally, but it seems that old boys from Rugby popularised the 'carrying game', founding the first football club in the world at Guy's Hospital in London in 1843.

The growth of football beyond schools led to the founding in 1863 of the Football Association (FA) at the Freemasons Tavern in London—the term 'association football' has now been shortened to 'soccer'. Much argument ensued over the rules, which still varied from club to club, particularly whether a player could catch the ball and run with it, and whether opposing players could try to stop him by holding, tripping or kicking him in the shins.

The FA decided to ban such practices and stick strictly to a 'kicking game'. A number of clubs promptly

In 1823 a student at Rugby school, William Webb Ellis, 'took the ball in his arms and ran with it,'

withdrew, led by the Blackheath Club, whose representative complained that the 'pluck and courage' would be drained from the game. In 1871 these other clubs met at the Pall Mall Restaurant, London, to found the 'Rugby Union'. Soon after, England and Scotland met in the first rugby international, and the divorce between the two forms of the game was complete. Today only the goalkeeper in a soccer game may catch the ball, and there are strict rules on tackling.

In early rugby games, the objective remained to kick a goal during play, but carrying the ball beyond the goal line entitled the attacking team to 'try' to kick a goal unimpeded. Over time 'tries' became more valuable than goals.

Rugby was split into two codes with different governing bodies and some distinctive rules in 1895 in England and in Australia in 1908, driven by the growing practice of some clubs, especially in the north of England, to pay their players. For the next century, 'rugby union' remained an amateur—and therefore mostly middle-class—game while 'rugby league' became professional, so was popular among the working classes. Some of that atmosphere remains, though elite rugby union players are now paid.

Technology made its impact. 'Pig skins' were eventually replaced by bladders made from the new 'vulcanised' rubber and enclosed in leather for durability. Balls remained round for the 'kicking game' but were elongated for the 'carrying game', in which the ball was passed hand-to-hand.

American football—often dubbed 'gridiron' from the array of lines marking the field—began as rugby. It was well established by the 1870s in the leading Ivy League colleges of the north-eastern United States. Two differences emerged early: greater importance was given to the 'touchdown' (try) in the scoring and sides were limited to 11 players, following the practice at Eton school in England.

Innovations soon transformed the game. Rule changes in 1880 gave possession of the ball to one team, which was free to play it forwards from the 'line of scrimmage'. To enliven the game, the team with the ball was required to carry the ball forwards at least 5 yards in three 'downs' (attempts) or lose it; today four downs are allowed to advance 10 yards.

From 1906 players could pass the ball forwards—strictly forbidden in the rugby codes—and in time the ball was reshaped to augment its aerodynamic qualities. From early days, the ball-carrier could be shielded from defenders by other players running in front; the notorious 'Flying Wedge' was invented at Harvard in 1892 and caused many injuries. Players not carrying the ball could be tackled—also not allowed in rugby. Solid body contact through blocking or tackling remains a feature; players are protected by helmets and padding. Over time liberalised substitution rules encouraged extreme specialisation, and development of separate offensive and defensive teams.

Soccer, in contrast, is now the 'world game', played in at least 200 countries, largely as a consequence of colonialism and the presence of British professionals in developing countries. The American game dominates in the United States. The rugby codes are popular in a number of countries, but other variations of football have followers, most notably the Gaelic code in Ireland and the not-unrelated 'Australian Rules', devised during the late 19th century to keep cricketers fit in winter. Played on very large grounds, these versions feature a combination of kicking, carrying and passing.

Get your skates on
James Plimpton and the Olsen Brothers

Skating on ice, for necessity or for pleasure, has a long history in climates where lakes and rivers freeze over. Its origins are lost in time. More recently, adventurous people tried replacing the blades with rows of wheels so you could skate on any smooth surface.

What we call the roller-skate today was invented by American James Plimpton in 1863. Seeking both stability and manoeuvrability, he used two pairs of wheels, one under the heels, and one under the ball of the foot—they were properly known as quad-skates. For the first time, skaters could move in a smooth curve and even skate backwards. The popularity of roller-skating began to boom. The use in the 1880s of ball bearings, then being incorporated into bicycles, made skating still easier and skates lighter. New technology impacted again in the 1960s, with tough plastic wheels—often made of polyurethane—replacing wood and rubber. Over the century the popularity of roller-skating waxed and waned, from frenzied to all-but-forgotten.

The in-line skate, as used by Joseph Merlin in 1760, was revived by the ice hockey-playing Olsen brothers of Minnesota after they came across an antique pair of pre-Plimpton skates. Thinking initially of providing off-ice training for their team, they incorporated modern materials and ideas from roller-skates, and roller-blading was born. Their first model was launched in 1983, but it had a number of technical problems and the Olsens soon sold Rollerblade Inc to someone better equipped to make a go of the innovation. New materials and better design cut the weight and increased safety. Within a few years, sales were enormous.

Roller-skating comes and goes in popularity, although in-line skates now outclass the old 'quad' format in terms of manoeuvrability and speed. Its most visible manifestation now is the skateboard, conceived in California as a surfboard on wheels.

> **Stranger than fiction**
>
> According to a possibly apocryphal story, London-based inventor Joseph Merlin wanted to make a grand entrance at a party in 1760. To achieve this feat, he devised 'in-line' skates, which would allow him to glide in smoothly, playing his violin. Sadly his skates were very hard to steer and he came to grief before the packed ballroom.

Meanwhile the quad-skate had morphed into something new, the skateboard with pairs of roller-skate wheels attached under the ends of a short plank. It appears this originated in the 1950s among Californian surfers who needed something to do when the waves and weather would not cooperate. Early models were hand-built. Devotees soon found that the skateboard needed the same 'oneness' with the board as surfing did. Again new materials and better technology moved the pastime forwards, bringing the capacity for trick-riding and cementing the skateboard as a key component of youth culture.

All the fun of the fair
George Ferris and James Miller

Visitors to the 1893 World's Fair in Chicago, celebrating 400 years since Columbus discovered the New World, blinked at a cascade of technological marvels, including the very new electric lights, powered courtesy of George Westinghouse. Another George exhibited something equally revolutionary. George Washington Ferris, who ran a steel-testing firm, reputedly invented the 'Ferris Wheel' while dining with fellow engineers. He drew a sketch and scribbled the specification on a table napkin. The basic design has never changed since.

Whatever the truth of that, the Ferris Wheel, designed purely for exciting entertainment, was a wonder, rivalling—as it was intended to—the impact of the Eiffel Tower erected in Paris four years earlier. Hundreds of passengers were lifted simultaneously in wooden cars to a height of 80 metres. Turned by two 1000 horsepower engines, the 15-metre, 70-tonne main axle was the largest single piece of steel forged up to that time. The original wheel was destroyed by fire in 1906, but its descendants enliven amusement parks across the planet, and include the immense London Eye on the banks of the Thames.

Gaining prominence at much the same time was the 'roller coaster' or 'Big Dipper'. The idea of a 'switch-back', in which carts were pulled up a slope and allowed to run down again, went back several decades in both France and the United States, before LA Thomson built one at New York's Coney Island in 1884. An early dipper simply went up and down a series of hills in a straight line but it was, no doubt, exciting enough in its day.

Prolific American inventor John Miller, holder of 100 patents, brought us the possibility of the modern roller coaster. His key innovation, patented in 1912 and still used today, ensures a passenger car remains firmly adhered to the track, in whatever direction it moves. Three sets of wheels are involved; one set above the track carries the weight of the car, one inside the track guides the car around turns, one underneath the track locks the car to it.

Miller was initially looking for greater safety, so that the car would not jump the track, even as the 'dips' were made deeper and steeper. He soon realised that more was possible and by the 1920s was building roller coasters with twists and turns, increasing excitement

Opposite: The turning of a Ferris Wheel is not as frenzied as this long-exposure image suggests. While they remain a popular attraction today, their impact does not match that of their first outing at the 1893 World's Fair in Chicago.

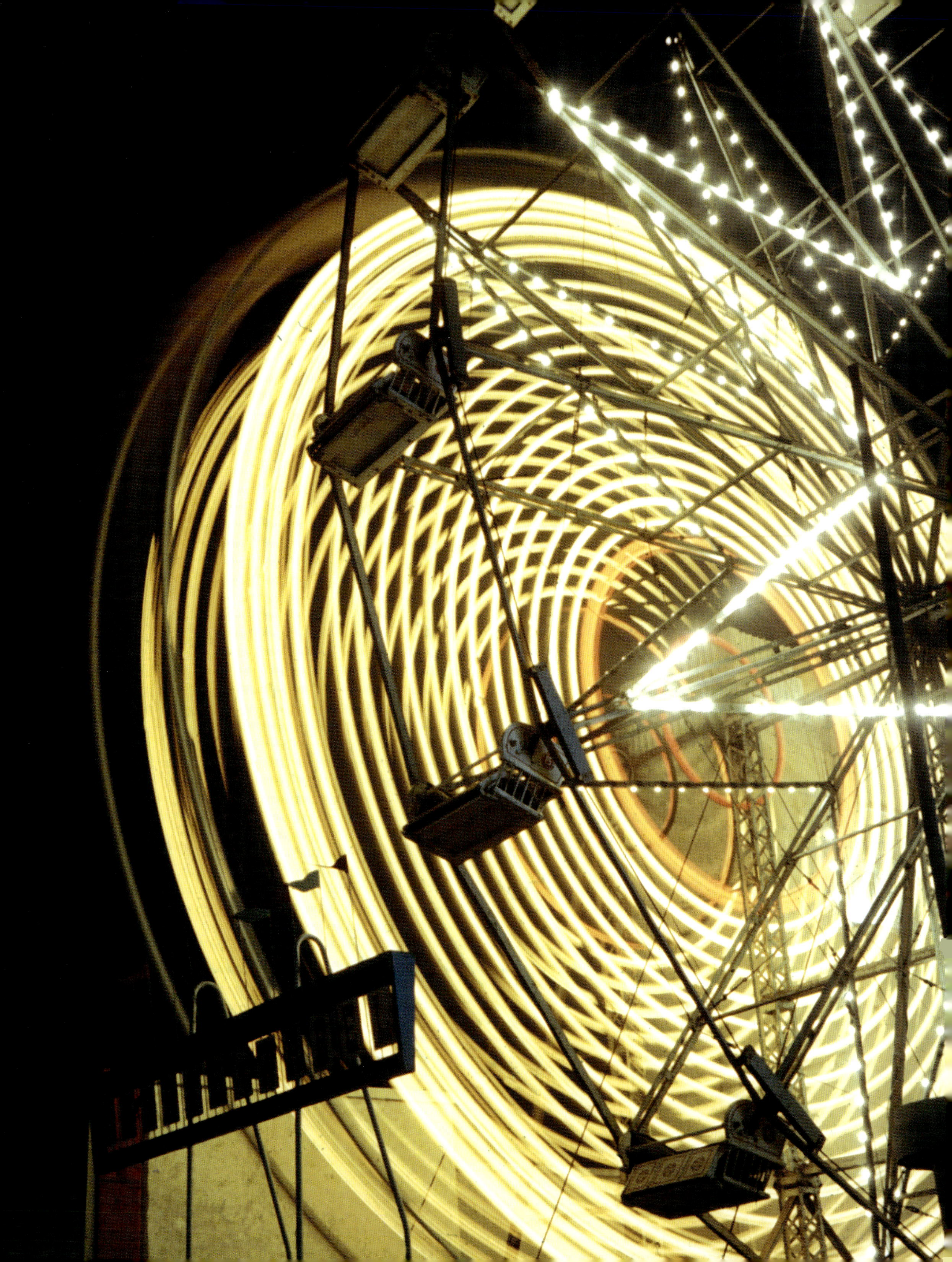

The Ferris Wheel ... was a wonder, rivalling—as it was intended to—the impact of the Eiffel Tower erected in Paris four years earlier.

and popularity from the heightened sense of danger and forming the basis for today's breathtaking rides.

Reinventing ancient toys
Richard Knerr, Arthur Melin, Pedro Flores and Donald Duncan

The hula hoop and the yoyo are ancient entertainments, if not as old as the doll. In the 20th century, they were rediscovered and re-engineered, and marketed so well they became 'crazes', enjoying recurring periods of intense popularity, if ignored in between.

Many cultures played with hoops of wood, bamboo or metal for millennia, but the modern resurgence began in the 1950s. Australian school children twirled cane hoops around their waists in exercise classes. Hearing the news, Americans Richard Knerr and Arthur 'Spud' Melin, who sold the Wham-O slingshot, grabbed the opportunity. Their hoops of brightly coloured plastic, often containing ball bearings, bells or other noise-makers, sold 25 million in two months and 100 million during 1958. It was the biggest toy fad in American marketing history.

The 'hula hoop'—so called because the hip-movement needed to keep the thing in motion resembled Hawaiian dances—was cheap and inclusive. People of any age could use one and devise their own style. Hoop-spinning contests were held at local fairs. An accomplished performer once spun 82 hoops simultaneously. It made *The Guinness Book of Records*.

Stone versions of the yoyo date back at least 3000 years. Many cultures used the toys, made from wood or bone, and they were popular, under another name, among 18th-century French nobility. The man responsible for starting the first modern craze appears to have been Pedro Flores. Moving to the United States from the Philippines, he started manufacturing yoyos in 1928; the name is Filipino. In a significant technical advancement, he looped the string around the axle, rather than affixing it. This let the yoyo spin at the end of the string, making possible feats like 'walking the dog' and 'around the world'. By 1929 yoyos were

Left: Most people find twisting a single hoop around their waists enough of a challenge, but skilled performers can do much better. Like the yoyo, such hoops were found in many cultures and many eras.

Stranger than fiction

The yoyo did not make Flores rich. He had already sold out for $25,000 to Donald Duncan, who later claimed 85 per cent of the market. Duncan had a genius for publicity, sponsoring contests and convincing movie and sports stars to pose yoyo in hand. In the 1960s he employed dozens of professional demonstrators across the country and his factory churned out 3000 yoyos an hour. He was too successful; 'yoyo' was declared a generic name. Loss of trademark protection bankrupted Donald, though he kept himself going marketing parking meters.

selling by the million, and hobbyist magazines carried articles on 'how to make your own'.

Fun indoors
Arthur Wynne and Charles Darrow

Crossword puzzles, popular everywhere, are a recent invention. Some very simple ones appeared in the 19th century, but the puzzles' profile today stems from 1913, when English-born journalist Arthur Wynne published one in the *New York World*, a Sunday newspaper. As today, clues were given to identify words that fitted together vertically and horizontally, but Wynne's first puzzles had no black squares. The outbreak of war diverted attention to other matters, but the 1920s saw an explosion of interest, with crossword puzzles—initially called word-cross puzzles—appearing in British newspapers from 1922 and in *The Times* from 1930.

The arrival of electronic games might have seen the end of the crossword, but in fact information technology now supports the pastime, with computer programs available from the late 1990s that allow anyone to construct a puzzle for themselves. An estimated 40 million people in the United States alone wrestle with a crossword daily.

If the crossword is the most popular pen-and-paper game—though now challenged by Sudoku—Monopoly remains the best-selling board game, despite the rise of Trivial Pursuit and many other pretenders. With 200 million sets sold in twenty languages, Monopoly has made a fortune for its publishers Parker Brothers, but it nearly was not so.

Charles Darrow, an unemployed salesman from Germanton, Pennsylvania, devised the game in the early 1930s for his own amusement, based on Atlantic City and drawn on his tablecloth. The game became popular among Darrow's family and friends, and he hand-made copies to sell, but an approach to Parker Brothers was surprisingly rejected. They claimed to find 52 'design errors', such as that the game took too long and the rules were too complex.

Undeterred, Darrow continued to market Monopoly himself, having a printer produce 5000 copies. Parker Brothers soon saw their mistake and offered Darrow a royalty deal. In its first year, 1935, Monopoly was the most popular game sold in America and Darrow was soon the first board-game millionaire.

Darrow was not, in fact, the first to design such a game, just the first to make a real success of it. Half a dozen board games involving the buying or selling of shares or property, such as Lizzie Magie's The

500
500
CHANCE
OLD KENT ROAD
£60
COLLECT
£200
AS YOU PASS
GO
SUPER
TAX
PARK LANE
£350

Chapter 22: Having fun

Above: Pinball machines were the first big attraction at the penny arcades where the poor sought cheap entertainment during the Depression. Despite decades of innovation, the venerable pinball lost out in the end to the video game, in the arcade and at home.

Landlord's Game, were being played somewhere; some had been patented. Some may have influenced the game Darrow devised, even its title. Parker Brothers quickly bought up the rivals, some very cheaply, to protect their investment.

Fun in the arcade
Raymond Maloney and Harry Mabs

During the dark days of the Depression, the unemployed flocked to the 'penny arcade' in search of cheap, time-filling entertainment. Popular among the attractions was the 'pinball' machine. A metal ball was fired across a sloping, glass-shrouded board studded with metal pins and equipped with a few holes into which the ball could drop as it rattled among the pins. A ball in the right hole could earn the player a useful prize.

Opposite: Monopoly has many rivals now, but it remains the archetype of board games and widely played. It is hard to believe that Parker Brothers, who made a fortune from publishing Monopoly, originally rejected it.

The pinball machine was based on a 19th-century entertainment known as a bagatelle, but was coin-operated like the earliest jukeboxes based on Edison cylinders. The first such machine was built in 1931 by Raymond Maloney from the Bally company, later better known as a maker of slot machines.

From the start, makers were constantly innovating, as players were always in search of something new to hold their attention. Once machines were electrified, bells, buzzers and flashing lights added to the excitement. Machines were soon equipped with a 'tilt mechanism'; sensors detected if a player tried to alter the path of the ball—and their chances of winning—by lifting, shaking or thumping the machine. Other versions saw the pins and holes replaced by spring-loaded bumpers. Hitting those scored points, which were automatically totalled.

Pinballs were more a matter of chance than of skill. Indeed, in 1942 the colourful Mayor of New York Fiorello LaGuardia banned them, on the grounds that the games actually constituted gambling. The ban lasted until 1976. In response to the ban, or perhaps just to provide something new, Harry Mabs from the Gottlieb company added electrically-powered 'flippers' to the new Humpty Dumpty machine in 1947. These were operated by the player to keep the ball in play, prolonging the enjoyment and adding a substantial measure of skill to the luck of the game. Mayor LaGuardia should have been satisfied.

Innovation continued with new layouts, more exciting lights and sounds, machines that played several balls at once, digital readouts, moving displays using LEDs or video screens, workings based on integrated circuits rather than electromechanics, machines that talked. But pinball machines and similar pastimes could not keep pace with the growth of purely electronic 'video games' and the number of manufacturers in the United States is now down to one. Interestingly, one popular video game is a computer simulation of pinball.

CHAPTER 23:
Making music

Any sort of music, other than singing, needs an instrument of some kind, be it as simple as a drum or a cut reed. Many music-making inventions took some inspiration from nature, but then added layers of human innovation, resulting in a form of machine. Early music machines were powered by human effort, by plucking, bowing, blowing or banging. More recently, help might come from electricity or compressed air, but human creativity remains in command.

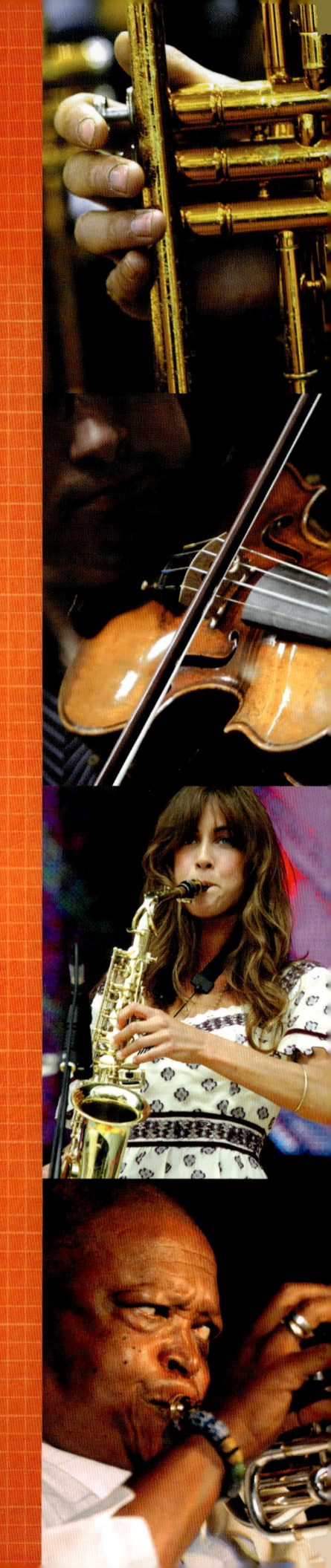

The loud and the soft
Bartolomeo Cristofori

Making music on 'keyboard' instruments goes back a long way, at least to the invention of the organ, perhaps 1000 years ago. Pressing keys on an organ let air into pipes of various lengths to make the different notes.

Not quite as ancient are the harpsichord, the related virginal and the spinet. In these the keys caused 'plectra', made from feathers or points of leather, to pluck strings, one for each note. The drawback with these instruments is that they are not 'sensitive to touch'; no matter how hard you press or bang the keys, the resulting notes have the same volume.

Bartolomeo Cristofori was a harpsichord maker from Florence. In 1709 he published plans for a new type of keyboard instrument which could be played loud and soft; 'forte' and 'piano' in Italian, hence the name 'pianoforte', now shortened to 'piano'. Instead of plectra, pressing a key would send a small leather-covered hammer upward to strike a string. The more firmly the keys were struck, the more vigorously the hammers struck and the louder the resulting note. The player now had full command over loud and soft.

There were other cunning innovations as well, an 'escapement' which immediately let the hammer fall back to its starting point, ready to be used again, and 'dampers', small cloth-covered blocks which pressed upon vibrating strings and stopped the sound.

Cristofori may not have been the first to experiment with such a mechanism; records survive of similar designs from elsewhere. But not until Cristofori did anyone follow-through consistently with the idea, and the oldest surviving instruments—from the 1720s—are his. The pianoforte was not immediately popular; as composers like Bach pointed out, the early instruments were heavy in touch and weak in the high notes.

So the harpsichord remained the keyboard of choice through most of the 18th century. Technical development of the piano continued, starting in Germany and England

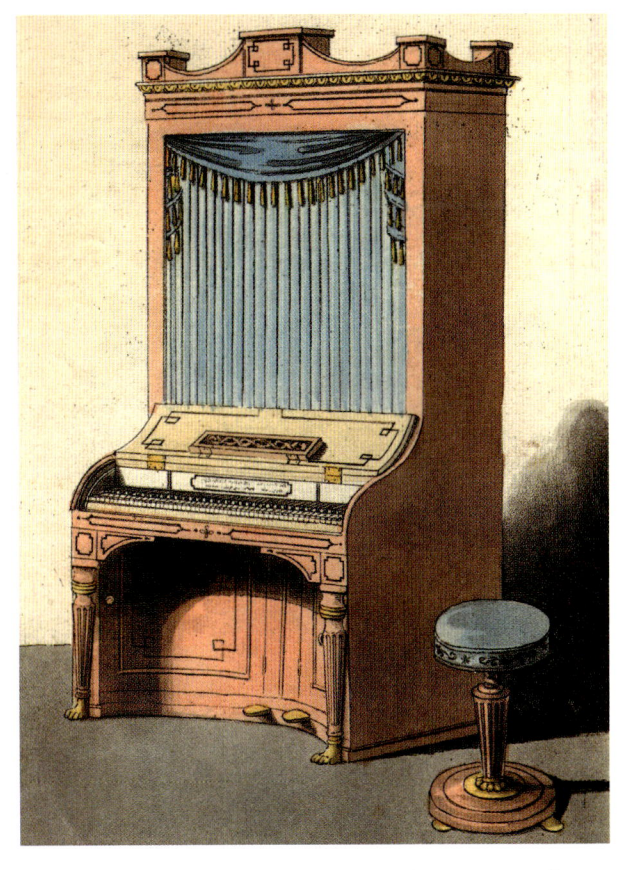

Above: The organ is the oldest keyboard instrument, the sound coming from vibrating air in pipes. Over the centuries it found its place not only in great cathedrals but (in scaled-down versions) in small churches and music rooms.

in the 19th century, leading to the advanced instruments of today. The most sophisticated 'grand pianos', costing hundred of thousands of dollars, use the principles pioneered by Cristofori. Until recently, every middle-class household in the Western world had a piano.

Many instruments together
Andrea Amati, Adolph Sax and many others

The modern symphony orchestra was not 'invented' in a single stroke; it is the result of half a millennium of innovation, beginning with diverse groups of musicians and instruments accompanying dancing or devotion. A symphony orchestra today is built around a core of stringed/bowed instruments, to which an increasing number and variety of blown instruments has been added over the centuries.

The violin and its relatives evolved from the earlier viols, so no one really invented them, but credit as the first great violin-maker usually goes to Andrea Amati from Cremona in northern Italy in the mid-16th century. He applied his lute-making methods to the crude violins of the day, using a mould to ensure precision in construction. He also enlarged the sound box, increasing volume while maintaining sweetness of tone. The oldest known survivor among the Amati violins dates from 1564.

The immediate popularity of Amati's violins began the great tradition of violin-making in Cremona, which included the Stradivarius and Guarneri families, active for nearly 200 years. More innovation came in the early

The oldest known survivor among the Amati violins dates from 1564.

19th century, with many changes to the structure of the violin in search of greater range and volume to suit the virtuosity of the players and increasing size of concert halls.

The typical orchestral complement of 'woodwinds' also evolved over time. The ('transverse') flute displaced the recorder only from around 1750, joining early oboes and bassoons with their double reeds. The latter had evolved from the more primitive 'bombard' of 200 years earlier.

We can say, however, that the single-reed clarinet was 'invented', the first such being made late in the 16th century by the instrument-maker Denner of

The valve: how does it work?

Operated by a key or piston, the valves on brass instruments enable the length of the column of vibrating air inside the instrument to be changed at will. The length of the column of air affects the pitch of the note being sounded. This was already possible with the sliding tubes of the trombone, but with a combination of valves other brass instruments could be developed that could play all the notes in a scale. The previous technology of 'crooks' to change pitch always left some notes unreachable, just as a bugle has many notes missing.

Nuremberg. It too had its ancestors, including the low-voiced 'chalumeau', but Denner's innovations, such as the 'speaker key' that allowed the upper range to be reached, made it essentially a new instrument. A century was to pass before the clarinet found a regular place in orchestras.

All the woodwind instruments enjoyed innovation in the 19th century and later, with a move to metal construction, changes in the shape of their 'bore' (the tube from the mouth-piece to the bell), and more complex sets of keys, such as the 'Boehm system', to ease the playing of intricate music.

For brass instruments such as cornets, trumpets and horns, the key invention was the 'valve', introduced around 1815 by the Germans Blumel and Stotzl. Through the use of valves, brass instruments gained the capability to play melodies like the strings or woodwinds—or rather 'regained' that ability. It had been possible a century earlier on 'natural instruments', but only through the special technique and extraordinary skill of the players.

Finding a bass instrument for the brass section needed further innovation, with various options such as the 'serpent' and the 'cimbasso' being tried before the valved 'tuba', first made around 1835, took over.

A memorable 19th-century name in instrument making was the Belgian Adolph Sax, who invented and named a new instrument in the 1840s. The 'saxophone' was a hybrid, made of brass, but using a single reed—as in the clarinet—and with a conical bore like an oboe. Composers were not very interested, but saxophones became popular in French military bands, linking the sound of the woodwinds with that of the brass. They soon moved into American military bands and then into jazz as it emerged, where they found an enduring place. Sax created a family of instruments, with the alto, tenor and baritone in common use, along with the soprano and occasionally the bass.

Sax also created a family of seven brass instruments to produce a uniform sound from highest to lowest. The 'saxhorns' are the mainstay of today's brass bands, going under a variety of names including flugel horn, alto or tenor horn and baritone.

The 'tuning fork' is not really an instrument. It plays only one note, but so reliably sounded that a number of instruments can agree on pitch and play in consonance. Now made of steel, it was invented in 1711 by John Shore, a trumpeter at the Court of St

Jukeboxes gave black recording artists a chance to be heard; for decades, radio was for white musicians only.

James in London. A notable player, he had solo parts written for him by Purcell and Handel.

Music from a machine
The player piano and the jukebox

Not everyone can create music for themselves, hence the market for machines producing music at the turn of handle or the press of a button. In the long-loved music-box, the tune is coded as an arrangement of spikes on a drum. As the drum turns, the spikes touch metal strips of varying lengths to create the different notes.

The 'player piano' was a step up in complexity and sophistication, drawing music from the familiar piano without human aid—other than to pump the pedals and drive the works. The first such, the 'pianola', was marketed by the Aeolian Company of New York in 1895, and was the creation of Edwin Votey. There were soon many manufacturers, including the Tonk brothers, whose pianos had a distinctive sound later dubbed 'honky-tonk'.

To control the performance, a long roll of perforated paper passed through the mechanism, the holes indicated when a particular note should be sounded by the aid of air pressure. The rolls were constructed, not by musicians, but by technicians who simply transferred the content of sheet music.

'Pianola' later came to mean almost any type of automatic piano, including the advanced 'reproducing pianos' invented a decade later in Germany. These added nuances of dynamics (variations in volume), rubato (variations in tempo) and pedalling, and many pianists and composers recorded on them, including Jan Paderewski, George Gershwin and Percy Grainger. Player-pianos let music-lovers recreate those performances in their living room, and were a major source of home entertainment until the 1930s, before being superseded by the wireless.

The purpose of the 'jukebox' was similar: to produce music from a machine on demand. The difference was that the music was completely pre-recorded. The first of these appeared in 1889, playing Edison cylinder recordings, with the customer putting a nickel in the slot and listening to the 'Nickelodeon' through a tube. Into the new century the demand for both recordings and 'coin in the slot' players boomed. Newer machines could hold more recordings, allowing a greater choice of music. As record players proliferated in the home, jukeboxes concentrated more on dance music, but the rise of radio took its toll here too.

The name 'jukebox', from the 1930s, is African-American; 'juke' is thought to mean 'to dance'. Appropriately, jukeboxes gave black recording artists a chance to be heard; for decades, radio was for white musicians only.

Electrons make music
Laurens Hammond and Leon Theremin

Electronic music did not begin with the 'synthesiser' or 'keyboard', which is an application of digital information technology. Following the invention of valves to amplify and oscillate, bright minds found ways to use them in musical instruments, rather than relying on vibrating strings or air columns. Two such devices, invented at much same time, were the Hammond organ and the theremin.

The Hammond organ, patented in 1934, is more familiar, with a keyboard and a recognisable 'theatre organ' sound, much associated with American gospel music. Its method of making music was altogether new. Soon after service in World War I, American engineer Laurens Hammond invented the synchronous electric motor, which kept in precise step with the 60-cycles-per-second alternating current used to distribute electricity. He first used this motor in a clock that did not tick.

He soon realised that the motor's unwavering rotation could generate precisely-pitched musical notes by spinning wheels with varying numbers of teeth through electromagnetic fields. The Hammond organ was, therefore, an electromechanical device, like some early computers. Loudspeakers using a revolving disc produced the unique trembling sound. Hammond organs could be a one-man band, with attachments to recreate the sound of a bass instrument or of percussion. They have recently undergone a revival among devotees of funk music, but the massive old-style Hammond organs are no longer made.

The theremin, created by Russian physicist and cellist Leon Theremin, was very different. With no keyboard, its notes were controlled entirely by the movement of the player's hands, effectively part of the electronic circuits. One hand determined the pitch of the sound, the other its volume. Theremin visited the United States in 1927 to demonstrate his instrument. It generated much interest, largely through the mesmerising skill of his young protégé Clara Rockmore. Her advanced sense of pitch and precise control of movement allowed her to master an instrument unlike any other.

Despite its ingenuity, and its place in the development of electronic music, the theremin was not a commercial success. The electronics manufacturer RCA put it on the market but sold only a few hundred. Yet it is likely you have heard its unforgettable, eerie tones in the soundtracks of classic movies like *Spellbound*, *The Lost Weekend* and *The Day the Earth Stood Still*.

CHAPTER 24:
Lighting the way

Providing a useful light in dark places has always been a major human concern. For much of the history of our race, we went to bed when the sun went down because we had no way to light up the night, other than by a campfire which was not portable and needed tending.

Until the early 19th century, we still needed fire to give us light, coming from burning candles, oil lamps, naphtha flares, and soon gas lights as well. But the 19th century also brought us electricity. While this would prove useful, indeed vital, in many ways, its first large-scale use was to provide light. Its success in this area made the other manifold uses of electricity cheap enough to be practical. In advanced countries, most of the light we use now comes from electricity, though we have developed much more efficient ways to make the conversion.

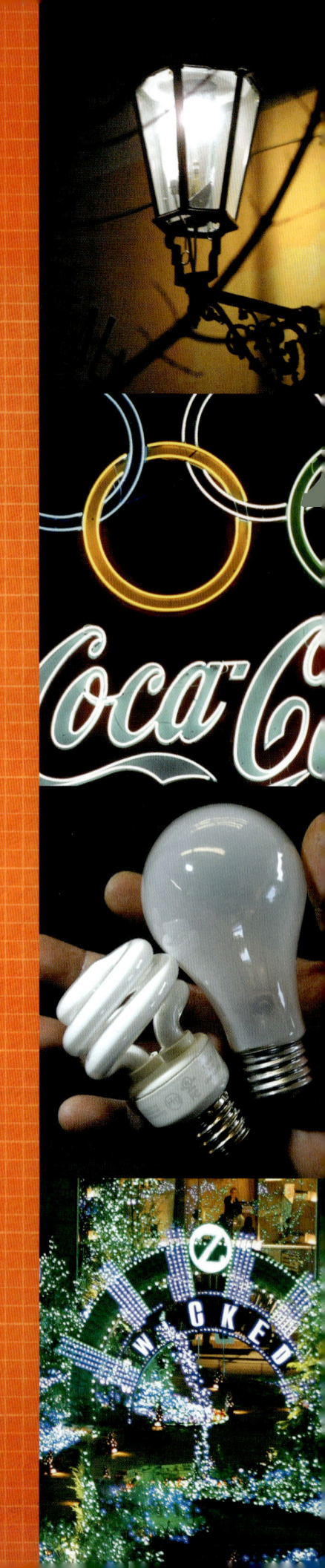

Chapter 24: Lighting the way

What happened when?

1792	William Murdoch lights a house using coal gas
1802	Humphry Davy uses electricity to make light
1820	Warren de la Rue creates an electric light bulb
1878	Joseph Swan improves the light bulb's design, making one that burns for 14 hours
1882	Thomas Edison develops a complete electricity supply system
1886	George Westinghouse and William Stanley install the first AC power system in Massachusetts
1888	Nikola Tesla patents the induction motor
1895	Westinghouse uses hydroelectric power from Niagara Falls
1902	Georges Claude patents the neon tube
1937	Fluorescent tubes displayed at the New York World's Fair
1962	Nick Holonyak produces the first light-emitting diode (LED)
2007	First cars come onto the market using LED headlamps

Lighting streets with coal
William Murdoch

When Abraham Darby first heated coal to make coke for smelting iron ore, he would not have known how much value was going up the chimney. Coal is a complex mixture of organic compounds. Heating it in the absence of air—so the coal does not burn away—releases many of these, and with care they can be trapped and collected for use.

The first compound found useful was coal gas, colourless if smelly. Composed mostly of hydrogen, methane and carbon monoxide—all of them flammable and the last poisonous—coal gas had been made and burned in small amounts previously. Not until the 1790s was it made in large amounts for distribution through buried pipelines to light houses and streets.

Existing domestic lighting came from candles and oil lamps; house interiors were dim at night. Streets were lit by oil lamps, if at all. In the absence of moonlight, careful pedestrians carried their own lamp. In 1792 Scottish engineer William Murdoch showed a new way. He lit a house in Redruth, Cornwall, with flames from a burning jet of coal gas.

Below: Gas lamps lit the streets of cities like London for a century from around 1800, though not very well. Before then the streets were mostly dark. By the early 20th century, electric lamps were putting the 'lamp lighters' out of a job.

Above: Some old-style gas lamps are still used for special effect, like this one in Prague's old town square. They are considered romantic, due to their soft, warm glow.

Over coming decades, coal gas lighting spread to the streets of major towns and the houses of the well-to-do. Gas heating and cooking followed soon after. To even out supply and demand, Murdoch devised large expandable storage tanks that he called 'gasometers'. His countrymen like to claim Murdoch as 'the Scot who lit the world'.

Murdoch had once walked 500 kilometres to Birmingham to get a job at the steam engine factory run by James Watt and Matthew Boulton. He became a long-serving and much respected member of the firm and later a director. While attending to the firm's interests among the Cornish tin miners he rediscovered the illuminating power of coal gas. He lit the Birmingham factory with gas in 1803.

He also pursued the technology of high-pressure steam engines, but Watt and Boulton were not interested, mostly because they thought them unsafe. This was often true and dissuaded Murdoch from further work.

Like Boulton and Watt, Murdoch was a member of the Lunar Society of Birmingham, sometimes dubbed the 'think tank' of the Industrial Revolution. The Society met on nights of the full moon, which would light the path homewards through dark streets. After Murdoch's innovation, they could have met on any night

Who invented the light bulb?
Humphry Davy, Thomas Edison or Joseph Swan

Whatever the books say, the prolific American inventor Thomas Alva Edison was not the first to make light from electricity. That honour belongs to the English scientist Humphry Davy in the first decade of the 19th century. Using the newly discovered 'voltaic battery', he passed a heavy current between two contacting pieces of carbon. When he slightly separated them, a brilliant white light burst forth—the first 'carbon arc lamp'.

It was, however, too bright and inconvenient for indoor use. Practical electric lighting had to wait more than half a century for the invention of the 'light globe'. Edison did not invent that either.

Credit for the light bulb as a concept belongs elsewhere, such as with Warren de la Rue in 1820. He placed a platinum coil inside a glass bulb from which the air had been evacuated, and made it glow with electricity from a battery. The high cost of platinum,

Opposite: A handful of light globes. The wasteful incandescent globe on the right is slowly giving way to the four-times-more-efficient compact fluorescent lamp. Both these are fitted with the venerable 'Edison screw'.

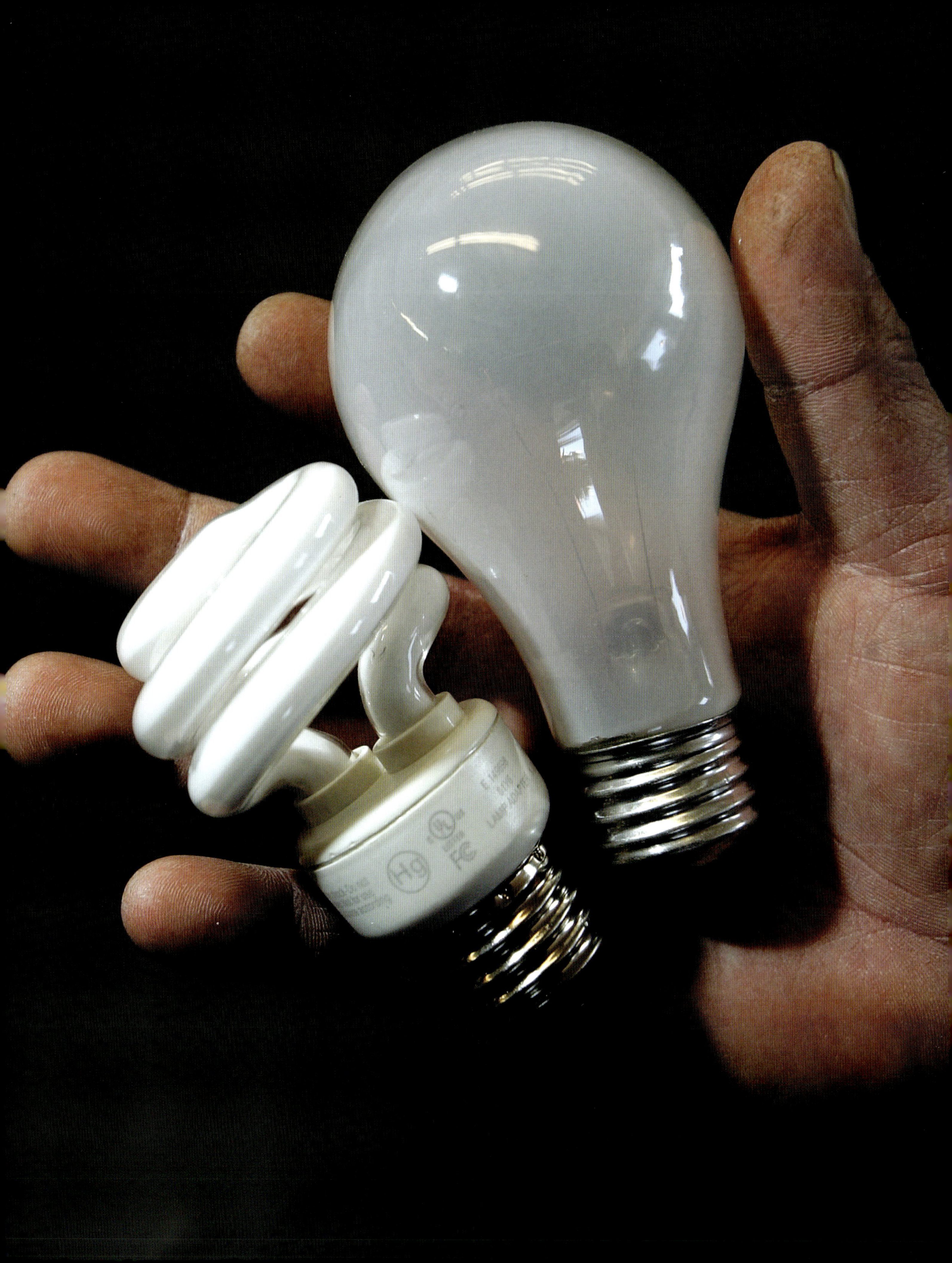
Hg

however, precluded mass use. From the 1850s inventors in England and Germany experimented with 'filaments' of carbon—cheap but also an electrical conductor—made by charring bamboo or paper impregnated with tar. In 1878 English physicist Joseph Swan patented a light bulb that used a filament of blackened cotton. This glowed for 14 hours before burning out, but mass production of his lamps did not begin until 1881.

Edison was now busy too, exploiting a patent bought from Canadian inventors James Woodward and Matthew Evans. Their lamp had a shaped carbon rod, held between metal struts in a hand-blown glass bulb filled with nitrogen to stop the filament burning away. Edison added his resources and ingenuity to the hunt for a cheap but durable filament. Carbonised cotton thread proved best. By 1879 his bulbs would burn for 40 hours, by 1880 for 1200 hours, long enough to be commercial.

So Edison's glory in the matter of the light bulb lies not with the idea or the invention but in its successful application on a wide scale. In this he was the innovator rather than the inventor. He pioneered the mass production of such bulbs and the electricity to power them.

The success story

The scale of Edison's achievement in relation to electricity should not be understated. His undertaking needed much insight and enterprise: a network of buried copper wires; parallel circuits so that each bulb enjoyed the same voltage; sockets with on/off switches; safety fuses; more efficient dynamos to make the electricity—he relied on the work of others there—and equipment to hold the voltage steady. Meters to measure how much power each customer bought came later.

Above: The old and the new: a wind turbine seen through a traditional incandescent globe. Edison created a glowing filament that could last long enough to be cost effective, but such globes remain very inefficient.

Edison opened the first American generating plant in Pearl Street, Manhattan, in September 1882, supplying 50 customers and 1300 lamps. However, this was not the world's first power station; Edison had set one up in London nine months earlier. By 1890 many cities were enjoying electricity from their own plants, with the Edison power companies becoming General Electric soon afterward. Edison could claim to have founded the electricity industry, though he shared that honour with George Westinghouse among others.

The battle of the currents
Thomas Edison versus George Westinghouse

American inventor Thomas Edison was first to develop a complete system to generate and distribute electricity, but he was not unchallenged. George Westinghouse, inventor of the air brake among other things, was also in the game. He lit the city of Buffalo in 1886.

The two had rival technologies. Edison used direct current (DC), which moved only in one direction along the wire. Westinghouse was exploiting alternating current (AC), in which the current reversed its direction many times a second. As Westinghouse saw it, AC had distinct advantages. A 'transformer' could raise the voltage, lowering the current and reducing losses in transmission over long distances. DC could not be transformed, and voltages low enough to be safe needed thick copper wires for acceptable losses. The cost of copper was rising; Edison feared he could not afford to run cables to customers more than a kilometre or two away.

The New York legislature was then considering new ways to execute criminals; hanging was thought too slow and painful. By 1888 they had decided on electrocution in an 'electric chair', with both DC and AC as possible power sources. Edison campaigned for the AC version; surely customers would not buy current from a man who also powered the electric chair. The committee formed to decide which system to use was headed by Dr Fred Peterson, who was on the Edison payroll. The committee chose AC.

Westinghouse refused to supply the generators. He also funded the appeal made on behalf of the first man sentenced to be so executed, on the grounds that the Constitution prohibited 'cruel and unnatural punishment'. The appeal failed; the first man to sit in the electric chair died from AC current supplied by an Edison-made generator.

The ploy did Edison no good in the long run. After powering the 1893 Exposition in Chicago, Westinghouse brought power from the first hydroelectric station in Niagara Falls in 1895, clearly demonstrating the benefits of AC in hauling electricity over long distances. By this time Edison was changing over, and AC has remained the basis of electricity distribution ever since.

Stranger than fiction

To scare customers away from Westinghouse, Edison claimed that AC was unsafe. In a bizarre act in 1887, he set up a 1000-volt Westinghouse AC generator and executed a dozen animals on an electrified metal plate in front of the press and the public. The term 'electrocution' was coined.

The high voltage wizard
Nikola Tesla

Hundreds of people attended the funeral of Croatian-born Nikola Tesla in New York in 1943, mourning an inventive genius. The 87-year old Tesla held 700 patents, but died an eccentric recluse, a misanthrope with few friends—humorist Mark Twain was one—a poetic visionary with a practical touch.

Arriving in America in 1884 almost penniless, Tesla worked briefly for Thomas Edison. The pair fell out over Edison's excessively methodical approach. Tesla later commented, 'a little theory and calculation would have saved him 90 per cent of his labour'. They also disagreed fundamentally over electricity, with Edison committed to direct current (DC) and Tesla to alternating current (AC), alongside George Westinghouse.

Tesla knew well the merits of AC: generation was simpler; and it could be 'transformed' to higher voltages,

The first man to sit in the electric chair died from AC current supplied by an Edison-made generator.

The 87-year-old Tesla held 700 patents, but died an eccentric recluse, a misanthrope with few friends, a poetic visionary with a practical touch.

reducing transmission losses over long distances. AC lacked a motor to convert it back into movement, but Tesla supplied that by inventing the 'polyphase' or 'induction' motor, soon put into use on the first electric trams and then into factories to drive machinery. It is fair to say that Tesla's genius made possible Westinghouse's electrical triumphs.

Lacking business skills and always short of funds for research, Tesla became a master of high voltages, generating artificial lightning, illuminating lamps without wires, staging spectacular shows in his laboratories where he passed high-voltage AC over his body to demonstrate its safety. Today his influence is pervasive. Every petrol-driven car uses a Tesla coil, a modified transformer, to generate the spark. Cathode ray tubes in television sets rely on similar technology.

Tesla was several years ahead of Marconi in devising technology to transmit and receive 'radio' waves (a later term). Marconi secured the patents—including American ones—and enduring fame, but Tesla remained active, building a massive tower on Long Island in 1900 to transmit messages, weather and stock-market reports, even pictures worldwide. He proposed to flood electric power into the air from similar towers, illuminating light globes at great distances.

Tesla's one great failure was the collapse of his project when his principal backer, banker JP Morgan, withdrew support. The dispute with Marconi over precedence in American patents was later reviewed by the US Supreme Court, which decided in Tesla's favour, two years after his death. To this day, the strength of magnetic fields is measured in 'teslas'.

Light from a tube
Georges Claude and Peter Hewitt

Modern electric light bulbs operate much as they did when pioneered into mass use by Edison and Swan, though filaments are now made of tungsten, and the machine-blown bulbs are argon-filled. They remain inefficient; only 5 per cent of the electricity consumed by most electric lamps emerges as light, the rest being wasted as heat.

Even when introduced in the 1880s, light globes did not satisfy everyone. 'Too small, too hot, too red,'

Left: The brilliant electrical pioneer Nikola Tesla, honoured by this statue in his native village in Croatia. Tesla's genius made the Age of Electricity possible. His technology allowed alternating current to be generated and distributed on a massive scale.

Above: Two easily recognisable symbols. Their outlines are formed by tubes of gas set glowing by electrical discharges. Neon signs once used just neon and made red light, but the generic term now applies to any colour.

sniffed a rival inventor, hoping to do better. An alternative was already to hand. For decades physicists in Germany and elsewhere had been experimenting with discharge tubes, sealed glass tubes with electrodes (metal plates) at each end. With most of the air pumped out and an electric current passing between the electrodes, the tubes glowed.

Replacing air with other gases produced light of various colours. French physicist Georges Claude used the rare, newly discovered neon gas. This generated red-orange light, the archetypal 'neon sign'. He patented his tubes in 1902, publicly displayed them in Paris in 1910 and began selling into the booming American market in the 1920s. The tubes, now available in many colours, can be drawn out and shaped into letters and symbols as we see every day in advertising. The Claude Neon company remains a manufacturer, though the industry grew mostly by infringing Claude's patents.

American Peter Hewitt began using mercury vapour in the 1890s. His lamps were less wasteful of electricity than incandescent globes and produced lots of light. It was, however, an unattractive blue-green which distorted other colours. This was less of a problem outdoors, and mercury vapour lamps first reached the market in 1902, backed by George Westinghouse.

By now the phenomenon of fluorescence had been discovered. Certain chemical compounds—known as phosphors—glow when bathed in ultraviolet light. In time the technologies merged. A mercury vapour tube, coated on the inside with a phosphor, became the first

The LED: how does it work?

Semiconductors are 'doped' (impregnated) with impurities to create excesses of negatively-charged electrons (in an N-type semiconductor) or positively-charged 'holes' (in a P-type). As a current passes across the junction between N-type and P-type material, holes and electrons combine to liberate energy as light. The colour of the light depends on the material used, and many compounds have been tried. The easiest colours to produce, and the first to reach the market, were red and infrared (for heat). Yellow, green and blue light, even ultraviolet, came much later—and still cost more. White light comes from mixing various colours or by making suitable materials fluoresce (glow) by bathing them in ultraviolet from a LED.

fluorescent lamp. Their greater efficiency provides 'light without heat'—or very nearly.

Pioneered by General Electric and first displayed at the New York World's Fair in 1937, fluorescents are now available in various flattering colours. Since the 1980s they have been wound into convenient coils to replace ordinary electric globes, using only 20 per cent as much power and lasting much longer. Though 'cool tubes' now produce most of the world's artificial light, their days too are numbered.

LEDs to light the way
Nick Holonyak

In our quest to make the darkness light, we have tried all sorts of technologies. The ideal light source would be cheap, efficient, robust, long-lasting and versatile. In light-emitting diodes (LEDs) we may have found that ideal.

In the history of LEDs, the first name we encounter is American Nick Holonyak. He produced his first LED, which made red light, in 1962 while working for the big firm General Electric. An LED is a cousin of the transistor.

LEDs seem to have all the advantages. Individual units are small but intensely bright and can be clustered for increased output. LEDs are long-lived; a typical LED will last ten years, far longer than other light sources. LEDs give off much less heat than incandescent globes, so are twice as good at turning electricity into light. They are also more compact and robust than fluorescent tubes—although the latter are more efficient. The impediment is mostly cost, and that is coming down rapidly as manufacturing techniques improve.

You are already encountering LEDs every day, in your television remote control, your optical mouse, the dashboard (and soon the headlamps) of your car, flashlights, traffic lights and in the increasingly light-weight message displays—the one in Time Square, New York is 36 metres high. The page opposite shows an LED display in the Shiodome business district in Tokyo. In time they are likely to replace all other forms of lighting for everyday purposes. New generations of technology, such as light-emitting transistors and transistor lasers, are still to make their mark.

A typical LED will last ten years, far longer than other light sources.

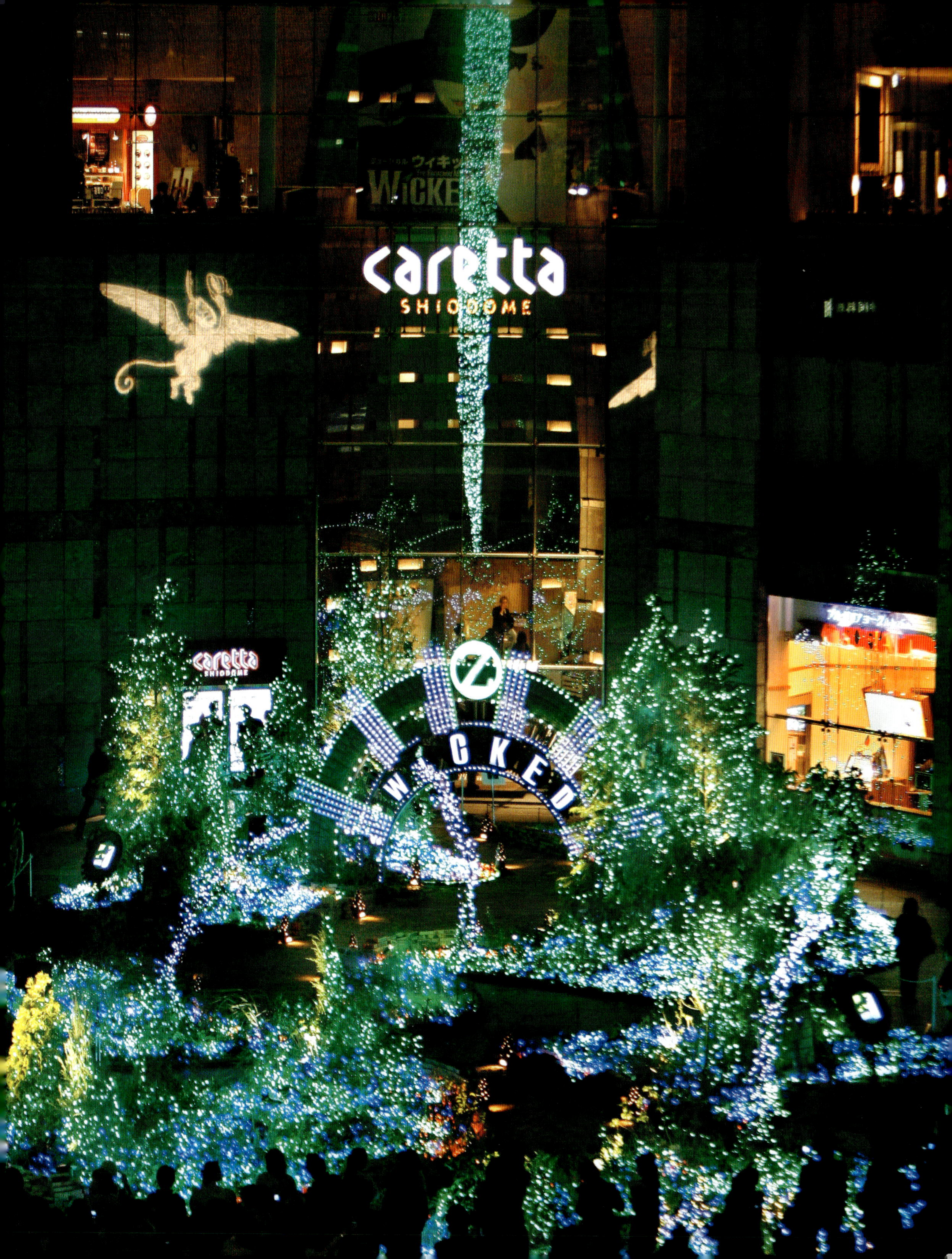
caretta
SHIODOME

Information sources

Virtually all of the information used in writing this book has been obtained via the Internet/World Wide Web, an immense and expanding source of documents, easily and quickly accessible and commonly much more up-to-date than printed sources, but one which requires some caution in its use. Typing the name of an inventor or an invention into a search engine will generate almost instantaneously hundreds or even thousands of responses. Given the ease of 'cutting and pasting', the majority of these will be derivative, reproducing without significant editing and usually without any acknowledgement material available elsewhere.

It is therefore crucial to assess as much as possible the originality and reliability of the information presented. I apply several criteria here. One is the independence of the source. Many of the responses will be from websites with a commercial association with the invention. While these are often sources of detail not available elsewhere, their vested interest may lead to some imbalance in the way information is presented. More broadly, I look for significant differences in presentation or detail between rival sources to ascertain whether they are independent, and therefore can be used to crosscheck information or whether one is derived wholly or in part from the other.

Much of the information supplied is in the form of anecdote, stories which have been passed down and have no original document on which they are based. These give a lot of enjoyable flavour to the accounts of inventors and inventions in this book but their reliability may be questioned. We commonly read slightly varying versions of these stories, with no real way of which if any is most reliable. My approach here has been to acknowledge the nature of this material, and preface it with phrases like 'one version of the story says that ... '

One of the most widely used, and in my judgment reliable, online resources, is *Wikipedia* (*http://en.wikipedia.org/wiki/Main_Page*). This is a remarkable and astonishingly diverse and comprehensive collaborative online encyclopaedia, established with protocols designed to build confidence in the accuracy and objectivity of material published. Entries are commonly compiled by individuals who appear to have a close association with, and detailed knowledge of, information presented and are replete with embedded links.

Wikipedia provides alternative ways to reach particular information, such as listings of inventors at *http://en.wikipedia.org/wiki/List_of_inventors,* and of inventions named after individuals at *http://en.wikipedia.org/wiki/List_of_inventions_named_after_people.* The article *History of Technology* (*http://en.wikipedia.org/wiki/History_of_technology*) is very useful usual and comprehensive, as is the *Timeline of Inventions http://en.wikipedia.org/wiki/Timeline_of_inventions.*

Another significant source dealing with inventions and inventors forms part of the comprehensive *About* site *(http://about.com)*. Listings of inventors and inventions arranged alphabetically can be found on *http://intentors.about.com*, with short informative pieces and hyperlinks to other sources. There are also useful and reasonably comprehensive timelines, again with embedded links to information sources.

The sources at *Ideasfinder http://www.ideafinder.com/history/index.html* again list by inventors, invention and times, and likewise at *Factmonster http://www.factmonster.com/ipka/A0004637.html*. However the need to assess the reliability of information found by the various routes, as discussed above, remains.

As to printed resources, I found a great deal of value in *The Inventions that Changed the World* (Readers Digest, London, 1982), *The Timetable of Technology* (Hearst Books, New York, 1982), dealing with 20th Century inventions, *Inventions* in the *Collins Eyewitness Guides* series (HarperCollins, London 1991) and *Eureka; an Illustrated History of Inventions from the Wheel to the Computer*, edited by Edward de Bono (Thames and Hudson, London 1979).

Fascinating perspectives on how important new technologies were view much closer to their time of invention have come from books like *The Wonderful Century* by AR Wallace (EW Cole, Sydney, 1906), *Science Remaking the World* by Otis Caldwell and Edward Slosson (Garden City Publishing, New York, 1923), *The Science Year Book for 1945*, edited by John Ratcliffe (Doubleday, Doran and Company, New York 1945) and *A Treasury of Science*, edited by Harlow Shapley, Samuel Rapport and Helen Wright (Angus and Robertson, London and Sydney, 1954).

My understanding of historical contexts and parallel events against which the history of invention can be assessed has been enriched by books like *Phillips World History; People, Dates and Events* (George Phillip Limited, London, 1999) and a much-loved and all-but-disintegrating *Encyclopedia of Dates and Events*, edited by LC Pascoe (Hodder and Stoughton, London, 1968). JD Bernal's iconic *Science in History* (Penguin Books, Melbourne, 1969) has been a constant companion.

David Ellyard

About the author

David Ellyard is one of Australia's best-known and most prolific science and technology communicators, with more than three decades of experience interpreting developments and discoveries in these fields. His insights, coupled with his contagious enthusiasm, help people with little science and technology background to assess the impact of new knowledge and invention on their everyday lives—and on the wider world. David enjoys exploring the human dimension of new ideas and inventions. He recognises that they are the products of human ingenuity and persistence, driven by curiosity and material need.

David's fascination with science began in childhood. Trained in both science and education, David has been involved in science and technology all his working life, as a researcher (including a year in Antarctica studying cosmic rays), high school and university teacher, radio and TV broadcaster, writer, communication consultant, public speaker and government policy and program officer.

After six years in the ABC Radio Science Unit, where he pioneered science commentary on 'popular radio', David worked as a researcher, producer and presenter in the ABC TV science programs *Towards 2000*, *Quantum* and *Skywatch*, and was a weekend weather presenter on ABC TV for nearly 20 years.

David is a former Governor of the University of New England and a former President of the Australian Science Communicators. He holds the NSW Director General's award for services to public education.

He has published more than a dozen books on science and technology topics including weather, climate and astronomy. His *Starwheel* map of the night sky has been in print for 20 years and has sold more than a hundred thousand copies. *Oliphant*, a biography of leading Australian physicist and public figure, Sir Mark Oliphant—co-authored with Stewart Cockburn—won the 1982 South Australian Premier's Prize for Non Fiction.

More recently, *Who Discovered What When* (New Holland Press, 2005) won the Eureka Awards Science Book Prize and was soon followed by *Who Invented What When*, released in December 2006. Like many of David's books, these combine his love of history with his concern for the impact of science and technology.

Index

A

acid rain, 73
aerosols, 261–3
air pressure, 47
aircraft, 209–12
　black box, 217
　helicopters, 212–13
　jet airliners, 214
　jet engine, 211–12
airships, 208
alkali, 72–3
Allen, Paul, 173
aluminium, 66–7
AM radio broadcasting, 147
Amati, Andrea, 280
ammonia, 76–7
anaesthetic, 231
aniline, 73–4
anthrax vaccine, 234–5
antibiotics, 237–8
antiseptic surgery, 233
Aplanalp, Robert, 262
Appert, Nicholas, 40
Apollo 13, 32
aqualung, 204
Archer, Frederick, 116
Armat, Thomas, 119
Armstrong, Edwin, 147
ARPANET, 108–9
Arthur, David, 197

Aspdin, Joseph, 60
aspirin, 235
atomic bomb, 26
audio cassettes, 132
automobile, 188–92, 196–7
　hybrid, 196–7
　mass production, 191–2
Aveling, Thomas, 187

B

Babbage, Charles, 162–3
Bacon, Francis, 32, 45
Baekeland, 84
Baird, John Logie, 156
bakelite, 84
Ballistic Missile, 27
bandaids, 259
barbed wire, 39–40
baseball, 267
bathyscaphe, 200, 201
bathysphere, 200
battery, 19
Bayard, Hippolyte, 115
bazooka, 195
Becquerel, Henri, 154
Beeb, William, 200
Bell, Alexander Graham, 106, 108
Bentz, Melitta, 260
Benz, Karl, 189–90
Berger, Hans, 248
Berliner, Emile, 130

Berners-Lee, Tim, 113
Bessemer, Henry, 62–4
Bezzera, Luigi, 259
bicycle, 187–8
Bidwell, Shelford, 154–5
Birdseye, Clarence, 45
Biro, Laszlo and Georg, 99
birth control, 242–3
Bissell, George, 20
Black, Joseph, 18
Blanchard, Pierre, 215
bleaching powder, 71
boats, 200, 202–4
bone china, 58
borer, 50
Bosch, Carl, 77
Bottger, Johann, 57
Boulton, Matthew, 18
Bourne, Gary, 169
Bowen, Edward, 148
Branca, Giovanni, 22
Brandenberger, Jacques, 84
Branley, Eduard, 139
brass instruments, 281
brassiere, 257
Braun, Karl, 153
Brearley, Harry, 67–8
Brebera, Stanislav, 81
Broadwick, Charles, 215
Brunel, Isambard, 51, 179, 203
Bullet Train, 182–3

Bullock, William, 96–7
Bush, Vannevar, 166
Bushnell, David, 201
Butler, Edward, 187

C

calculator, 161
camcorder, 124
cameras, 116–17
  digital, 124–6
  Polaroid, 120–3
camera obscura, 115
Campbell, John, 223
Campbell-Swinton, Alan, 154–6
can opener, 40–1
canned food, 40
carbolic acid, 74
Carlson, Chester, 98
cars *see* automobile
Caselli, Abbé, 154
Cassini, Jean-Dominique, 221
cast iron, 50
CAT scans, 249–51
cathode ray tube, 153, 246
cavity magnetron, 148–9
Cayley, George, 209
CD-ROM, 136
CDs, 134–6
cellophane, 84
cellular phone, 110–13
celluloid, 83, 116
cement, 60, 65
centrifuge, 44–5
Chain, Ernest, 238
champagne, 253

Chanute, Octave, 209
Chappe, Claude, 101
Charles, Jacques, 207
chemicals, 70–81
  cures in medicine, 237
chemotherapy, 237, 238–40
china pottery, 57
chips, potato, 41–2
chlorine, 71
cinematograph, 119
clarinet, 280–1
Clarke, Arthur C, 149–50
Claude, Georges, 291
clocks, 219, 223–4
coal, 20–1
coal gas lighting, 285–6
coal tar, 73–4
coffee, 259–60
coherer, 139
coke (fuel), 50
colour film, 119–20
combine harvester, 36, 38–9
communications, 100–113
  electric telegraph, 102
  email, 109–10
  Internet, 108–10, 113
  mobile phones, 110–13
  Morse code, 102–5
  satellite, 149–51
  telegraph lines, 101
  telephone, 106–8
  television, 152–9
  wireless, 139–51
compact disc, 134–6
compass, 225–6
computer chip, 169–70

computers, 160–73
  Apple, 173
  central processing unit, 169
  Commodore 64, 173
  Electronic Numerical Integrator and Calculator, 164
  hard drive, 167
  hypertext, 166
  IBM, 173
  mouse, 166, 172
  PCs, 172–3
  punch-card system, 163
  supercomputers, 170–1
  Universal Automated Computer, 164–6
comsats, 149–51
concrete, 60, 65
contraceptive pill, 242–3
Cooke, William, 102
Cooper, Martin, 110
Coover, Harry, 263
Cormack, Allan, 250
Cornish engines, 19
correction fluid, 99
corsets, 257
Cort, Henry, 50
cotton gin, 37
Cousteau, Jacques, 204
Crapper, Thomas, 255–6
Crawshay, Richard, 54
Cray, Seymour, 170–1
cream separator, 22, 44
cricket, 267
Cristofori, Bartolomeo, 279
crossword puzzles, 275
crude oil, 20–2

Crum, George, 41–2
crystal, 59
Cugnot, Nicolas, 49, 185–6
Cummings, Alexander, 255
Cunard, Samuel, 202–3

D

da Vinci, Leonardo, 22
Daguerre, Louis, 115
Daimler, Gottleib, 189–90
Damadian, Raymond, 251
Darby, Abraham, 4
Darrow, Charles, 275
Davy, Humphry, 66, 115, 178, 231, 286
de Forest, Lee, 142, 144
de Laval, Carl, 22
de Mestral, Georges, 261
de Rivaz, Isaac, 185–6
'Devil's Rope', 39, 40
DDT, 77–8
deep-sea diving, 199–200, 204
dialysis machine, 240
Dickson, Earle, 257, 259
Diesel, Rudolf, 22–3
Digges, Leonard, 220
digital photography, 124–6
diode, 141
disposable razor, 264
diving bell, 199
Donald, Ian, 249
double acting engines, 19
Drais, Karl, 187
Drake, Edwin, 20, 21
Dresden china, 58

Duncan, Donald, 274–5
Dunlop, John, 190
Durand, Peter, 40
DVDs, 127
dynamite, 76

E

Eastman, George, 116
Eckert, J Presper, 164–5
Edison, Thomas, 106, 117–18, 129–30, 141, 286–9
Eddystone Lighthouse, 60
Ehrlich, Paul, 237, 240
Einthoven, William, 247
elastic, 61
electric chair, 289
electric lighting, 288, 290
electric telegraph, 102
electricity, 19, 26–7, 288–90
electrocardiograph, 247–8
electroencephalograph, 247–8
electrolysis, 66
Electronic Numerical Integrator and Calculator, 164
electronics, 141, 142
Elswood, Isaac, 39, 40
email, 109–10
energy, 16–33
  coal, 20
  crude oil, 20–2
  diesel engine, 22–3, 25
  electricity, 19–20
  fuel cells, 32
  fusion, 30–2
  nuclear, 25–30

  petroleum, 20–2
  solar, 33
  steam engine, 17–19, 48
  turbines, 22
Engelbart, Douglas, 166, 172
espresso coffee machine, 259
explosives, 74–6, 81

F

Faraday, Michael, 19
farm machinery, 34–40
Farnsworth, Philo, 157–8
Fermi, Enrico, 25–7
Ferris, George, 272
ferris wheel, 272
Fessenden, Reginald, 141–2, 145
fibreglass, 59
film, 116–17
Fleming, Alexander, 237–8
Fleming, John Ambrose, 141
Flores, Pedro, 274
Florey, Howard, 238
fluorescent light, 291–2
flush toilet, 254–5
flying machines, 209–111
FM radio broadcasting, 147
food preservation, 40, 45
football, 269–70
Ford, Henry, 191–2
fountain pen, 94, 99
Fourdrinier machine, 95
Frisius, Gemma, 220
frozen food, 45
fuel injection, 25
Fulton, Robert, 200–3

G

Gagnan, Emile, 204
Galilei, Galileo, 223
Galvani, Luigi, 19
Garbor, Denis, 123
Garnerin, André, 215
gasoline, 20
Gates, Bill, 17
Getting, Ivan, 226–7
Gibbon, John, 240
Gillette, King, 264
glass, 58–60, 68
Glauber, Johann, 72
Glidden, Carlos, 97
Glidden, Joseph, 39, 40
Global Positioning System (GPS), 226–7
Goldmark, Peter, 134
golf, 268
Goodyear, Charles, 61–2
Gore, Wilbert, 89
Gore-Tex, 89
Graham, Bette, 9
grain stripper, 39
gramophone, 130
graphite, 93
Gray, Elisha, 106, 108
Greatbatch, Wilson, 240–2
Gregg, David, 134
grid iron, 270
Groves, William, 32
guncotton, 74
gunpowder, 74
gyrocompass, 226

H

Haas, Earle, 259
Haber, Fritz, 76–7
Hadley, John, 222
Haish, Jacob, 39
Hall, Charles, 66
Halley, Edmond, 199
Hammond, Laurens, 283
Hammond organ, 283
Hancock, Thomas, 61–2
hard drive, 167
Harrington, John, 254–5
Harrison, John, 225
heart-lung machine, 240
heart transplant, 240
helicopter, 212–13
Henlein, Peter, 219
Hennebique, Francois, 65
Herault, Paul, 66
Herschel, John, 116
Hertz, Heinrich, 139
Hewitt, Peter, 291
hockey, 267
Hoff, Ted, 169
Hoffman, Albert, 78–81
Hoffman, Felix, 236
Hofmann, August Wilhelm von, 74
Holland, John, 202
Hollerith, Herman, 163
holography, 123
Holonyak, Nick, 292
Hounsfield, Geoffrey, 250
hula hoop, 274
Hulls, Jonathan, 200
Hunt, Walter, 254

Huygens, Christiaan, 223–4
Hyatt, John and Isaiah, 83
hydraulic cement, 60
hydrogen bomb, 30–2
hypertext, 166–7
hypodermic syringe, 232–3

I

ice hockey, 267
ice-skating, 271
Industrial Revolution, 46–55
instant coffee, 260
internal combustion engine, 23, 32, 188
Internet, 109, 113, 137
iron, 49–55, 62–3
iron bridge, 49
Irvin, Leslie, 217

J

Jacobs, Mary, 257
Jacquard, Joseph, 161–2
Jenkins, Charles, 156
jet engine, 211–12
Jobs, Steve, 173
Johnson, Reynold, 167
Judson, Whitcomb, 260
jukebox, 282

K

Kalmus, Herbert, 119–20
Kato, Satori, 260
Keller, Friedrich, 94
Kelly, William, 63

kerosene, 20
Kevlar, 89
kinetoscope, 117
Knerr, Richard, 274
Koch, Robert, 234
Koenig, Friedrich, 94
Kohler, Georges, 243
Kolff, William, 240
Kwolek, Stephanie, 89

L

lacrosse, 267
Laennec, René, 245
Land, Edwin, 122
Lauterbur, Paul, 251
Laval, Gustav de, 44
Lavoisier, Antoine, 71
Lay, Herman, 41–2
LEDs, 292
Leary, Timothy, 78
Leblanc, Nicholas, 72–3
Leibniz, Gottfried, 49, 161
Lenoir, Etienne, 23, 186
Lenormand, Sebastien, 215
Lesage, Georges, 102
Levassor, Emile, 191
light bulb, 286–8, 290
light-emitting diodes (LED), 292
lighting, 284–93
Lilienthal, Otto, 209
Liquid Paper, 99
Lister, Joseph, 43, 233
longitude, calculating, 225
LSD, 78–81
Lumière, August and Louis, 119
Lundstrom, Johan and Carl, 257
lycra, 89

M

Mabs, Harry, 277
Macmillan, Kirkpatrick, 187
Magdeburg hemispheres, 47
magnetic resonance imaging (MRI), 249–51
Maloney, Raymond, 277
Manhattan Project, 25–7
Mansfield, Paul, 251
map making, 220–1
Marconi, Guglielmo, 139–40, 153
Marker, Russell, 242
Mason, James, 260
matches, 256–7
Mauchly, John, 164–5
Maybach, Wilhelm, 189
McAdam, John, 186–7
McCormick, Cyrus, 38–9
McGee, Carl, 195–6
McGill, George, 254
McIntosh, Charles, 61–2
McKay, Hugh, 38–9
medicine, 231–51
Meikle, Andrew, 36
Melin, Arthur, 274
Michaux, Pierre, 187
microchip, 169
microwave ovens, 149
microwaves, 148–9
milk pasteurisation, 43
Miller, James, 272
Milstein, Cesar, 243
Miner's Friend, The, 17, 18
mobile phone, 110–13
Monier, Joseph, 65
monoclonal antibodies, 243
Monopoly, 275
Montgolfier, Joseph, 207
Morgan, Garrett, 193
Morse code, 102–5
Morton, William, 231–2
motion pictures, 117–18
 'talkies', 132–3
Mouchout, August, 33
MP3, 136–7
Muller, Paul, 77–8
Murdoch, William, 285
music recordings, 133–7
musical instruments, 278–83
mustard gas, 238–9
Muybridge, Eadweard, 117

N

Nasmyth, James, 54–5
neon lights, 291
Neumann, John von, 166
Newcomen, Thomas, 17, 18
Niepce, Joseph, 115
Nipkow, Paul, 154
nitroglycerine, 76
nitrous oxide, 231
Nobel, Alfred, 74, 76
nuclear fission, 25–7
nuclear fusion, 30–2
nuclear reactors, 26–30
nuclear submarine, 27
nylon, 85

O

ocean liners, 203
Oersted, Hans, 19, 20
Ohain, Hans, 211–12
Ohr, Russell, 33
oil, crude, 20–2
Oldendorf, William, 250
Oliphant, Mark, 30–1, 148
Olsen, Scott and Brennan, 271
oscilloscope, 153
Otto, Nicolaus, 23, 188–9
Oughtred, William, 161

P

pacemaker, 240–2
paddle-steamer, 200–1
pain relief, 231–2, 235–7
Panhard, René, 191
paper, 94–7
paperclip, 254
Papin, Denis, 19, 48
paracetamol, 236–7
parachute, 215–17
paraffin, 20
Parkes, Alexander, 83
parking meters, 195–6
Parsons, Charles, 22
Pascal, Blaise, 161
Pasteur, Louis, 40, 43, 234–5
pasteurisation, 43
pen, 93–4, 99
pencil, 93
penicillin, 238
penny-farthing bicycle, 188
percolated coffee, 260
Perignon, Dom, 253
Perkin, William, 74
personal computers (PCs), 172–3
petroleum, 20, 22
phenol, 73–4
phonograph, 129
photocopier, 98–9
photography, 115–17, 119–20
   digital, 124–5
piano, 279
pianola, 282
Piccard, Auguste and Jacques, 200
Pilkington, Alistair, 68
pinball machine, 277
Pincus, Gregory, 243
plane table, 220
plastics, 83, 85–9
plate glass, 68
Plimpton, James, 271
plutonium, 25–30
Polaroid instant camera, 120
polymers, 86, 89
polystyrene, 86
polythene, 86
porcelain, 57–8
Porsche, Ferdinand, 196
Portland cement, 60
Post-It Notes, 263–4
potato chips, 41–2
Poulsen, Valdemar, 130, 132
Pravaz, Charles, 232
preserving food, 40
pressure cooker, 19, 48
printing press, 95–6
pump, steam-driven, 17
punch cards, 163
PVC, 86

Q

quill pen, 93–4
quinine, 74

R

rabies vaccine, 235
radar, 142, 145–8
radio, 139, 142–8
railways, 177–80
Ravenscroft, George, 58
rayon, 84
razor blades, 264
RDX, 81
reaping machine, 38–9
records, 130, 134
reinforced concrete, 65
Rennie, John, 37
Rickover, Hyman, 27
Ridley, John, 38–9
road construction, 186–7
Robert, Nicholas, 95
Roebuck, John, 71–2
Roentgen, Wilhelm, 246
roller-blades, 271
rollercoaster, 272
roller-skates, 271
Rosenthal, Ida, 257
Rosing, Boris, 157
Rotheim, Erik, 262
rubber, 61–2
   eraser, 93
   tyres, 190

rugby, 270
Runge, Friedlieb, 73

S

safety matches, 256–7
safety pin, 254
safety razor, 264
Sakharov, Andrei, 30, 32
Sarnoff, David, 145
Sasson, Steven, 125
satellites, 149–51, 227
Savery, Thomas, 17, 18
Sax, Adolph, 281
saxophone, 281
Scheele, Karl, 71, 115
Schickard, Wilhelm, 161
Scholes, Christopher, 97
Schonbein, Christian, 74
Shore, John, 281
scuba diving, 204
seed drill, 35
semaphor, 102
Semtex, 80, 81
sextant, 221–3
Shinkansen, 182–3
Shivers, Joseph, 89
Sikorsky, Igor, 212–13
Silliman, Benjamin, 20
Silver, Spencer, 263
Simpson, James, 231
skateboard, 272
skim milk, 44
slide rule, 161
Smeaton, John, 60
Sobrero, Ascanio, 76

soccer, 269, 270
soda ash (sodium carbonate), 72
solar energy, 33
Solvay, Ernest, 72–3
sound recording, 128–37
spandex, 89
Spencer, Percy, 149
Sperry, Elmer, 225
spray-can, 261–3
sport, 266–72
stainless steel, 67–8
staples, 254
Starley, John, 188
steam engine, 17–19, 48, 185–6
steam hammer, 54–5
steam locomotive, 177–8
steam ships, 202–4
steam wagon, 185
steamboat, 200
steamroller, 187
steel, 62–4, 67–8
Stephenson, George, 177
stethoscope, 245
submarines, 201–2
  nuclear, 27, 202
sulphuric acid, 72
sulphuric ether, 231
Sundback, Gideon, 260
Super Glue, 263
supercomputers, 170–1
surgery, antiseptic, 233
surveying, 220
Swan, Joseph, 288
Swinton, Ernest, 194
synoptic charts, 105
synthetic dyes, 74

synthetics, 82–9
syringes, disposable, 233

T

tabulating machine, 163
Talbot, William Henry Fox, 116
talking pictures, 132–3
tampons, 259
tank warfare, 194–5
tape recorders, 130–2
Technicolor, 119–20
telegraph lines, 101–5
telegraphic money order, 105
telephone, 106–8
television, 153–9
  cable, 158
  colour, 159
  high definition, 159
Telford, Thomas, 187
Teller, Edward, 30
Tenderich, Gertrude, 259
Tennant, Charles, 71
tennis, 268–9
Tesla, Nikola, 289–90
theodolite, 220–1
Theremin, Leon, 283
Thompson, Robert, 190
threshing machine, 36–7
TNT, 76
toilet, 254–6
Tomlinson, Ray, 109–10
Torricelli, Evangelista, 47
toys, 274
traffic lights, 192–4
trains, 176–83

air brake system, 180
magnetic levitation, 183
Shinkansen, 182–3
TGV, 183
transport, 176–227
Trevithick, Richard, 19, 177
triode, 142–3
Tucek, Marie, 257
Tull, Jethro, 35–6
tuning fork, 281
turbine, 22
typewriter, 97–8
tyres, 190

U

UHT pasteurisation, 43
Ulam, Stanley, 30
ultrasound, 248–9
underwater salvage, 199
uranium, 25–6

V

Vaaler, Johann, 254
vaccination, 234–5
vacuum, 47
Velcro, 261
video cassette, 124
videotape recorder, 123–4
violin, 280
Volta, Alessandro, 19
von Guericke, Otto, 47
vulcanised rubber, 62

W

Walker, John, 256
Walkman, 132
Warren, David, 217
watches, 219
waterproof fabric, 61
Watson-Watt, Robert, 146
Watt, James, 18
Westinghouse, George, 180, 288–9
Wheatstone, Charles, 102
Whitney, Eli, 37
Whittle, Frank, 211–12
Wilkinson, John, 49–50
windmill shutters, 37
window glass, 68
winnowing machine, 36
Wire, Lester, 193
wireless, 139
woodwind instruments, 280–1
World Wide Web, 113
Wouk, Victor, 196–7
Wozniak, Steve, 173
Wright, Wilbur and Orville, 209–11
wrought iron, 50, 52
Wynne, Arthur, 275

X

X-rays, 246–7

Y

yoyo, 274

Z

Zech, Jacob, 219
Zeppelin, Ferdinand von, 208–9
Zinn, Walter, 27, 29
zipper, 260–1
Zworykin, Vladimir, 157